GATEWAY
Science

David Acaster | Mary Jones | David Sang

CAMBRIDGE
UNIVERSITY PRESS

CAMBRIDGE UNIVERSITY PRESS
Cambridge, New York, Melbourne, Madrid, Cape Town, Singapore, São Paulo

Cambridge University Press
The Edinburgh Building, Cambridge CB2 2RU, UK

www.cambridge.org
Information on this title: www.cambridge.org/9780521685405

First published 2006

Printed in the United Kingdom at the University Press, Cambridge
Cover and text design by Blue Pig Design Ltd
Page layout and artwork by HL Studios, Long Hanborough

A catalogue record for this publication is available from the British Library

ISBN-13   978-0-521-68540-5 paperback
ISBN-10   0-521-68540-0 paperback

Contents

Introduction   iv

Biology

B1a Fit for life   1
B1b What's for lunch?   8
B1c Keeping healthy   16
B1d Keeping in touch   28
B1e Drugs and you   37
B1f Staying in balance   46
B1g Who am I?   54
B1h New genes for old   58

B2a Ecology in our school grounds   66
B2b Grouping organisms   75
B2c The food factory   88
B2d Compete or die   94
B2e Adapt to fit   102
B2f Survival of the fittest   110
B2g Population out of control?   121
B2h Sustainability   131

Chemistry

C1a Cooking   147
C1b Food additives   155
C1c Smells   161
C1d Making crude oil useful   167
C1e Making polymers   174
C1f Designer polymers   182
C1g Using carbon fuels   188
C1h Energy   193

C2a Paints and pigments   200
C2b Construction materials   205
C2c Does the Earth move?   211
C2d Metals and alloys   218
C2e Cars for scrap   227
C2f Clean air   232
C2g Faster or slower (1)   239
C2h Faster or slower (2)   249

Physics

P1a Heating and cooling   257
P1b Keeping houses warm   265
P1c How insulation works   269
P1d Cooking with radiation   275
P1e Communicating with infra-red
     radiation   279
P1f A wireless world   284
P1g Light waves   290
P1h Earth waves   297

P2a Solar cell technology   305
P2b Generating electricity   313
P2c Fuels for power   322
P2d Nuclear radiations   330
P2e Magnetic Earth   338
P2f Exploring the solar system   344
P2g Threats to Earth   350
P2h Beginnings and endings   356

Answers to SAQs

Biology   362
Chemistry   369
Physics   376

Glossaries

Biology   379
Chemistry   385
Physics   390

Physics formulae   392

Periodic Table   393

Index   394

Acknowledgements   403

MUFC

Introduction

To the pupil

This book is divided into three sections – Biology, Chemistry and Physics. Each of these sections is then arranged by Item, as in the exam specification. Each Item has the following features.

- **Self-Assessment Questions (SAQs)** are placed within the text and refer to material that has gone before. Answers to these are provided at the back of the book.
- **End-of-chapter questions**, including exam-style questions. Answers to these are not provided in this book.
- **Summaries** at the end of the chapter that show the information you need to know in order to do well in the exam.
- **Higher-level** text, questions and summaries are shown by a side bar marked with the letter 'H'.
- **Context boxes** that give you the opportunity to read about the history behind the discoveries and to learn about real-world applications.
- **Worked example boxes** (where relevant).

The book also contains glossaries, a list of physics formulae, a Periodic Table and an index at the back of the book.

To the teacher

The *Cambridge Gateway Sciences* series has been written to cover the new Gateway Specification (B) developed by OCR.

This text contains materials for the Science specification. It is accompanied by a *Science Teacher File* CD containing adaptable planning and activity-sheet resources, as well as answers to the end-of-chapter questions, and by *Science* CDs of interactive e-learning resources, including animations and activities for whole-class teaching or independent learning, depending on your needs.

The Additional Science specification is supported by a book and CDs in the same way.

For further information on all accompanying materials, visit the dedicated website at www.cambridge.org/newsciences

High altitude, low oxygen

Up on top of the world, in the high Himalayas, oxygen is hard to come by. At Everest Base Camp, a height of 5240 metres above sea level, the air is much thinner than at the kinds of altitudes where people usually live. However fit a person is, they struggle to get enough oxygen into their body.

Reinhold Meissner and Peter Habeler were the first people to climb Everest without breathing any oxygen from a cylinder. Meissner wrote:

'Now, at a height of 8800 metres, we can no longer keep on our feet while we rest. We crumple to our knees, clutching our axes ... Breathing becomes such a strenuous business that we scarcely have the strength to go on. Every ten or fifteen steps we collapse into the snow to rest, then crawl on again. My mind seems almost to have ceased to function ... I crawl, I cough, but I am drawn on.'

Figure 1a.1 At high altitude.

Respiration

All the energy that our bodies use is provided by **respiration**. Respiration is a chemical reaction that takes place inside every body cell. This reaction releases energy from **glucose**. The energy can then be used by the cell for whatever is needed. For example, if it is a muscle cell, it can use the energy for contracting (getting shorter). If it is a cell in the digestive system, it can use it for making enzymes. If it is a nerve cell, it can use it for sending electrical impulses to the brain.

Each cell must release its own energy, so every living cell respires.

Aerobic respiration

When cells get enough oxygen, they respire **aerobically**. In **aerobic respiration**, glucose reacts with oxygen. The **word equation** for this reaction is:

glucose + oxygen → carbon dioxide + water

This is an example of a **metabolic reaction**. A metabolic reaction is a chemical reaction that takes place inside a living organism.

SAQ

1 This is the word equation for another metabolic reaction:

carbon dioxide + water → glucose + oxygen

What is this reaction? Where does it take place?

H   This is the balanced symbol equation for aerobic respiration:

$$C_6H_{12}O_6 + 6O_2 \rightarrow 6H_2O + 6CO_2$$

In this process, glucose is oxidised. In many ways, aerobic respiration is similar to combustion (burning) – a fuel (glucose) reacts with oxygen and energy is released. However, burning cannot be allowed to happen in a cell because a rapid temperature rise would destroy the proteins and other molecules in the cell.

In a living cell, the oxidation of the glucose takes place in a series of small, carefully controlled steps. The energy is released gradually, not all at once as happens during burning.

SAQ

2   Write down two similarities between aerobic respiration and combustion.

3   Write down two differences between aerobic respiration and combustion.

Anaerobic respiration

Cells are also able to release energy from glucose *without* using oxygen. This process is called **anaerobic respiration**. The glucose is broken down to produce **lactic acid**:

glucose → lactic acid

Anaerobic respiration is not very efficient. That means that you don't get as much energy from a certain amount of glucose as you would with aerobic respiration. This is because anaerobic respiration does not break down the glucose completely. It is really a 'last resort' for cells when their oxygen supply runs out but they still need to use energy.

Imagine you are running fast. Your leg muscles are working flat out. Your heart rate has increased and so has your breathing rate, to supply oxygen at a much faster rate than usual to the muscles. But their best efforts are not enough and your muscles need more energy than they can release through aerobic respiration. So they use anaerobic respiration as well.

The lactic acid that is made in anaerobic respiration is toxic (harmful) to cells. As it accumulates (builds up) in your muscle cells, they begin to hurt. If a lot of lactic acid accumulates in them, they can no longer carry out anaerobic respiration. You have reached your limit; no matter how much you want to keep on running fast your muscles simply cannot do it.

Figure 1a.2 A top-class sprinter may not breathe at all during the 10–11 seconds of a 100 m race but relies heavily on anaerobic respiration.

The lactic acid made during anaerobic respiration diffuses out of the cells where it is made and into the blood. It is carried to the liver. This is helped by the rapid beating of the heart, which keeps blood moving swiftly through the arteries and veins.

When you stop running, you keep on breathing hard so you are still getting lots of oxygen into the body. Some of this oxygen is used by the liver cells to break down the lactic acid. You keep on breathing faster and deeper than usual until all the lactic acid has been broken down. The extra oxygen that you need to do this is called your **oxygen debt**.

Aerobic respiration	Anaerobic respiration
Similarity	
energy released by breakdown of glucose	energy released by breakdown of glucose
Differences	
uses oxygen	does not use oxygen
no lactic acid made	lactic acid made
carbon dioxide and water made	no carbon dioxide or water made
releases a large amount of energy	releases a small amount of energy

Table 1a.1 A comparison between aerobic and anaerobic respiration in humans.

Keeping fit

Fatigue

In June 2005, Centre Court at Wimbledon was brought alight by a young Scottish tennis player who had seemingly appeared from nowhere. 18-year-old Andrew Murray took the first two sets from David Nalbandian, playing brilliant tennis. But as they went into the third set, Andrew began to suffer from muscle cramp and fatigue. He had all the tennis know-how and ability, all the agility, flexibility and speed. But he lacked the stamina to take a world-class player to the end of a five-set match. Nalbandian took the last three sets and the match.

Andrew went away from Wimbledon knowing that he had the ability to get into the top ten in the world. His priority was to work on his physical fitness, training his heart, lungs and muscles so that they work together to supply his legs with the energy that they need to keep going to the end of a hard-fought match.

Figure 1a.3 Andrew Murray playing David Nalbandian.

Andrew may not have been fit enough to get through a gruellingly long and tough match but he is probably much fitter than almost everyone in your class. Most of us don't need to be fit enough to play five sets of hard tennis on a hot day. But being fit is a good thing to aim for. It makes you feel good and helps your general health. It won't stop you from getting infectious diseases such as colds and flu, but it does help you to enjoy physical activity. And that can improve your health, because it can make it less likely that you will suffer from heart disease when you are older.

Measuring fitness

If you ask a trainer at a gym to devise a fitness programme for you, they will probably want to start off by measuring your **fitness**. There are many different aspects of fitness that can be measured and lots of different ways of measuring them. Here are a few examples.

- **Strength** of a particular set of muscles can be measured by making them work as hard as they can and measuring the force they produce. For example, the strength of the hand and arm muscles can be measured with an instrument where you squeeze as hard as you can on a hand-grip monitor, which gives you a readout of the force you have generated.
- **Stamina** is how long you can keep going with a particular exercise. You could measure how long you can keep running at a particular speed or how many press-ups you can do without a break.
- **Flexibility** is how easily and how far your joints can allow you to bend. For example, you could do a sit-and-reach test, in which you sit on the floor with your legs straight in front of you and then measure how far you can reach with your arms.
- **Agility** is how quickly and easily you can turn and twist while you are moving. One test for agility is called the 'hexagonal obstacle' test. You stand in the centre of a hexagon with 66 cm sides, marked out on the floor. You then have to jump as quickly as possible from the centre over one side, back to the centre and over another side, and so on all around the hexagon.

Figure 1a.4 Measuring stamina.

Figure 1a.5 Measuring flexibility.

H

- **Speed** is how fast you can run. You can measure this on a treadmill.
- **Cardiovascular efficiency** is how well your heart manages to supply oxygen to your muscles. There are a great many ways of measuring this. One method is to measure **VO₂ max**. This is short for 'maximum volume of oxygen'. VO₂ max is the maximum rate at which your muscles can use oxygen before they have to go over to anaerobic respiration (because the oxygen has run out). The better your heart and blood are at getting oxygen to your muscles, the higher your VO₂ max. Cardiovascular efficiency affects your stamina.

SAQ

4 Which of the above types of fitness do you think let Andrew Murray down in his five-set match against David Nalbandian?

5 Choose two of the methods described above and discuss how useful they are in giving information about someone's fitness.

Blood pressure

Your heart is made of cardiac muscle, a kind of muscle that naturally contracts and relaxes rhythmically, all the time. Each time it contracts, it squeezes inwards on the blood inside the heart, increasing its pressure and squirting it out of the

heart into the big arteries that carry the blood to other parts of the body.

You can feel this **blood pressure** when you take your pulse. If you hold two fingers gently over the tendon on the inside of your wrist, or at the front of your neck, you can feel the arteries expanding and recoiling every time the heart beats. The arteries have thick walls with a lot of elastic tissue in them. This allows them to be stretched outwards by the surge of high-pressure blood each time the heart contracts, and then recoil back to their normal diameter when the heart relaxes.

Figure 1a.6 Measuring blood pressure.

If you have your blood pressure measured, you will be told two numbers. The first is the highest pressure that blood in the big arteries in your arm reaches. This happens when the heart is contracting and it is called the **systolic pressure**. When the heart relaxes, the pressure in the arteries drops and this is called the **diastolic pressure**.

In science, pressure is measured in units called **pascals**. In medicine, though, doctors still use old units for measuring pressure. They are **millimetres of mercury**, written as mmHg for short. This refers to the height that a particular pressure can push a column of mercury up a tube. Now we have digital blood pressure meters, but many doctors still use tubes of mercury to measure blood pressure.

What affects blood pressure?

A good blood pressure for a 16-year-old to have is somewhere around 120 over 80. This means that your systolic pressure is 120 mmHg and your diastolic pressure is 80 mmHg.

So what affects your blood pressure? We all vary a bit from each other, just naturally, but there are several things that definitely have an effect.

As you **age**, your blood pressure will probably increase a bit. By the time you are 60, it is fairly normal to have a blood pressure of around 135 over 89.

Blood pressure is also affected by **diet**. A diet containing a lot of salt can increase your blood pressure. This is because the high concentration of salt in your blood means that the kidneys have to allow there to be a lot of water in the blood, to dilute the salt. So you have a greater volume of blood squeezed inside your blood vessels, increasing the pressure.

It is also affected by the amount of **exercise** that you do. While you are exercising, the heart beats harder and faster and your blood pressure increases. When you relax, the pressure goes down to normal. If you exercise regularly, your resting blood pressure will probably be lower than if you are unfit.

Your **weight** is also important. Overweight people tend to have higher blood pressure than people of normal weight.

Alcohol affects blood pressure. Regularly drinking large amounts of alcohol increases blood pressure. We don't really understand why this happens.

Stress is yet another factor that increases blood pressure. Some stress is good for you – it means that you have challenges in your life and can make your life more interesting and rewarding. But some kinds of stress – the kinds you can't control or escape from – are not good, especially if they go on for a long time. Stress can have several damaging effects on your body, including increasing your blood pressure.

Figure 1a.7 Stress can cause an increase in blood pressure.

High and low blood pressure

High blood pressure can cause considerable harm to the circulatory system and to other organs.

● It puts extra strain on the heart, which can increase the likelihood of burst blood vessels. If this happens in the brain then part of the brain may be damaged – either because the leaked blood builds up and presses dangerously on the brain tissue, killing brain cells, or because the burst vessel no longer supplies oxygen to the brain cells, so they die. This is called a **stroke**.

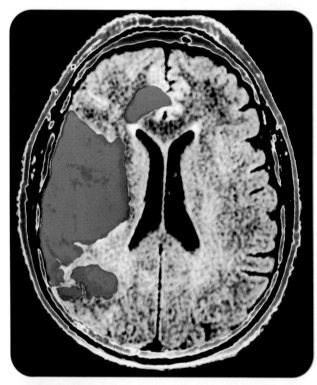

Figure 1a.8 This is a scan of the brain of someone who has had a stroke. The orangey-red area shows where blood has flooded into the brain tissue.

● It can damage the kidneys. Blood pressure is a little higher in the kidneys than in other parts of the body. This helps them to filter the blood efficiently. But if the blood pressure is high already then the extra-high pressure in the kidneys can damage them. Kidney failure is much more common in people who have high blood pressure.

Although most people know that high blood pressure is a health risk, not so many ever think about **low blood pressure**. But this is dangerous too. If you have low blood pressure then blood isn't moving through your blood vessels as fast as it should. So, for example, less oxygen gets carried to your brain and you may get dizzy spells or faint, especially when you stand up quickly after sitting down. Poor blood circulation means that your fingers and toes may not get enough blood to them, which can harm the cells.

6 If someone has high blood pressure, do you think a high diastolic or a high systolic pressure is the more dangerous? Explain your answer.

7 Explain fully why low blood pressure can make someone faint.

Summary

You should be able to:

◆ describe how energy is released from glucose in aerobic respiration and in anaerobic respiration, and write the word equations

Ⓗ ◆ write the balanced equation for aerobic respiration

◆ explain why your heart beats faster, and you breathe faster, when you are exercising

Ⓗ ◆ explain how lactic acid is removed as you recover from exercise

◆ describe what is meant by *diastolic* and *systolic* blood pressure, and know that they are measured in mmHg

◆ know that diet, age, exercise, weight and alcohol intake affect blood pressure

Ⓗ ◆ explain the consequences of high blood pressure

◆ explain the difference between being fit and being healthy

Ⓗ ◆ describe some ways of measuring fitness

Questions

1   Copy and complete these sentences to compare aerobic and anaerobic respiration.

   a   Anaerobic respiration releases .... energy than aerobic respiration.

   b   Anaerobic respiration makes .... acid.

   c   Carbon dioxide is made in .... respiration but not in .... respiration.

2   Jenny is told her blood pressure is 160 over 90.

   a   Which number is her diastolic pressure and which is her systolic pressure?

   b   What are the units her blood pressure was measured in? Choose from: pascals, millimetres of mercury, newtons, joules.

   c   Jenny's blood pressure is too high. Suggest two changes she could make to her lifestyle that might bring her blood pressure down.

3   The graph shows the percentage of men and women in different age groups who had high blood pressure in 2003.

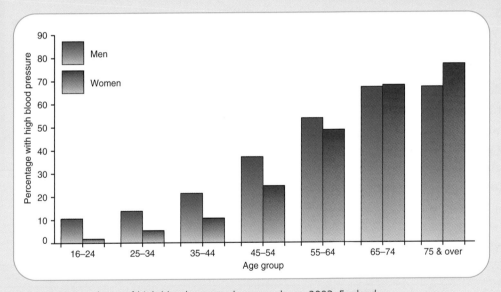

Figure 1a.9 Prevalence of high blood pressure by sex and age, 2003, England.

   a   Describe the relationship between age and high blood pressure.

   b   Describe the relationship between a person's sex and their risk of having high blood pressure.

   c   Explain why high blood pressure is dangerous.

A balanced diet

You have seen that every living cell must respire, to provide the energy that it needs to stay alive. For aerobic respiration, oxygen is required. Equally importantly, cells need a supply of glucose as their energy source.

The glucose used in respiration comes from your **diet**. So the food you eat provides the source of energy for your body. Food also provides the **raw materials** that you need to produce the structures that make up your body. A diet that provides all of your energy needs plus all of the raw materials for growth and maintenance of the body is called a **balanced diet**.

Table 1b.1 summarises the functions of the different nutrients you need in your diet and **deficiency diseases** caused by lack of them.

Individual diets

Your diet is what you eat each day. It should contain some of all the different nutrients listed in Table 1b.1.

It should provide you with enough energy or your cells will not be able to function well and you will feel tired. However, it should not contain more energy than you use or your body will store the extra as fats, and you may become overweight. Remember that the energy-containing nutrients are carbohydrates, fat and proteins.

The amount of energy you need will depend on how much you use each day. This varies hugely between different people. It depends on your age, your sex and how much exercise you do. Figure 1b.1 shows some examples.

Your diet should also contain some of each kind of **vitamin** and mineral. Each one has its own particular function in the body. Fibre is also essential. It stimulates the muscles in the walls of the alimentary canal, which ensures that food keeps moving through. This prevents constipation, and is also thought to protect against diseases such as bowel cancer.

SAQ

1   Which types of nutrients provide energy?

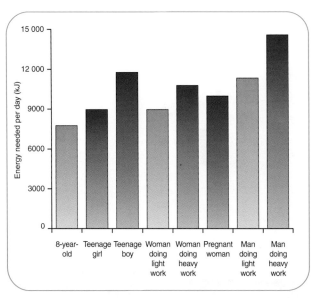

Figure 1b.1 Daily energy requirements.

Nutrient	Functions	Deficiency disease
carbohydrate (e.g. glucose, starch)	provides energy	
fat	provides energy; used as energy store (e.g. under the skin); provides heat insulation; protects internal organs	
protein	is used to build cells, so needed for growth and repair; can also be used as an energy source	kwashiorkor
vitamins (e.g. vitamin C)	vitamin C strengthens skin	scurvy
minerals (e.g. iron)	iron is needed for making haemoglobin, which is the red pigment inside red blood cells that transports oxygen	anaemia
fibre	keeps the muscles of the alimentary canal working well; also reduces the risk of bowel cancer	constipation

Table 1b.1 Nutrients.

Protein in the diet

Foods that contain protein tend to be more expensive than those made mostly of carbohydrate and fat. In many parts of the world, especially in some developing countries, people do not get enough protein in their diet. This is especially dangerous for young children, who are growing fast and need protein to make new cells. **Kwashiorkor** is most often seen in children between the ages of 9 months and 2 years, after they have stopped feeding on their mother's breast milk.

Children with kwashiorkor are underweight for their age. However, they may look quite fat,

Figure 1b.2 This child has kwashiorkor.

because the diet may contain a lot of carbohydrate and because fluid often accumulates in their abdomens. They are weak, because their muscles have not developed properly. If they are given a high-protein diet their health will improve, but they will probably never attain the height that they would have done with a good diet during their growing years.

Figure 1b.3 on page 10 shows the structure of a protein molecule. It is made of a long chain of smaller molecules called **amino acids**. There are 20 different kinds of amino acid. Any of these 20 can be joined together in any order to make a protein molecule. Each protein has its amino acids joined in a precise order. Even a small difference in the order of amino acids makes a different protein, so there are millions of different proteins that could be made.

So, to make all the proteins our cells require, we need all 20 amino acids. Cells are able to change some kinds of amino acids into others. However, there are eight amino acids that we can't make like this. They are called **essential amino acids**. We have to eat these in food.

Some of these essential amino acids are only found in proteins in foods that have come from animals. These are called **first class proteins**.

To help people to know how much protein they should eat, **recommended daily allowance** protein intakes can be calculated. They are often

War and famine

There is more than enough food in the world for everyone to eat and no-one to go hungry. So why do so many children suffer from deficiency diseases like kwashiorkor?

Often, fighting is to blame. In the rural areas of Kitgum in northern Uganda, four different groups of 'warriors' regularly rustle cattle and terrify people. They have AK47 rifles. The insecurity this causes has had severe impacts on education and health care. Money, drugs and food are often stolen by the 'warriors'. In 2000, at least 562 children under 5 died from kwashiorkor. Their mothers have done their

best. In a report made for Oxfam, Esther Luk is quoted as saying: 'The warriors' cattle eat all the cassava and the potato leaves, no one should complain or she will be shot.' Oscar Ogwang says: 'Even with relief food we sometimes get, the armed groups waylay the vehicles and disrupt the distribution.'

Oxfam knows that just pouring food aid into this area won't solve the huge health problems there. Something has to be done to alter the underlying situation. Maybe the guns can be got rid of. Maybe a political solution might persuade people to behave differently.

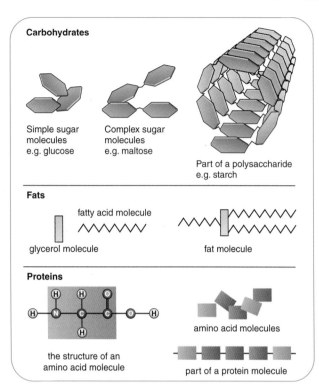

Figure 1b.3 The structures of some nutrient molecules.

known as **RDAs**, and you may have seen them listed on the side of cereal packets. As a rough guide, your RDA (in grams) for protein is 0.75 multiplied by your body mass in kilograms.

SAQ

2 A man has a body mass of 70 kilograms. What is his RDA for protein?

3 The RDA for protein for a teenage boy is often given with a multiplier of 0.9 rather than 0.75, and the RDA for a teenage girl is often given with a multiplier of 0.8. Suggest why these values are larger than the RDAs for adults.

4 Proteins aren't normally used to provide energy unless a person does not have enough carbohydrate or fat in the diet. Explain why.

Different diets

What you have for lunch today is probably not the same as everyone else. Maybe you don't care about your health, so you are eating salty chips, a fat-loaded burger or a chocolate bar. Maybe you are keen on sports and play a lot of football; you have a high-protein meal with plenty of complex carbohydrates, say pasta. Maybe your religion doesn't allow you to eat pork, so there's no bacon or sausages on your plate. Maybe you are allergic to peanuts, eggs or fish, so those are missing from your lunch. Maybe you are a vegetarian (you don't

The obesity time bomb

In Britain, it is estimated that 75% of adults are either overweight or obese. And things are getting worse. More and more young people are obese; childhood obesity has tripled in the past 20 years.

Why is this, and why does it matter? It probably happened because people take less exercise now (they travel by car rather than walk or bike, watch television or use computers rather than playing football in the park) and because we eat a lot of pre-prepared and fast food, which is often very high in fats. Doctors are talking about the 'obesity time bomb'. They are referring to the effect they expect this to have on people's health as they grow up. More people are developing type 2 diabetes (the kind that starts later in life – type 1 begins in young children), and at a much younger age. They will probably be more vulnerable to heart disease, arthritis and many other weight-related illnesses. England has the fastest growing obesity problem in Europe.

Figure 1b.4 This person is very overweight.

H eat meat) or a vegan (you don't eat anything that has come from animals, including milk or eggs) because you don't like the idea of animals being killed or used for food.

Whatever your religion, medical condition or personal choice, you can still choose a balanced diet that has all the right nutrients in it for your personal needs, in suitable proportions. If you are a vegan, you may be missing out on some essential amino acids, so you need to find out about those and take nutrient supplements. For most of the rest of us, supplements aren't necessary. Most research shows that supplements don't improve people's health at all, unless they have a specific medical condition that requires them. And sometimes they can be positively harmful. For example, vitamin A (which we need to help us to make a substance called rhodopsin, which our eyes need for night vision) shouldn't be taken by pregnant women because too much can harm their unborn child.

SAQ

5   Make a list of what you have had, or will have, for lunch today. (No cheating!)

   a   Is it a balanced meal? If not, why not? Can you make up the balance by what you eat at other meals today?

   b   Compare your meal with those of two other people in your class. What differences are there? What are the reasons for the differences?

Obesity

We have seen that, if a person eats more energy-containing foods than they need, the excess will be stored as fat in the body. If this amount becomes large, then the person may become very overweight, or **obese**.

Obesity is dangerous, because it increases the risk of getting various diseases. These include:

●   arthritis, in which the joints are damaged and may become so swollen and painful that movement is difficult

●   heart disease, in which the blood vessels supplying the heart muscle with oxygen may

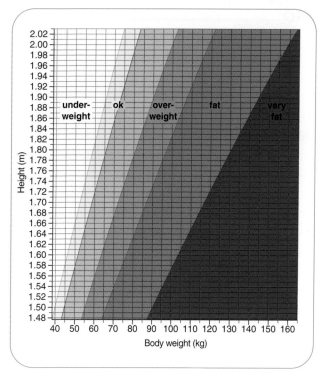

Figure 1b.5 Chart of body weight against height.

become narrow, increasing the risk of a heart attack

●   diabetes, in which the body loses control of blood sugar levels

●   in women, an increase in the risk of developing breast cancer.

To check if your body weight is roughly correct, you can calculate your **body mass index** or **BMI**. You need to know your height in metres and your body mass in kilograms.

BMI = mass in kg ÷ (height in m)^2

It has been found that the ideal BMI is somewhere between 20 and 24. If your BMI is less than 20, you are underweight. If it is between 25 and 30, you are overweight. If it is above 30, you are obese.

Worked example

Adam is 1.7 m tall. His body mass is 78 kg.

$$BMI = 78 \div 1.7^2$$
$$= 78 \div 2.89$$
$$= 27$$

Adam is overweight.

SAQ _____

6 Ingrid is 1.6 m tall. Her body mass is 57 kg. Calculate her BMI. Is she overweight, underweight or about right?

Dangers of dieting

Many people would like to lose weight. Some of them need to lose it, because their health will be harmed if they don't. But thousands of others worry about being too fat when they are exactly the opposite – they should be worrying about being too *thin*.

Some jobs put a great deal of pressure onto people to have a body weight well below what is normal for them. Jockeys have to make low weights in order to get rides on the best horses. Fashion models have to be skinny before they can get work or become famous. People who appear regularly on television may feel they need to be very thin in order to look good.

This obsession with being thin has spread out to many other people, too. A young person may have a strong desire to be perfect, to be liked and loved, and to feel they are in control of their lives. Sometimes this can lead to obsessive dieting. Not eating is a way of being in control. They look in the mirror and see a fat person, not the skinny one that everyone else sees.

Figure 1b.6 What does the mirror really show?

It is not difficult for this attitude to lead to a health-threatening weight loss. If the body does not get the nutrients that it needs, then it will raid them from parts of the body that aren't immediately essential for survival. So proteins will be taken from muscles and other organs, so that the heart can keep beating and the brain can keep working. Gradually, organs and systems in the body start to fail. Without treatment, the person will die.

The first stage of preventing all of this from happening is to recognise the problem. Usually, the dieter can't see it themselves, not even when someone else points it out to them. The second stage is to begin to slowly increase the quantity of nutrients taken into the body, to try to stop or reverse the damaging effects of starvation. The third and most important stage is to help the person to understand what has been happening, and to improve their self-image so that they are once more in control of their diet, but this time in a healthier way.

SAQ _____

7 More young women than young men suffer from this kind of health-threatening weight loss. Suggest possible reasons for this.

Digestion

Look back at Figure 1b.3 (page 10). Molecules of starch, fats and proteins are each made up of smaller molecules linked together. These big molecules cannot get through the walls of the **digestive system** and into the blood. So they have to be broken down into their smaller components. This is called **digestion**.

Physical and chemical digestion

In humans, digestion takes part in two stages. First, we break down large pieces of food into smaller pieces. This is called **physical digestion**. It starts in the mouth, where we chew food to crush it and mix it with saliva. Physical digestion continues in the stomach, where muscles in the stomach wall churn the food around and break it up even more.

Next comes **chemical digestion**. This is on a much, much smaller scale, because it involves breaking big molecules into small ones. This is controlled by **enzymes**. The enzymes are secreted (made and released) by cells lining the digestive system and by the pancreas. Figure 1b.7 shows how enzymes are involved in the digestion of starch (a carbohydrate), fats and proteins.

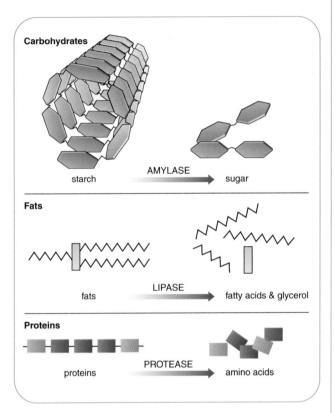

Figure 1b.7 Digestion of carbohydrates, fats and proteins.

The digestive system

Figure 1b.8 shows the structure of the human digestive system.

SAQ

8 Put your finger just outside the mouth of the person shown in Figure 1b.8. Trace the pathway food would move along as it travels through the digestive system.

 a Write down the parts of the digestive system that the food moves through, in the correct order.

 b The abdomen is the lower part of the body, below the diaphragm. Name two organs in the abdomen through which food does *not* pass.

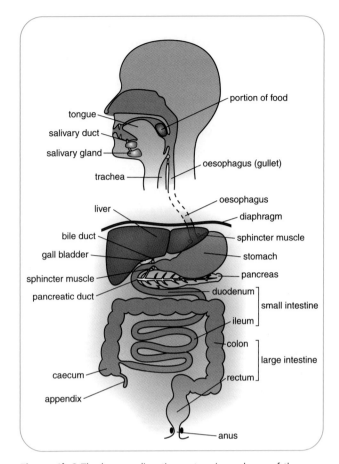

Figure 1b.8 The human digestive system is made up of the alimentary canal – the long tube along which food passes – plus the liver and pancreas.

In the **mouth**, teeth break down the food into smaller pieces and mix it with saliva. Saliva contains the enzyme **amylase**, and so some starch molecules are broken down to sugar in the mouth. Amylase is a carbohydrase, an enzyme that breaks down carbohydrates.

The food then travels down the oesophagus (gullet) to the **stomach**. Here, the churning movements of the stomach walls help to break down any lumps of food and mix them up well. The stomach wall secretes **protease**, so protein molecules start to get broken down to amino acids. It also secretes **hydrochloric acid**. This is needed to create the very low pH in which protease works best. The acid also helps to kill any bacteria in the food.

Food can be stored for several hours in the stomach. It then moves on into the **small intestine**. Here, juices made by the **pancreas** pour in. They contain amylase, protease and also **lipase**, so now fats are digested to fatty acids and glycerol.

Bile also pours in along the **bile duct**. Bile is made in the **liver** and stored in the **gall bladder**. Bile helps with the digestion of fats and it also helps to neutralise the acid coming through from the stomach. There are no enzymes in bile.

Bile and fat digestion

Fats are insoluble in water. This means that they form big globules when they are mixed with the water-based fluids inside the digestive system. This reduces the surface area available for lipase to get into contact with them.

Bile helps to solve this problem. When food arrives in the small intestine, bile flows down the **bile duct** and mixes with the food. It acts rather like washing-up liquid, helping to break the fat into little globules so it can mix into the watery fluids. This is called **emulsification**. It provides a large surface area that lipase can work on, breaking the fat molecules into glycerol and fatty acids.

Absorption

When food has been broken down into small molecules, these can diffuse through the wall of the digestive system. This is called **absorption**.

Most absorption happens in the small intestine. Here, blood and lymph (a clear fluid that flows in lymph vessels) are very close to the inside edge of the intestine wall. This makes it quick and easy for the small food molecules to diffuse from inside the intestine and into the blood or lymph. Sugars and amino acids go into the blood. Fatty acids and glycerol go into the lymph.

Water is also absorbed in the small intestine. However, there is still a lot of water left, and most of this is absorbed later on, in the colon.

Summary

You should be able to:

◆ explain that food is the energy source for living organisms

◆ describe what is meant by a *balanced diet*, and how a balanced diet varies according to a person's age, sex and activity

◆ discuss the ways in which religion, personal choice and medical conditions can influence a person's diet

◆ explain why you need protein in your diet, and why kwashiorkor is common in children in some parts of the world

◆ explain what is meant by *first class proteins*

◆ calculate recommended protein intake using the formula RDA in g = 0.75 × body mass in kg

◆ explain factors that can cause a person to become dangerously underweight

◆ explain what is meant by *obesity* and why it is dangerous

◆ calculate body mass index using the formula BMI = mass in kg ÷ (height in metres)^2

◆ know what is meant by *physical digestion* and *chemical digestion*

◆ explain why digestion is necessary and describe how carbohydrases, proteases and lipases digest carbohydrates, proteins and fats in the mouth, stomach and small intestine

◆ describe how bile helps with fat digestion

◆ know that small molecules are absorbed into the blood from the small intestine by diffusion

Questions

1 Copy and complete these sentences, to explain why you need different nutrients in your diet.

   a Carbohydrates and .... are the main sources of energy in the diet. We can also use .... for energy in an emergency.

   b Vitamin .... prevents scurvy. The mineral .... is needed to make the red blood pigment called ....

2 Copy this diagram. Draw lines from each kind of food to show the part of the digestive system where it is digested, and the enzyme that digests it. One line has already been drawn for you. Starch and protein are each digested in two places, so you will need to draw extra lines for them.

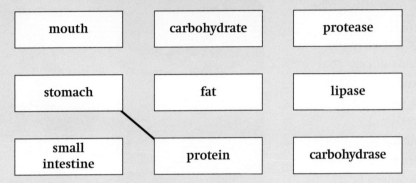

3 Table 1b.2 shows the nutrients in a tin of baked beans.

   a If you have a 200 g serving of baked beans, how many kilojoules of energy does that contain?

   b How much of the carbohydrate in 100 g of baked beans is not sugars? What kind of carbohydrate could this be?

Typical values	Per 100 g	Per serving (207 g)
Energy	306 kJ	633 kJ
Protein	4.6 g	9.6 g
Carbohydrate (of which sugars)	12.9 g (4.8 g)	26.8 g (9.9 g)
Fat (of which saturates)	0.2 g (trace)	0.4 g (trace)
Fibre	3.7 g	7.6 g
Sodium	0.3 g	0.7 g
Salt equivalent	0.9 g	1.8 g

Table 1b.2

   c Baked beans are high in fibre. Explain why this is an important part of a balanced diet.

   d Sally weighs 50 kg. Calculate her recommended daily intake (RDA) of proteins, using the equation

   RDA in g = 0.75 × body mass in kg

   e If Sally ate only baked beans, what mass would she need to eat to get her RDA for protein? Show how you worked out your answer.

   f Explain why eating only baked beans would not give you a balanced diet.

Diseases

Most of us get ill sometimes. Some illnesses come and go quickly, while others may last for a long time. Some diseases can be caught from someone else, while others just start up in one person's body and can't be passed on. Some illnesses are easily treated, while others can't yet be cured.

So what makes someone ill? There are many different reasons.

- A **deficiency** in the diet. For example, scurvy is caused by a lack of vitamin C. Anaemia can be caused by a lack of the mineral iron.
- A disorder arising in the **body**. For example, **cancer** is the result of mutations in cells that cause them to divide uncontrollably. Diabetes is the result of the pancreas not secreting insulin (type 1 diabetes) or the body cells not responding to it (type 2 diabetes).
- A problem with a person's **genes**. For example, red–green colour blindness is caused by a faulty gene. So are sickle cell anaemia and cystic fibrosis (see Item B1h).
- Disease-causing **micro-organisms**. For example, colds, measles, mumps, flu and AIDS are caused by viruses. Cholera and **tuberculosis** are caused by bacteria. Malaria and dysentery are caused by protozoa.

SAQ

1 Which of the types of diseases listed on the left do you think you could catch from someone else?

Cancer

Cancer can be caused when a cell's **DNA** (the material that genes and chromosomes are made of) changes or **mutates**. Some of our genes control

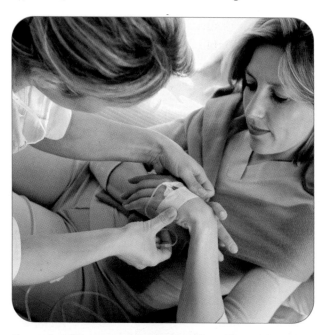

Figure 1c.1 This woman is being given chemotherapy for cancer.

Tuberculosis

Tuberculosis (TB) used to be a common disease in Britain. It is caused by a bacterium that infects cells in the lungs. The bacteria stay there for years and the person with TB gets progressively weaker. Without treatment, they often die.

When antibiotics were first used against TB in the 1940s, the effects were miraculous. People could be completely cured of TB. What's more, fewer people were getting it, mostly because living conditions had improved.

So most people forgot about TB and thought that it had gone away. Now, though, it is making a real comeback. The number of cases of TB in Britain is rising, and this is happening in many countries all over the world. It is estimated that 3 million people in the world die from TB each year.

Why is this happening? There are two main reasons. Firstly, the TB bacteria are becoming resistant to the antibiotics that are used against them. Secondly, more and more people in the world are becoming infected with HIV. This virus, which causes AIDS, damages the immune system. The result of this is that people with HIV/AIDS are open to infection from all kinds of other viruses and bacteria. As the number of people with HIV increases, so does the number of people with TB.

cell division – they tell a cell when to divide and when not to divide. If a mutation happens in one of these control genes then a cell may begin to divide repeatedly. These dividing cells form a lump called a **tumour**.

Some tumours are **benign**. The cells that form them just stay in one place. A benign tumour may cause problems if it gets so big that it harms the tissues around it. But most benign tumours are not dangerous and many can be removed by surgery, if necessary.

Other tumours are **malignant**. These tumours invade other tissues around them. They contain cells that are liable to break away and settle down in other parts of the body, where they start up other tumours. This is what cancer is.

What causes cancer? There are many different causes, although they always involve mutations in DNA. It seems that there probably has to be more than one mutation in a cell before it becomes cancerous. What's more, our own immune system is very good at detecting and destroying these cells before they get anywhere near causing cancer.

So when we think about reducing the risk of getting cancer, we need to consider how we can reduce mutations in our DNA. Some of the things we need to do are positive ones (that is, we ought to do something) while others are negative (that is, we ought *not* to do something). Three of the most important ways to reduce the risk of getting cancer include the following.

● Not smoking. Smoking cigarettes hugely increases the risk of getting cancer. More than 80% of cases of lung cancer happen in smokers. Smoking also increases the risk of many other kinds of cancer. One-third of all cancer deaths are related to smoking.

● Not getting sunburnt. Sunlight contains ultraviolet light. This can cause mutations in the DNA in skin cells. Many years later, these cells may become cancerous. If detected early, most skin cancers can be cured but one type, called melanoma, is very dangerous.

● Eating a good range of different foods in your diet. There is a lot of evidence that eating a diet containing plenty of fresh fruit and vegetables really helps to protect against cancer. It seems

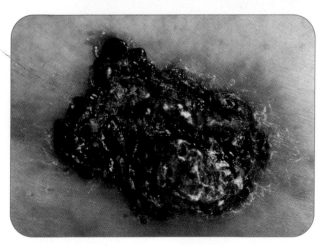

Figure 1c.2 A melanoma.

that taking lots of vitamin and mineral supplements doesn't have the same protective effect.

Records are kept of the numbers of people out of every 100 000 who get cancer (the **incidence rate**) and the number of people out of every 100 000 who die from it (the **mortality rate**).

Figure 1c.3 shows the incidence rates for men and women in 1971 and 2000. You can see that this has gone up for both. The graph also shows that more men get cancer than women, but that the increase in cancer for women is greater than that for men. The increase for women is largely because of increases in breast cancer and lung cancer.

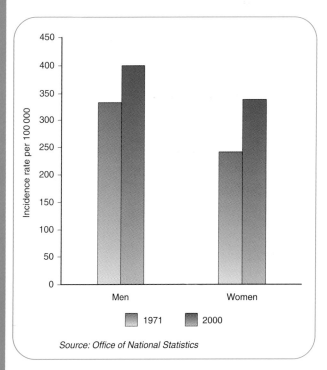

Source: Office of National Statistics

Figure 1c.3 Incidence of cancer in England in 1971 and 2000.

H *SAQ*

2   Figure 1c.4 shows the mortality rates for cancer for men and women in 1971 and 2000.

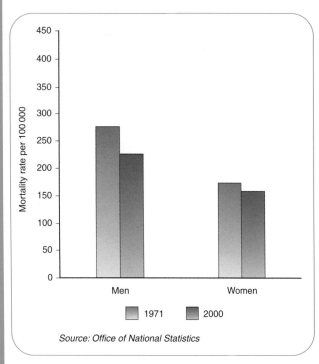

Source: Office of National Statistics

Figure 1c.4 Mortality rates for cancer in England in 1971 and 2000.

a   What has happened to mortality rates from cancer in men?

b   Is the trend for women the same as your answer to **a**?

c   Although more men still smoke than women, the number of men who smoke has fallen greatly since 1971, while the number of women who smoke has only begun to fall more recently. Suggest how this might explain some of the data shown in Figure 1c.4.

d   There have been great improvements in the treatment of breast cancer in recent years. How might this explain some of the data shown in Figure 1c.4?

H

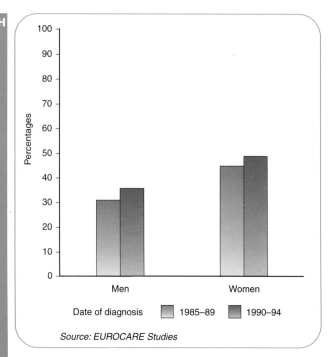

Source: EUROCARE Studies

Figure 1c.5 Five-year cancer survival rates for England in 1971 and 2000.

3   Figure 1c.5 shows the percentage of cancer patients who survived for five years or more after they had been diagnosed.

a   What percentage of male cancer patients diagnosed in 1985 survived for at least five years?

b   By how much had this improved for patients who were diagnosed in 1990?

c   Summarise the trends shown in Figure 1c.5.

d   Lung cancer is very difficult to treat. How might this be related to the trends in Figure 1c.5?

Infectious diseases

An **infectious disease** is one that you can catch from someone else. Infectious diseases are caused by **pathogens**. A pathogen is a micro-organism that causes disease. Figure 1c.6 shows the four kinds of micro-organism and an example of a disease caused by each of them.

Malaria

Malaria is caused by a protozoan (called *Plasmodium*). The protozoan lives in liver cells and red blood cells. It causes fevers that keep coming and going. Malaria causes millions of deaths each year, mostly in tropical, developing countries.

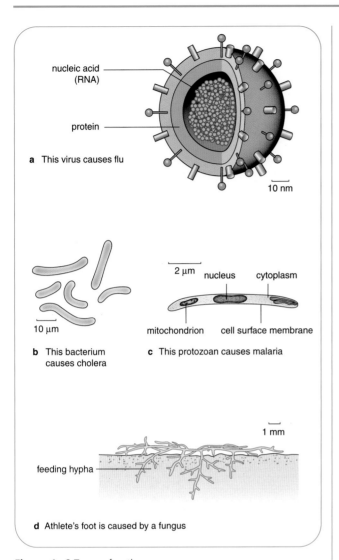

a This virus causes flu

nucleic acid (RNA)

protein

10 nm

b This bacterium causes cholera

10 μm

c This protozoan causes malaria

2 μm

nucleus   cytoplasm

mitochondrion   cell surface membrane

d Athlete's foot is caused by a fungus

1 mm

feeding hypha

Figure 1c.6 Types of pathogen.

The protozoan is a **parasite**. A parasite is an organism that lives in or on its **host**, feeds on it and does it harm. In this case, the host is the person it has infected.

You can get malaria if a mosquito bites you. Mosquitoes are the **vector** for malaria – they pass on the protozoan from one person to another. The kinds of mosquito that can transmit malaria live only in hot countries, so you are unlikely to get malaria in the UK. But if you go abroad for a holiday, it is important to check if there is a risk

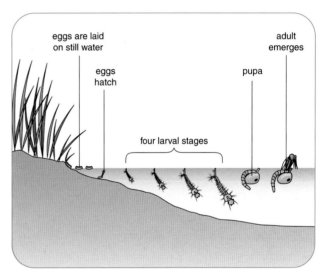

eggs are laid on still water

eggs hatch

four larval stages

pupa

adult emerges

Figure 1c.8 The life cycle of *Anopheles*, a mosquito which transmits malaria.

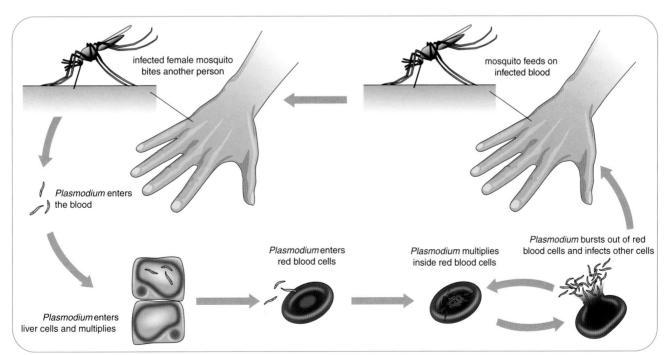

infected female mosquito bites another person

mosquito feeds on infected blood

Plasmodium enters the blood

Plasmodium enters liver cells and multiplies

Plasmodium enters red blood cells

Plasmodium multiplies inside red blood cells

Plasmodium bursts out of red blood cells and infects other cells

Figure 1c.7 How malaria is transmitted.

Figure 1c.9 Sleeping under a mosquito net is a good protection against malaria. The mosquitoes which are vectors for the disease are most active after dark.

of getting malaria. If there is, you can take drugs to reduce the risk of getting it.

SAQ

4 As global warming happens, it is possible that malaria could become commoner in Britain. Explain why this might happen.

H 5 People in countries where the mosquitoes that transmit malaria live try to avoid being infected.

Look at the list of precautions they might take. Explain how each method works.

- sleeping under a mosquito net
- keeping mosquitoes away from anyone who has malaria
- spraying a house with insecticide
- wearing long sleeves and long trousers in the evenings
- putting insect-eating fish into ponds
- clearing up any rubbish like old tyres that could fill up with rainwater

Keeping them out

Your body has many different defences that stop pathogens getting in. Table 1c.1 summarises them. Figure 1c.10 shows how the tubes leading down to the lungs help to keep bacteria and

viruses out. They have a lining called a **mucous membrane** – a layer of cells that makes sticky mucus. Bacteria in the air passing over these cells get stuck in the mucus. The beating **cilia** sweep the mucus upwards to the back of the throat, and you swallow it.

Method of entry	Natural defences
through skin	• the outer layers of skin are thick and strong, making it difficult for pathogens to get through • when skin is damaged, blood forms a clot to seal the wound and stop pathogens getting in • tears contain **lysozyme**, which kills bacteria on the surface of the eye
into the gaseous exchange system	• cilia and mucus in the trachea and bronchi trap pathogens and sweep them away from the lungs • white blood cells patrol the surface of the alveoli and ingest any pathogens they find
in food or water	• we don't like to eat food that smells or looks as though it might have pathogens in it • hydrochloric acid and protein-digesting enzymes in the stomach kill bacteria

Table 1c.1 How the body prevents infection.

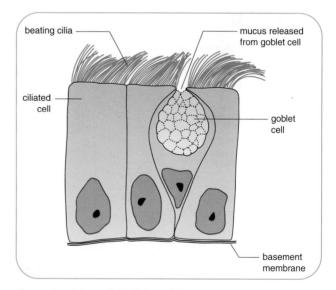

Figure 1c.10 Part of the lining of the respiratory passages.

Symptoms of an infectious illness

If pathogens do get into the body, they may breed inside you. They can breed really quickly, so just a few pathogens can become a million in a few hours. Sometimes they invade your cells and kill them. So, for example, you get a sore throat because viruses are damaging the cells there. Sometimes they produce harmful chemicals called **toxins** that spread around the body and damage cells in many different places. For example, the toxin produced by the cholera bacterium gives you really bad diarrhoea.

Immune system

Your **white blood cells** attempt to destroy pathogens inside your body. They are part of your **immune system** – the system in your body that protects you from invading pathogens.

Some white blood cells engulf (take in) and destroy pathogens (Figure 1c.11). Others produce chemicals called **antibodies** (Figure 1c.12). These chemicals go into the blood and are carried all around the body. Antibodies are a special kind of protein molecule.

Antibodies are produced when the white blood cells detect **antigens**. An antigen can be a substance on the surface of a pathogen, or it might be a toxin that a pathogen has made. Each white blood cell has just one antigen that it responds to. If it detects that antigen, then it starts dividing to form hundreds of white blood cells just like itself.

These white blood cells secrete antibodies that exactly fit that particular antigen. The antibody

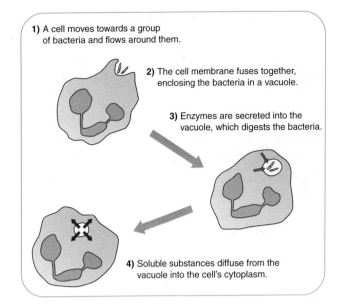

1) A cell moves towards a group of bacteria and flows around them.

2) The cell membrane fuses together, enclosing the bacteria in a vacuole.

3) Enzymes are secreted into the vacuole, which digests the bacteria.

4) Soluble substances diffuse from the vacuole into the cell's cytoplasm.

Figure 1c.11 A white blood cell engulfing bacteria.

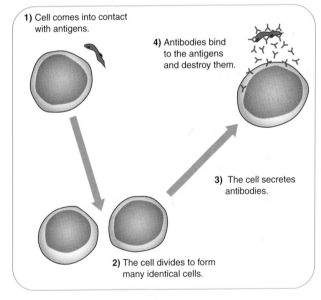

1) Cell comes into contact with antigens.

2) The cell divides to form many identical cells.

3) The cell secretes antibodies.

4) Antibodies bind to the antigens and destroy them.

Figure 1c.12 How white blood cells respond to antigens.

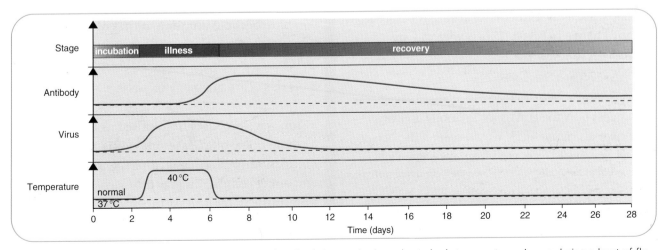

Figure 1c.13 How the number of viruses and amount of antibody in your body, and your body temperature, change during a bout of flu.

locks on to the antigen. If the antigen is on a bacterium then the bacterium may be killed. If it is a toxin then it will be neutralised. This is called the **immune response**.

SAQ

6 You will need to look at Figure 1c.13 to answer these questions.

    a How long after the person was infected by the virus did they become ill? Use the information on the graph to explain why they didn't feel ill straight away.

    b Suggest why the level of antibodies didn't start to rise until 4–5 days after infection.

    c Explain how the rise in antibody concentration helped the person to get over the illness.

Immunity

Often, if you have an infectious disease once, you won't get it again. This is true for mumps and measles. You are said to be **immune** to that disease.

The first time a pathogen enters your body, it takes a little while for the immune response to swing fully into action. This gives the pathogen a chance to breed, so you get ill.

However, not all of the white blood cells that were made in response to the antigen take on the job of making antibodies. Some of them just stay in the blood and become **memory cells**. They stay in your blood for a very long time after an infection. If that same pathogen gets into your body again, it is likely to be recognised by a memory cell really quickly. The immune response to the pathogen is immediate, killing it before it has any chance to breed. This is why you are immune to that disease.

SAQ

7 Suggest how the curves for antibody and virus, shown on Figure 1c.13, would differ if the person had previously recovered from the same illness.

Active and passive immunity

The kind of immunity in which your body has reacted to a pathogen, made antibodies and built up a population of memory cells is called **active immunity**. Your body has done it all by itself and has actively made its own future protection.

You can also become immune in a different way. When a baby is first born, its immune system is not very well developed and it cannot easily fight off infections. But during pregnancy, the mother's antibodies cross the placenta and get into her baby's blood. There are also antibodies in her breast milk, so a breast-fed baby gets more antibodies this way. This is called **passive immunity**. The baby isn't making its own antibodies, but just 'borrowing' some from its mother. This kind of immunity doesn't last long, because the antibodies soon break down. It does protect the new baby at a time when it is very vulnerable to all the new micro-organisms to which it is now exposed.

H Immunisation

Getting ill isn't the only way to become immune. You can be **vaccinated** (or immunised) against some infectious diseases.

A harmless form of the pathogen that causes the disease is put into your body. Usually, you have an injection but polio vaccine is given by mouth. The pathogen has its normal antigens on it and your white blood cells respond just as they would if it was a real, live, disease-causing pathogen. They quickly multiply, secrete (make and release) antibodies and produce memory cells. The memory cells remain in your body, so you have become immune to that disease.

SAQ

8 Is this active immunity or passive immunity?

You were probably immunised as a baby against a whole range of diseases including polio, measles, mumps, rubella, diphtheria and whooping cough. You may also have been immunised later against tuberculosis (TB) and, if you are a girl, against rubella. This is done because, if a woman gets rubella while she is pregnant, her baby may be deformed.

The MMR controversy

In 1988, a group of 13 researchers published an article in the medical journal *The Lancet*. They had been investigating a possible link between bowel diseases in children and autism (a condition in which a child has difficulties in interacting with other people). They looked at 12 children who had both conditions to see if they could find any reason why bowel disease and autism might be linked. One researcher thought it was possible that the MMR vaccine – against measles, mumps and rubella – might have something to do with it. He had absolutely no evidence for this. It was just his own idea. All the same, it got a mention in the article.

The media pounced on it. Newspapers ran huge headlines and melodramatic stories describing examples of children who had become autistic after having an MMR jab. (They often didn't point out that the age at which autism first shows is about the same as the age when children have their MMR jabs, so it is hardly surprising that the two might happen together.) The government denied there was any connection and urged parents to let their children have the vaccination.

Parents were worried. Very naturally, they wanted to do the best for their children. Very naturally, they were confused by the conflicting information they were getting. Many decided to play safe and not let their children have the MMR jab. What would you do in this situation if you had a new baby?

But was this really 'playing safe'? Measles, mumps and rubella are all very nasty diseases.

Figure 1c.14 Vaccination against MMR.

Children still die from these diseases in some countries. People had forgotten how unpleasant they were, because so many children had been vaccinated against them that scarcely anyone had actually seen a child with one of these illnesses. Not surprisingly, cases of all three started to rise.

In 2004, ten of the scientists whose names had been on the original article wrote another article saying that there was absolutely no link between the MMR jab and autism. Many other doctors and scientists have looked hard for a link and none has been found. But the genie is out of the bottle. Many parents are still fearful of letting their children have the vaccine.

Antibiotics

Sometimes, your immune system can't control and kill a pathogen quickly enough. If you have an infectious disease caused by a bacterium or a fungus, you can be given drugs called **antibiotics** to help you to fight against them. Antibiotics are chemicals that kill bacteria, and sometimes fungi, but don't usually harm you if you take them as instructed. Penicillin is an example of an antibiotic.

SAQ

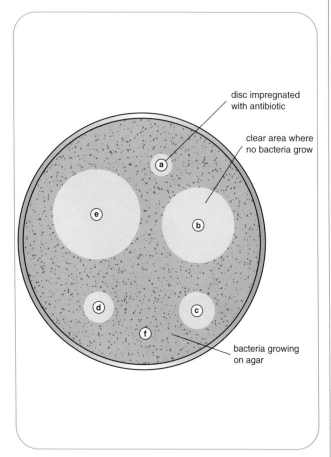

disc impregnated
with antibiotic

clear area where
no bacteria grow

bacteria growing
on agar

Figure 1c.15 Testing a range of antibiotics against a pathogen by disc diffusion.

9 Figure 1c.15 shows an antibiotic sensitivity test for a bacterium. The Petri dish has been filled with a jelly on which bacteria are growing. Little filter paper discs, each soaked in a different antibiotic, were placed on the jelly.

The light-coloured areas show where the bacteria weren't able to grow.

a Which antibiotic would you use to treat a disease caused by this bacterium?

b Suggest how the antibiotics spread out from the paper discs.

H Penicillin works by preventing bacteria forming cell walls. When a person infected with bacteria is treated with penicillin, the bacteria are unable to grow new cell walls and they burst open.

The population of bacteria in the person's body may be several million. By chance, one of them may mutate to a form that is not affected by penicillin. This mutant bacterium will have a

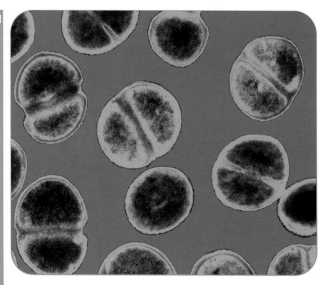

Figure 1c.16 MRSA.

tremendous advantage. It can reproduce while the other bacteria around it die. Soon, its descendants form a huge population of penicillin-resistant bacteria.

In hospitals, an especially dangerous strain of a common bacterium called *Staphylococcus aureus* has evolved. It is resistant to almost all antibiotics, including one called methicillin. It is called **MRSA** – methicillin-resistant *Staphylococcus aureus*. In most people, MRSA doesn't do any harm. But if someone is weakened by illness or is recovering from an operation, it can cause a dangerous infection. What's more, these people are often in hospital, and that is just where MRSA tends to be found – probably because lots of antibiotics have been used there. Many people who were already ill have picked up MRSA infections while in hospital and died as a result.

The more we use an antibiotic, the more likely it is that a strain of bacteria will evolve that is resistant to it. Doctors should not give you an antibiotic unless you really need it. Some kinds of antibiotics are kept in reserve, to be used only when all others have failed to cure someone who is seriously ill. And drug companies are constantly searching for new ones. A constant 'arms race' is on between us and pathogenic bacteria.

Testing new drugs

The big pharmaceutical companies are always working on producing new drugs. If they can make something new that really helps people in some way, they can make very large amounts of money from it. But it also takes very large amounts of money to develop and test a new drug.

Usually, the drug is first tested on animals or on human tissue that has been grown in a laboratory. Currently, there is a move away from testing drugs on animals, partly because many people don't think this is morally acceptable and also because you can't be sure that a drug will behave in the same way in a human as in another animal. Sometimes, computer models can be used to predict the effects of the drug. This is only useful if scientists understand thoroughly what the drug is doing, so that they can put all the relevant facts and information into the computer program.

If the drug passes all of those tests then it will be tried out on human volunteers. The aim of this test is not to find out if the drug works but to see whether it has any harmful side effects.

Figure 1c.17 Taking part in a drug trial.

Next, the drug will be tested on large numbers of people. Usually, this is done as a **double-blind test**. There are two groups of people. In one group, everyone is given the new drug. In the other, they are given a **placebo** – that is, a substance that has no effect on the body. They don't know which group they are in and nor do the doctors or other researchers who are giving them the drugs and recording the effects. The records of who is taking the drug and who is taking the placebo are kept secret from everyone participating in the trial.

The researchers can then analyse the results of the trial to look for any improvement in the

patients taking the drug compared with those taking the placebo. Quite often, there is no difference – and that will be the end of that drug. Sometimes, the drug will be shown to have a useful effect. This has to be balanced against any side effects that it has caused.

If the drug works, and if the side effects aren't too serious, then the drug will be released onto the market. Doctors will be allowed to prescribe it. All the time, though, doctors and researchers from the drug company will be watching out to see how the drug performs. Sometimes, serious side effects happen that weren't picked up during the trials. In that case, the drug will be withdrawn.

Summary

You should be able to:

- state that infectious diseases are caused by pathogens, and the types of pathogen – virus, bacterium, fungus or protozoan – that cause flu, cholera, athlete's foot and dysentery

- explain the meaning of the terms *host* and *parasite*, and describe how mosquito vectors spread malaria

- explain how we can control malaria

- describe how the body stops pathogens getting in

- explain how white blood cells in the immune system destroy pathogens

- explain how we can become immune to a disease, and the difference between active and passive immunity

- discuss the benefits and risks associated with immunisation

- explain why we should not use antibiotics unnecessarily

- describe how a new drug is trialled before it can be prescribed to patients

Questions

1   Copy this chart, and then draw lines to link each pathogen with the disease it causes.

bacterium		athlete's foot
fungus		dysentery
protozoan		flu
virus		cholera

2   Abbi had a bad cut on his hand. It had dirt in it. He went to the Accident and Emergency Department of the hospital to have it cleaned and stitched. The doctors were worried he might have got tetanus bacteria into it. They gave him an anti-tetanus injection, which contained antibodies that would stick to the toxin produced by the bacteria and make them harmless.

a   The injection made Abbi immune to tetanus. Was his immunity active or passive? Explain your answer.

b   A normal anti-tetanus injection, which you are given as a routine vaccination, contains weakened tetanus bacteria. Explain why it would have been no use to give Abbi this kind of injection.

c   Describe what Abbi's blood would do when he was cut, to help to stop bacteria getting into the wound.

3   A company has manufactured a new drug called fluticasone, which they hope will ease the symptoms of hay fever. They have carried out many different trials to find out if the drug works.

In one trial, they divided 200 volunteers – all of whom had hay fever – into two groups. One group was given fluticasone and the other was given a placebo. Neither the people collecting the results nor the volunteers knew which they were being given.

The volunteers were asked to score some of their symptoms using a scoring system where the higher numbers mean the symptoms were worse. Some of the results are shown in Table 1c.2.

continued on next page

Questions - *continued*

Measurement	Given fluticasone	Given a placebo
nasal congestion	1.43	1.63
sneezing	1.18	1.42
itching nose	1.14	1.35
runny nose	1.43	1.54

Table 1c.2

a   What is a placebo? Why is a placebo used in trials like this?

b   What is the name given to this kind of trial?

c   Suggest how the volunteers should have been divided into groups, to make sure the trial was fair.

d   Do the results suggest that fluticasone is a good treatment for hay fever? Use the data in the table to support your answer.

e   Suggest what other kinds of trials and measurements should be carried out on fluticasone before it is allowed to be prescribed to patients.

Sense organs

Most of us take our senses for granted, until something goes wrong with them. They are part of our **nervous system** – the cells and organs that transmit information between different parts of the body as **nerve impulses**. Our sense organs are the parts of the body that keep us in touch with our environment. They pick up information from the environment and transmit it to the central nervous system.

Changes in the environment that are detected by sense organs are called **stimuli** (singular: stimulus). Sense organs contain specialised cells that are sensitive to stimuli, called **receptor cells**.

The eye is a sense organ that is sensitive to light. It has receptor cells that respond to light by sending electrical impulses to the brain.

Sense organ	Stimulus detected	Name of the sense
eye	light	sight
skin	pressure, temperature and pain	touch
tongue	chemicals in food	taste
nose	chemicals in air	smell
ear	sound and movement	hearing and balance

Table 1d.1 Human sense organs.

Going blind

Figure 1d.1 shows the view that one grandfather has of his grandchildren.

He has an illness called retinitis pigmentosa. The sensitive cells in the retina, at the back of his eye, are gradually dying. As the disease progresses, his field of vision will get less and less, until eventually he may become totally blind.

Experiments have been taking place to try out an electronic way of helping blind people to see. Figure 1d.2 shows the retina of someone with retinitis pigmentosa who has volunteered to try out one of the very first retinal implants. It is a tiny chip containing 16 tiny electrodes, which stimulate his optic nerve. He can't see a clear image, but he can tell whether it is light or dark. This technology is in its very early days, but there is hope that eventually retinal implants may be able to restore useful sight to some blind people.

Figure 1d.1 This is how a grandfather with retinitis pigmentosa sees his grandchldren.

Figure 1d.2 A retinal implant.

SAQ

1 Many of the receptors listed in Table 1d.1 are in the head. Suggest why this is a good place to have them.

Structure of the eye

Figure 1d.3 shows what a human eye would look like if you sliced it in half.

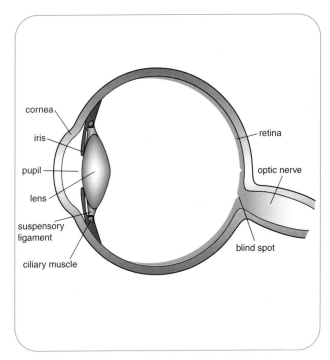

Figure 1d.3 Section through a human eye.

We have already seen that the part of the eye that is sensitive to light is the **retina**. When light falls onto the retina, the receptor cells are stimulated. They send electrical impulses along the **optic nerve** to the brain. These impulses travel at tremendous speeds, often as fast as 100 m per second. So it takes only a tiny fraction of a second for the impulse to get from your eye to the brain. This is vital if you are to respond quickly to what you see.

The receptor cells in the retina transfer the energy in the light into electrical energy in the optic nerve. Most people have three kinds of receptor cells that respond to different colours – red, green and blue. Some people have only two kinds of these cells and they can't tell the difference between red and green. This is called **red–green colour blindness** and you can read more about it on pages 60–61.

The parts of the eye in front of the retina – that is, on the left of Figure 1d.3 – must all be transparent, so that light can pass right through them. The coloured part of the eye, called the **iris**, does not let light pass through it. In the centre of the iris is a gap called the **pupil**. The size of the pupil can be controlled by the muscles in the iris. The pupil is made large in dim light and small in bright light.

Focusing light

Some of the parts of the eye in front of the retina have the function of focusing light onto it. You can think of the retina as a screen. If the light isn't sharply focused then the image on the retina is blurry and your brain 'sees' a blurred image.

Figure 1d.4 shows how parallel rays of light are focused onto the retina. The light rays are **refracted** as they pass into the eye. The **cornea** is responsible for most of the refraction. Its curved shape means that it acts like a convex lens, refracting (bending) the light rays inwards.

The rays then pass through the **lens**, which is where fine adjustments to the focusing are made.

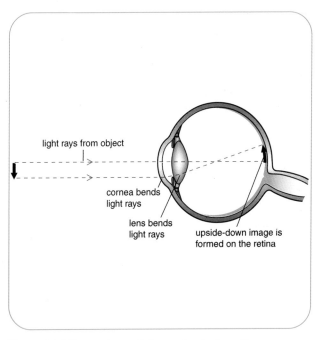

Figure 1d.4 How an image is focused onto the retina.

SAQ

2 The lens is a tissue made of living cells, but they are not supplied with blood through capillaries as most tissues are. Instead, they get their nutrients and oxygen from the fluids that fill the eye. Suggest why the lens does not contain blood vessels.

The image that is made on the retina is upside down. The brain sees the image the right way up. People have done experiments where they have worn special glasses that turn the image on the retina the right way up. For several days, the brain sees the world upside down. But eventually it learns that it has to deal with the image differently and stops turning it over. When a person takes the glasses off, the brain has to learn all over again to turn the image over.

Accommodation

Accommodation is the name for the way in which the eye adjusts the focusing of light, so that a sharp image is formed on the retina.

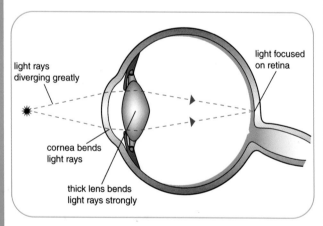

light rays diverging greatly

light focused on retina

cornea bends light rays

thick lens bends light rays strongly

Figure 1d.5 Focusing on a nearby object.

Figure 1d.5 shows an eye that is focused on a nearby object. The light rays falling onto the eye are spreading apart from each other, or diverging. If you look carefully at the diagram, you can see that the rays are refracted as they pass through the cornea and then again as they pass through the lens. The lens is very curved, which increases the refraction of the light rays. The curvature of the lens is exactly right to ensure that the rays are brought to a focus on the retina.

Figure 1d.6 shows an eye that is focused on a more distant object. The light rays are not diverging as much when they reach the eye, so less refraction is needed to focus them. The cornea is a fixed size and shape, so the adjustments have to be made by changing the shape of the lens. You can see that the lens is now less curved than when the eye is focused on a near object.

Ciliary muscle

The lens is held in position by a ring of **suspensory ligaments**. You can see these in side view and front view in Figure 1d.7. (You also have ligaments at your joints, holding the bones together.)

The ligaments are fixed to the lens and also to a circle of muscle called the **ciliary muscle**. This is the muscle that is responsible for changing the shape of the lens.

When the muscle is relaxed, it makes a wide circle around the lens. The suspensory ligaments are pulled taut. This keeps the lens stretched out, so that it is quite thin. This is how your eyes are when they are relaxed. In this state, they are focused on distant objects. Most people find it more

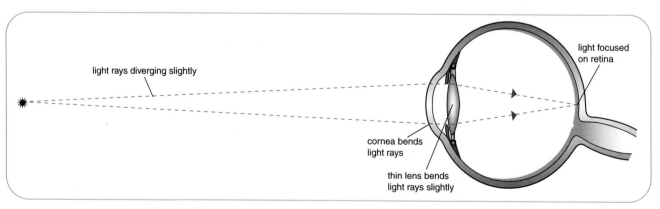

light rays diverging slightly

light focused on retina

cornea bends light rays

thin lens bends light rays slightly

Figure 1d.6 Focusing on a distant object.

restful to look into the distance than to concentrate on something that is close to their face.

When the muscle contracts, it gets shorter. This means that the ring of muscle gets smaller. As it squeezes inwards, there is less tension on the suspensory ligaments. Now they are not pulling outwards on the lens. The lens falls back into its natural shape, which is more spherical. In this state, the eye is focused on a nearby object.

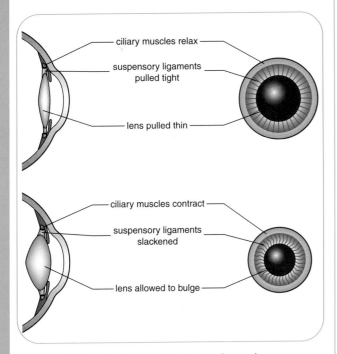

Figure 1d.7 How the shape of the lens is changed.

SAQ

3 Suggest why people who work at a computer screen for long periods are advised to look up and into the distance at intervals as they work.

Poor sight

How many people in your class wear glasses or contact lenses? In one year, more than 5 million people in England have their eyes tested and over 3 million pairs of glasses or contact lenses are prescribed.

Some people are able to focus on things that are far away but cannot focus on nearby objects. They have **long sight**. Long sight happens when the cornea and lens don't bend the light rays enough. The rays still have not met when they hit the retina. This may happen because the eyeball is too short, or because the lens can't be made fat enough.

In young people, **short sight** is a common problem. If you are short sighted, you can easily see things close to you, such as the writing on this page, but can't focus on things a long way away. Short sight happens when the cornea and lens bend the light rays too much, so that they come to a focus before they reach the retina (Figure 1d.9). This may happen because the eyeball is too long, or because the lens can't be stretched out thin enough.

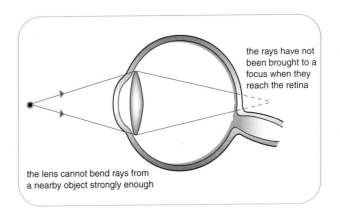

Figure 1d.8 Long sight.

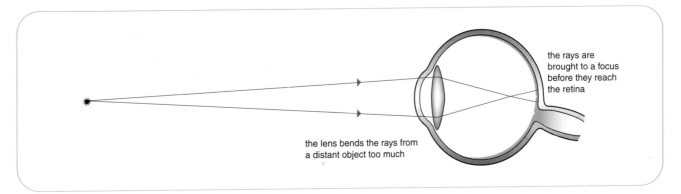

Figure 1d.9 Short sight.

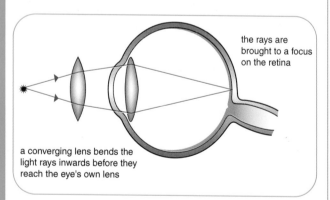

the rays are brought to a focus on the retina

a diverging lens bends the rays outwards before they reach the eye's own lens

Figure 1d.10 Short sight is corrected by wearing glasses or contact lenses that make light rays diverge more before they hit the cornea.

As people age, the lens becomes less elastic. When the suspensory ligaments pull outwards on it, it can't be stretched as much as when the person was young. When the ligaments are slackened, the lens doesn't go back to as fat a shape as it used to do. So older people often have problems in focusing on both distant and nearby objects.

Correcting poor sight

Short and long sight can be corrected by wearing glasses or contact lenses that bend the light rays before they reach the cornea and lens. Figures 1d.10 and 1d.11 show how this is done.

the rays are brought to a focus on the retina

a converging lens bends the light rays inwards before they reach the eye's own lens

Figure 1d.11 Long sight is corrected by wearing glasses or contact lenses that bend light rays inwards before they hit the cornea.

SAQ

4 Elderly people often wear bifocal glasses. These have lenses divided into two parts. At the top is a concave lens, and at the bottom is a convex lens. Suggest how this helps elderly people to see better.

Some people choose to have surgery to correct their sight. The surface of the cornea is reshaped with a laser so that it bends light rays less or more. This can mean that the person does not have to wear glasses or contact lenses.

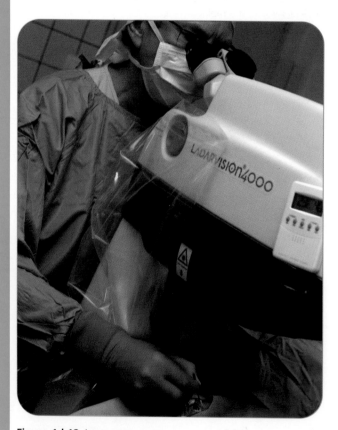

Figure 1d.12 Laser eye surgery.

Binocular vision

Binocular vision means having both eyes looking at the same object. Humans have excellent binocular vision. Our eyes both face forward (Figure 1d.13). They each give us a slightly

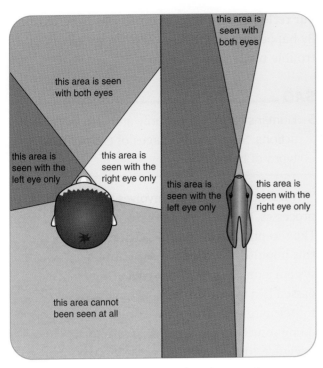

Figure 1d.13 Binocular vision. How does the area of binocular vision of a rabbit differ from that of a human? How does its overall field of vision differ?

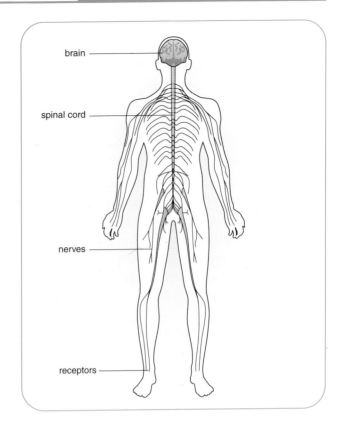

Figure 1d.14 The human nervous system.

different image of the object on which they are focused. The closer the object is to us, the more different the image is. The brain can use the differences to judge how far away an object is.

Many animals don't have binocular vision. A rabbit, for example, has eyes on the side of its head, rather than in front. Each eye therefore sees a completely different image. This is called **monocular vision**. Rabbits can't easily judge distances. But they can see much further to the side and rear of the head than we can. This is really helpful to them in spotting predators.

The nervous system

Figure 1d.14 shows the human **nervous system**.

The brain and spinal cord make up the **central nervous system**. The **nerves** and **receptors** make up the **peripheral nervous system**.

Neurones

Many of the cells in the nervous system are highly specialised cells called **neurones**. Figure 1d.15 shows a neurone that carries impulses from the central nervous system to a muscle or gland. It is called a **motor neurone**. The impulse is carried along the **axon**.

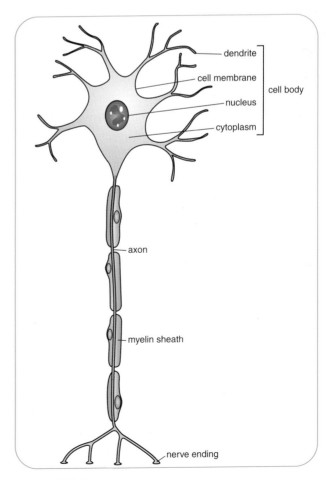

Figure 1d.15 A motor neurone.

Neurones are superbly adapted for carrying electrical impulses extremely rapidly from one part of the body to another. Some neurones have axons that are more than a metre long.

In the neurone shown in Figure 1d.15, the impulse is carried from the top to the bottom. The impulse will begin in one of the dendrites, which will receive it from another neurone. The many branching dendrites keep the neurone in touch with many other neurones around it.

The myelin sheath around the axon insulates it from electrical activity in any other axons nearby. It also helps the impulse to be carried up to 50 times faster than in an axon with no sheath.

Reflex actions

Much of what we do is intentional. We decide to stand up or sit down, to talk or to eat, to read a book or to dance. We call these actions **voluntary**. They are under the conscious control of the brain.

But some of our actions happen without us thinking about it. For example, if you unexpectedly touch a very hot object with your hand, you will quickly pull your hand away. It happens so fast that you have probably pulled your hand away before you realised what was happening.

An action like this – fast, automatic, happening without any conscious thought – is called a **reflex action**. Other examples include jerking the leg upwards when someone hits you on your kneecap and the change in size of your pupil when you move from dim light to bright sunshine. Many of

our reflex actions help to protect us from harm. By happening so quickly, they can get us out of trouble really fast.

SAQ

5   Humans can act voluntarily and also by reflex actions. Suggest advantages of each of these types of behaviour.

A reflex action always involves the same kinds of processes. First, a stimulus is received by a receptor. The receptor starts up an electrical impulse, and this impulse is carried to the central nervous system along a sensory neurone. The impulse is passed to a motor neurone, which carries it to a muscle or gland. The muscle or gland responds by doing something. It is called an **effector**.

Figure 1d.16 shows the pathway that a nerve impulse takes as you respond to touching a very hot object. It is called a **reflex arc**. This reflex passes through the spinal cord, so it is an example of a **spinal reflex**.

Synapses

If you look very closely at the neurones in Figure 1d.16, you will see that the three neurones do not quite connect with one another. There is a small gap between each pair. These gaps are called **synapses**. Figure 1d.17 shows a synapse in more detail.

Inside the sensory neurone's axon are hundreds of tiny **vesicles**. They contain a

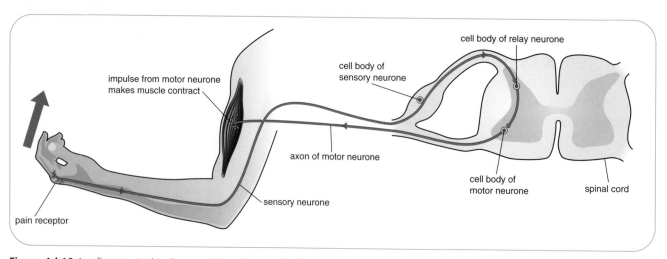

Figure 1d.16 A reflex arc. In this diagram, the spinal cord is shown as it would look if you cut it across and looked at it end on. Follow the pathway taken by the impulse as it begins at the receptor and ends at the muscle (the effector).

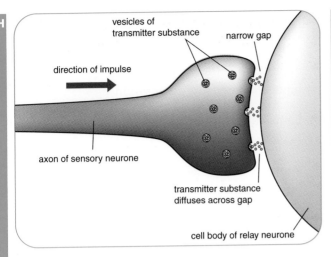

Figure 1d.17 A synapse.

chemical called a **transmitter substance**. When an impulse arrives, it makes these vesicles empty their contents into the space between the two neurones. The transmitter substance diffuses across the gap and slots into receptors in the cell membrane of the next neurone. This triggers an electrical impulse in that neurone.

SAQ

6 Using what you can see in Figure 1d.17, explain why a nerve impulse can only cross a synapse in one direction.

Snake and spider venoms

Many snake venoms contain chemicals that interfere with the way that synapses work. For example, the South American rattlesnake has a venom called crotoxin. Crotoxin reduces the release of a transmitter substance called acetylcholine. This transmitter substance is especially important for transferring nerve impulses to muscles, to make them contract. So crotoxin can cause paralysis.

The venom of the black widow spider does exactly the opposite. It enhances acetylcholine release. Black widow spider venom is 15 times more toxic than rattlesnake toxin. It causes severe pain, sometimes feeling like appendicitis. Even though the venom is so toxic, black widow spiders are small and only inject tiny amounts of it, so their bite does not usually kill.

Figure 1d.18 A black widow spider.

Summary

You should be able to:

- state the type of information gathered by each of the body's sense organs

- name and label the parts of the eye, and describe their functions

- describe the pathway of light through the eye

H - explain how the eye focuses light

- describe the causes of long sight and short sight

H - explain why older people cannot focus as well as young people and describe how long and short sight can be corrected

- describe the differences between binocular and monocular vision

- name and label the main parts of the nervous system

- know that neurones carry electrical impulses along axons

H - explain how neurones are adapted for their function

- explain what is meant by a *reflex action* and describe a reflex arc

H - explain how synapses work

Questions

1 Copy and complete these sentences, using Figure 1d.3 to help you.

   a The part of the eye that contains light-sensitive cells is the ....

   b The .... and the .... refract (bend) light rays as they pass through the eye.

   c The .... nerve carries electrical impulses from the eye to the brain.

2 Joe is padding around in bare feet when he steps on a nail. His leg pulls upwards so quickly that he nearly falls over.

   a What is the name for a fast, automatic reaction like this?

   b Write down the receptor and the effector in this reaction.

   c Write down the three kinds of neurones, in the correct order, that carry the electrical signals in this reaction.

Figure 1d.19

H 3 a Josh is short sighted. Explain what this means.

      b Describe how glasses can help Josh to see more clearly. Draw a diagram if it helps your answer.

      c Explain why older people often have trouble in focusing on objects at different distances.

Drugs and you

Types of drug

A **drug** is a substance that changes the chemical processes in your body. Many drugs are helpful. Many are used in medicine. But some are misused, damaging the health and lives of people, and of their families and their friends.

Drugs can be classified according to the general effects that they have on your body.

- **Depressants** slow down the activity of parts of the brain. They include alcohol, solvents and temazepam.
- **Painkillers** reduce sensations of pain, sometimes by blocking nerve impulses. They include aspirin and heroin.

Figure 1e.1 Drugs can be helpful.

- **Stimulants** increase activity in the brain. They include nicotine, ecstasy and caffeine.
- **Performance enhancers** increase an athlete's abilities to run quickly, throw a long way or keep going for longer. They include anabolic steroids, which increase muscle development.
- **Hallucinogens** produce strange pictures and ideas in someone's brain, distorting what is really seen or heard. They include cannabis and LSD.

Some of these drugs are easily available and it is legal to take them. Many of us use aspirin for headaches and drink cola or coffee (which contain caffeine), and many people drink alcohol. Some smoke cigarettes, which contain **nicotine**. Nicotine and alcohol are sometimes said to be **social drugs**. This means that it is acceptable to use them – even though we know that they can be harmful.

Other drugs are only available in hospital or on prescription. For example, heroin is a powerful painkiller, used in medicine to treat severe pain. It is illegal for anyone other than doctors to possess or use this drug. Temazepam is prescribed to people who have difficulty sleeping, or are very anxious.

How stimulants and depressants work

Stimulants and depressants change the way that synapses in the brain function. (If you have forgotten about synapses, look back to page 34.)

Stimulants increase the activity of the brain. Some of them do this by increasing the level of a transmitter substance called **dopamine** in the neurones in the brain. This can make a person feel happier, more talkative and more energetic. It also makes the heart beat faster and blood pressure increase.

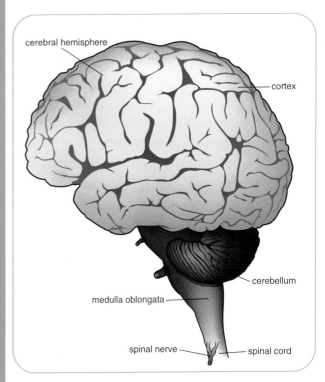

Figure 1e.2 External view of a human brain.

Depressants decrease the activity of the brain. They do this in many different ways. Alcohol increases the effect of a transmitter substance called **GABA**. GABA tends to slow down the activity of neurones. At the same time, it decreases the effect of another transmitter substance called **glutamate**. Glutamate tends to increase the activity of neurones. So, overall, alcohol decreases the activity of neurones in the brain. This happens especially in the cortex, the part of the brain used for logical thinking and decision making. It also affects the cerebellum, the part of the brain at the back of the head that is responsible for control of movement.

Misusing drugs

Drugs can be enormously helpful. Without antibiotics, thousands more people would die from bacterial infections. Without painkillers, there would be much needless suffering.

But many drugs can do harm, especially if they are not used carefully. Some drugs are powerful and can cause big changes in the brain. They can also affect other parts of the body, especially the liver, because this is the organ that has the role of breaking down harmful substances so that they won't harm other organs. As the liver works to destroy drugs and toxins, it can be damaged. Eventually, the damage can be so great that the liver cannot work properly.

Drugs that are misused are classified into three groups: Class A, Class B and Class C (Table 1e.1).

- **Class A drugs** are the most dangerous. Someone convicted for possessing them can be given up to 7 years in prison. Someone convicted for supplying them can be imprisoned for life.
- **Class B drugs** are still dangerous, but not quite as much as Class A drugs. Sentences for

possessing or supplying them are similar to those for Class A drugs.

- **Class C drugs** are still dangerous, but less so than Class B drugs. Someone can be sentenced to 2 years imprisonment for possessing them and 14 years for supplying them.

Addiction

Class A and some Class B drugs are **addictive**. This means that a user can come to depend on them, and cannot do without them. They may suffer **withdrawal symptoms** if they stop using the drug. These can be extremely unpleasant – they may feel sick and anxious, be unable to sleep, have constantly watering eyes and nose, and feel pain all over the body. This can go on for days.

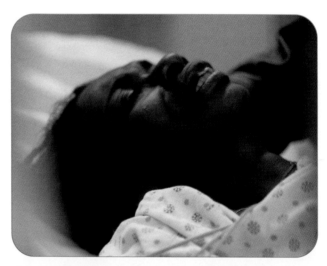

Figure 1e.3 Withdrawal symptoms can be extremely unpleasant.

Cocaine is an especially addictive drug. Some people become addicted to it after just one use.

Being addicted to an illegal drug can completely change a person's life. They can often think of scarcely anything other than where they will get their next dose from. Their habit costs money and they may have to resort to crime to get enough to be able to pay for their drugs.

These problems are made even worse because people often need to take more and more of a drug to get the same effect. This is called **tolerance**. The body learns how to deal with the drug more effectively, so an increasing dose is needed to achieve the same feelings as before.

It is really hard for someone who is addicted to a drug to stop taking it. They are likely to need a

Class	Examples
A	heroin, ecstasy, LSD
B	amphetamines (a kind of stimulant)
C	cannabis, anabolic steroids, temazepam

Table 1e.1 Classes of illegal drugs.

Classifying cannabis

For many years, cannabis was classified as a Class B drug. You could be sentenced to almost as long in prison for possessing cannabis as you could for possessing heroin.

Many people argued that cannabis should be reclassified as a Class C drug. There didn't seem to be all that much evidence that cannabis was harmful. Indeed, many people with long-term diseases, especially multiple sclerosis, insisted that cannabis helped them to feel better. Others argued that many young people tried cannabis, found that it didn't seem to do them any harm and they didn't become addicted to it, and so thought that people must also be exaggerating the danger of Class A drugs.

One of the arguments for keeping cannabis classified as Class B was that some statistics showed that if you smoked cannabis you were more likely to go on to using Class A drugs. Others argued that this didn't actually mean that smoking cannabis *caused* people to use Class A drugs. It might have been that the kind of people who used cannabis were simply people who were likely to misuse other drugs as well.

The British government eventually decided to reclassify cannabis as a Class C drug. This happened in 2004. As luck would have it, almost immediately afterwards, new and very convincing evidence emerged that young people who smoked cannabis were much more likely than non-cannabis users to develop the serious mental condition schizophrenia. Some people are worried that the reclassification to a Class C drug has given out the wrong message. It suggests that smoking cannabis is completely safe, which we now know is not true.

lot of help. They may need medical support to help them to avoid or to get through withdrawal symptoms. They may need more support to help them to rebuild their life, get back to work and find a lifestyle in which they feel more happy, secure and able to avoid drugs in the future. This is called **rehabilitation**.

Smoking

Tobacco smoke contains the drug **nicotine**. Nicotine is a stimulant and is addictive. It is addiction to nicotine that makes it so difficult to give up smoking tobacco once you've started.

Everyone knows that smoking is dangerous and can lead to fatal diseases. These include emphysema, bronchitis and cancer (of all kinds, but especially of the lungs, mouth, throat and oesophagus). Smoking also increases the risk of developing heart disease.

What is in tobacco smoke?

Tobacco smoke contains a very large number of different chemicals. Four are known to be causes of illness. They are nicotine, carbon monoxide, tar and particulates.

Figure 1e.4 Now that offices are non-smoking areas, people who have difficulty giving up the habit have to smoke outdoors.

We have already seen that **nicotine** is the addictive drug in tobacco. Nicotine also affects the circulatory system by making blood vessels get narrower. This increases blood pressure, which puts extra strain on the heart. Nicotine increases the risk of a heart attack or a stroke.

Carbon monoxide decreases the amount of oxygen that can be carried in the blood. Oxygen is transported by **haemoglobin** inside red blood cells. But haemoglobin combines much more readily with carbon monoxide than with oxygen. What's more, once it has combined, it is very reluctant to let the carbon monoxide go. So a lot of the haemoglobin of a smoker is taken out of action, permanently combined with carbon monoxide. This means that less oxygen is delivered to tissues. If a smoker exercises, they have to breathe faster and their heart must beat faster than that of a non-smoker to get the same quantity of oxygen to their muscle cells.

Both nicotine and carbon monoxide can harm the development of a foetus in a woman's uterus. Both of them can cross the placenta and get into the baby's blood. This has a similar effect to the fetus smoking a cigarette. The baby is likely to be smaller than average at birth and more likely to suffer several diseases.

Figure 1e.5 It's best not to smoke when you are pregnant.

Tar contains chemicals that are **carcinogens** – substances that can cause cancer. Carcinogens in tar cause genes to mutate, which can make cells begin to divide out of control (see Item B1c, page 16). Even low tar cigarettes contain enough carcinogens to seriously increase the risk of getting cancer.

Particulates are tiny pieces of carbon and other materials. They cause irritation in the lungs. White blood cells try to remove them. White blood cells and other protective cells secrete chemicals that are supposed to help to defend the person against pathogens, but in this case they often do a lot of harm to other body cells.

Bronchitis and emphysema

Your lungs have their own protection systems in place. We have seen that the mucous membrane in the trachea and bronchi helps to stop bacteria and other particles getting down into the lungs. Unfortunately, carbon monoxide and other chemicals in tobacco smoke stop this system from working. They stimulate the goblet cells to make even more mucus and, at the same time, stop the cilia from beating. The extra mucus just trickles down and accumulates in the lungs. This is why smokers cough – they are trying to get rid of this mucus.

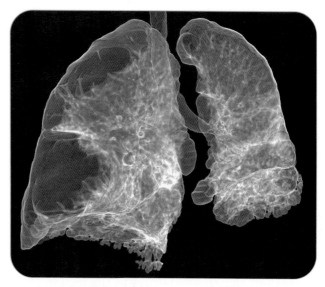

Figure 1e.6 Smoking damages your lungs.

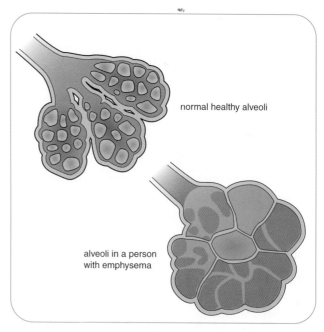

normal healthy alveoli

alveoli in a person with emphysema

Figure 1e.7 The alveoli on the left are healthy; those on the right have been damaged by smoking.

With these defences out of action, bacteria begin to accumulate in the lungs. They often breed in the mucus, both in the lungs and in the bronchi. They cause inflammation and the person has **bronchitis**. Smokers often develop chronic bronchitis – that is, they have bronchitis practically all the time.

The constant coughing damages the delicate walls of the alveoli. Other chemicals in the smoke cause the walls of the alveoli to break down. This reduces the surface across which oxygen and carbon dioxide can be exchanged with the blood. The smoker needs to breathe faster to get enough oxygen into the blood. This condition is called **emphysema**. The longer someone smokes, the worse

the emphysema gets. Many smokers end up unable to get enough oxygen unless they breathe it from a cylinder. Emphysema can eventually confine a person to bed – they just don't have enough energy to be able to stand up, let alone walk around.

Alcohol

We have seen that alcohol is accepted as a 'social drug'. Unfortunately, this has meant that many people aren't aware that it is dangerous if it is misused.

People drink alcohol at least partly because others drink it and it makes them part of the crowd. It can make people feel relaxed and able to talk and join in more easily.

SAQ

1 These graphs show the incidence of lung cancer in men and women, and the percentage who smoked, between 1951 and 2002.

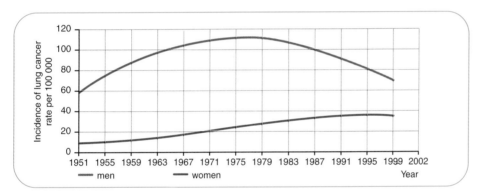

Figure 1e.8 Incidence of lung cancer.

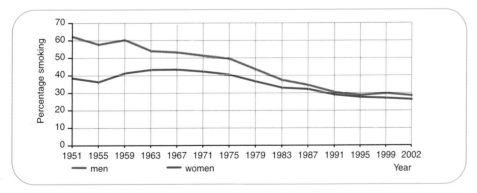

Figure 1e.9 Percentages of men and women who smoked.

   a Describe the change in lung cancer incidence in men between 1951 and 2002.

   b Describe the changes in smoking in men between 1951 and 2002.

   c Suggest how the two changes you have described in **a** and **b** could be linked.

   d Compare the smoking rates for women between 1951 and 2002 with those for men.

   e Use the data in the graphs to suggest reasons for the differences you have described in **d**.

Effects of alcohol

Alcohol is a depressant and slows down the parts of the brain that are involved in decision making and coordination of movement. Drinking a large quantity of alcohol can cause blurred vision and difficulty in speaking clearly. It can make you sleepy. It causes the blood vessels in your skin to dilate (get wider), which allows more heat to be lost from your skin. Each year, people die from hypothermia after getting drunk and falling asleep outside in the cold. Moreover, alcohol is a toxin, and someone who drinks a very large quantity of alcohol can fall into a coma and die, usually because the alcohol inhibits the function of the parts of the brain that control breathing movements.

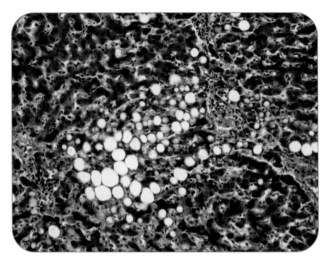

Figure 1e.10 Alcohol can cause liver damage. The white patches are fat globules inside the liver cells. The pink areas are bands of fibrous tissue. If either of these grow much larger, the liver will not be able to work properly.

Alcoholism

George Best was born in Belfast in 1946. He had an amazing talent as a footballer and joined Manchester United in 1963. Many people think he was perhaps the best footballer in the world.

Best was a great celebrity: he was good looking, a brilliant and exciting player and had an outgoing personality. He became a favourite with the crowds and the media. But he handled his fame badly. He led an extravagant lifestyle, and this involved heavy use of alcohol.

We don't know what makes some people likely to become addicted to alcohol while others don't, but Best was one of the unlucky ones. He came to be unable to manage without alcohol and drank so much that he rapidly destroyed his health.

For people who easily become addicted to alcohol, the only way to avoid ruining their lives is to stop drinking completely. But Best did not do this. At the age of 37, he was imprisoned for drink driving. By the late 1990s, his liver was so badly damaged by alcohol that he was told he would die within a few years. He was lucky – after an 8 month wait, a suitable liver was found for him and he had a successful liver transplant at the age of 56. But within months he was drinking heavily again, destroying the new liver that he had been given.

He died in 2005.

Figure 1e.11 George Best was a talented footballer but he suffered from alcohol addiction.

These are all short-term effects. Someone who drinks large amounts of alcohol on a regular basis also has a high risk of other, long-term effects. One of these is liver disease. One of the roles of the liver is to break down toxins in the body. In the process, the liver's own cells become damaged. The person has **cirrhosis** of the liver, which can kill.

Another long-term effect of alcohol is a heightened risk of heart disease. Alcohol increases blood pressure, so people who regularly drink large amounts of alcohol are more likely to have a heart attack or a stroke.

Drinking and driving

To be a good driver, you need to be alert, to be able to make quick decisions and to have fast reactions. A driver who has drunk alcohol has none of these qualities. In Scotland, it is estimated that one in five road accidents is due to drink driving.

When you drink alcohol, it quickly goes into the blood. The legal blood alcohol limit for a driver in Britain is 80 mg of alcohol in 100 cm^3 of blood. For an airline pilot, it is 20 mg of alcohol in 100 cm^3 of blood. It is even better to have no alcohol in your blood if you are driving. Many young people follow that rule. Convictions for drink driving are more frequent among middle-aged men than among young drivers.

SAQ

2   These tables show some statistics about alcohol and road traffic accidents between 1979 and 2002.

1979	1989	1990	1991	1992	1993	1994	1995	1996	1997	1998	1999	2000	2001	2002
1640	810	760	660	660	540	540	540	580	550	460	460	530	530	560

Table 1e.2 Fatal casualties in collisions where one or more driver or rider was over the legal limit.

1979	1989	1990	1991	1992	1993	1994	1995	1996	1997	1998	1999	2000	2001	2002
32%	19%	18%	19%	20%	19%	21%	19%	21%	17%	15%	17%	18%	18%	21%

Table 1e.3 Percentage of drivers killed in Great Britain whose blood alcohol level was known and who were over the legal limit.

1979	1989	1990	1991	1992	1993	1994	1995	1996	1997	1998	1999	2000	2001	2002
41	102	113	119	109	106	110	119	159	189	210	215	213	202	196

Table 1e.4 Breath tests administered following injury accidents in Great Britain (thousands).

1979	1989	1990	1991	1992	1993	1994	1995	1996	1997	1998	1999	2000	2001	2002
34%	10%	9%	8%	7%	7%	7%	6%	5%	4%	4%	4%	4%	4%	4%

Table 1e.5 Proportion of breath tests that were positive or refused.

a   Describe the trend in fatal casualties between 1979 and 2002 for collisions in which one or more driver or passenger was over the legal limit.

b   Compare the trend you have described in **a** with the figures for the percentage of drivers killed who were over the legal limit.

c   Use the data in the two tables about breath tests to suggest reasons for the trends you have described in **a** and **b**.

Summary

You should be able to:

◆ give examples of depressants, painkillers, stimulants, performance enhancers and hallucinogens, and describe their general effects

◆ explain the meanings of *addiction*, *withdrawal symptoms*, *tolerance* and *rehabilitation*

H ◆ explain how depressants and stimulants affect synapses

◆ describe the legal classification of drugs

H ◆ discuss the legal classification of drugs in the context of national policy

◆ explain how smoking affects health

◆ explain how drinking alcohol affects the liver and nervous system

Questions

1   Health authorities recommend that women drink no more than 14 units of alcohol a week, and men no more than 21 units. A unit is:

   one glass of wine

   one measure of spirits (e.g. gin, rum, whisky)

   one 330 ml bottle of alcopop

   half a pint of beer or lager.

a   Emma always drinks wine with her evening meal but doesn't drink alcohol at other times. How many glasses are the most she should have with each meal?

b   Jon drinks two pints of beer every evening and two measures of vodka about three times a week. Should he be cutting back?

The 'units' system is easy to remember, but it isn't very accurate because different drinks have different alcohol content and you don't always get the same volume of wine or beer in a drink. You can work out the number of units of alcohol in a drink like this:

units of alcohol = (% of alcohol in the drink × volume of drink in ml) ÷ 1000

c   Emma's wine has an alcohol content of 12% and her glass holds 300 ml. How many units of alcohol are there in one glass of wine? What does this suggest about the simple 'one glass of wine = one unit of alcohol' system?

continued on next page

Questions - *continued*

2 Table 1e.6 shows the smoking habits of mothers having their first baby in Scotland in the early 1990s. It also shows the length of their pregnancy and the birth weight of their babies.

Smoking habit	Birth weight of baby (g)			Number of babies
	1000 or less	1000–1500	1500–2500	
never smoked	31	126	986	1143
smoking now	24	70	700	794
given up smoking	5	19	147	171

Table 1e.6

a For mothers who never smoked, the percentage of babies who weighed less than 1000 g is:

$$\frac{31 \times 100}{1143} = 2.7\%$$

Do the same calculation for the percentage of babies who weighed less than 1000 g born to mothers in the other two smoking categories.

b Draw a bar chart to display these results.

c Use your bar chart, and the table, to discuss the effects of smoking on birth weight.

3 a Give two examples of drugs that act as stimulants.

b How do stimulants affect the body?

c Heroin is a Class A drug. Explain what this means and why heroin is placed in this class.

d Discuss the reasons for the reclassification of cannabis from a Class B drug to a Class C drug. Do you think this was a good decision? Explain the reasons for your decision.

Staying in balance

Homeostasis

In one day, your body can experience a wide range of different conditions. You might have to work in a very hot room and then go outside where it is cold and wet. You might get up so late that you have no time for breakfast and be so hungry at lunchtime that you eat an enormous meal. You might drink three cups of tea in the morning and then not drink anything else until the evening.

But the cells inside your body cannot be allowed to experience all of these changes. For them, it is important that their temperature, their food supply and their water content stay roughly constant. This allows them to function at their best.

Humans, like all mammals, keep the internal conditions in the body roughly constant, no matter what happens in their external environment. This includes:

- temperature
- blood glucose level
- carbon dioxide (CO_2) level
- water level.

Keeping internal conditions constant is called **homeostasis**. It is done using automatic control systems – we don't have to think about it, the body just gets on with it. As a general rule, it is done by balancing inputs and outputs. For example, if you have too much carbon dioxide in your blood, you will breathe faster to get rid of it through your lungs. If you drink more water than you need then the kidneys will get rid of more in the urine (Figure 1f.1).

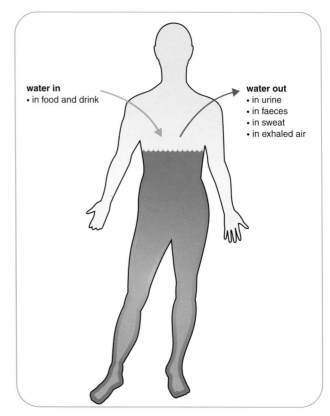

water in
• in food and drink

water out
• in urine
• in faeces
• in sweat
• in exhaled air

Figure 1f.1 Water balance. Water makes up around 65–70% of the body.

Hypothermia

Sadie was found at 9.00 a.m. on Boxing Day morning. She was unconscious and barely alive. Outside it had been snowing and the temperature overnight had fallen to −6 °C. Inside Sadie's house it didn't feel much warmer. The heating in her house was turned right down. Her neighbours said she worried about paying the gas bills and often turned it off completely.

Sadie couldn't get around very easily, so she spent most of her time sitting in her armchair. Lately, she had even found it difficult to put the kettle on so she hadn't been making her usual cups of hot tea.

Now she had hypothermia – a body temperature so low that her brain could no longer function normally. If she hadn't been found that morning, she would probably have died.

Figure 1f.2

 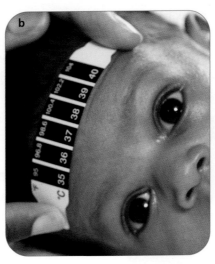 

Figure 1f.3 Ways to measure body temperature: **a** thermometer in the mouth; **b** sensitive strip on forehead or finger; **c** digital recording probe in ear.

Temperature control

One of the first things that would happen to Sadie once she was found would be measurement of her body temperature. Figure 1f.3 shows some of the different ways in which this can be done.

Your body temperature is kept at around 37 °C. This is the temperature at which your enzymes work best – their **optimum** temperature. Much lower than this and you can suffer from **hypothermia**, which is when metabolic reactions slow down so much that you may die. Much higher than this and enzymes inside the cells will be denatured (damaged). If the body can't cool itself down, you can suffer from **heat stroke**, which can also kill. This is especially likely to happen if you have been sweating so much that you have become seriously dehydrated.

We maintain our body temperature by balancing heat inputs and heat outputs (Figure 1f.4). If the body temperature begins to rise too high then heat inputs are lowered and heat outputs increased. If it begins to fall too low then heat inputs are increased and heat outputs lowered.

Animals such as humans that keep their body temperature constant are said to be **homeothermic**.

SAQ

1 Suggest some reasons why hypothermia is much more likely to happen in an elderly person than in a young and active one.

When temperature drops too low

If the body temperature drops below 37 °C then these responses will take place.

● Your muscles may start to contract and relax very quickly – they **shiver**. To do this, they have to release energy from glucose by respiring. This generates heat, which is transferred into your blood and carried all over the body. You may also decide to jump up and down, which has the same effect.

● You may put on extra clothing. Clothes trap a layer of warm air next to your skin. Air is a poor conductor of heat, so this helps to reduce the rate of heat loss from your body.

heat inputs
- *respiration* generates heat inside cells by breaking down glucose
- *radiation* from the sun or a fire
- *conduction* from hot objects in contact with the body, including hot food and drink

heat outputs
- *radiation* from the skin
- *conduction* for example from the body into cold water
- *evaporation* for example of water in sweat

Figure 1f.4 Heat inputs and heat outputs.

H ● The arterioles (small arteries) that deliver blood to the skin surface get narrower. This is called **vasoconstriction**. The blood is diverted to flow deeper so that less heat is lost from it to the air by radiation (Figures 1f.5 and 1f.6).

Figure 1f.5 This thermal image shows where infrared radiation is being given out from a person's body.

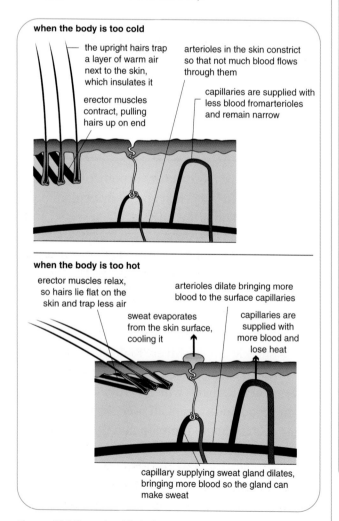

when the body is too cold

the upright hairs trap a layer of warm air next to the skin, which insulates it

erector muscles contract, pulling hairs up on end

arterioles in the skin constrict so that not much blood flows through them

capillaries are supplied with less blood fromarterioles and remain narrow

when the body is too hot

erector muscles relax, so hairs lie flat on the skin and trap less air

sweat evaporates from the skin surface, cooling it

arterioles dilate bringing more blood to the surface capillaries

capillaries are supplied with more blood and lose heat

capillary supplying sweat gland dilates, bringing more blood so the gland can make sweat

Figure 1f.6 How the skin helps in temperature regulation.

SAQ

2 When you are cold you may get goose pimples, as your hairs stand on end. This doesn't help to keep you warm but, in a furry animal, it can help a lot. Using what you know about methods of heat transfer, explain how it can do this.

When temperature rises too high

If the body temperature rises about 37 °C, these responses will take place.

● The sweat glands in your skin secrete extra sweat. This watery liquid lies on the surface of the skin. The water in it evaporates. It needs heat to do this, which it takes from the skin and therefore cools the skin down.

H ● The arterioles supplying the skin surface get wider. This is called **vasodilation**. This brings more blood close to the surface so that more heat can be lost by radiation.

Negative feedback

The part of the body that keeps track of your temperature is the **hypothalamus**. This is a tiny organ that hangs beneath the brain, almost exactly in the middle of the head. It constantly monitors the temperature of the blood that flows through it. If this temperature departs from normal then nerve impulses are sent to the arterioles, sweat glands and muscles to bring about the responses described above.

Figure 1f.7 outlines how this kind of control system works. It is an example of **negative feedback**. 'Feedback' refers to the fact that, when the hypothalamus has made your body take action to increase heat loss, information about the effect of these actions is 'fed back' to it as it senses the drop in blood temperature. It is called 'negative' because this feedback stops these actions.

SAQ

3 Make a copy of Figure 1f.7. Alter the words in the boxes and between the arrows so that they apply particularly to the control of body temperature.

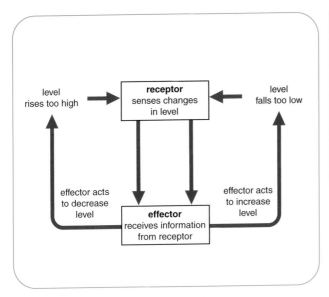

Figure 1f.7 A negative feedback control loop.

Hormones

Hormones are chemicals that are **secreted** (made and released) by glands in your body called **endocrine glands**. Like neurones, they carry information from one part of the body to another.

Figure 1f.8 shows the position of the main endocrine glands and some of the hormones that they secrete.

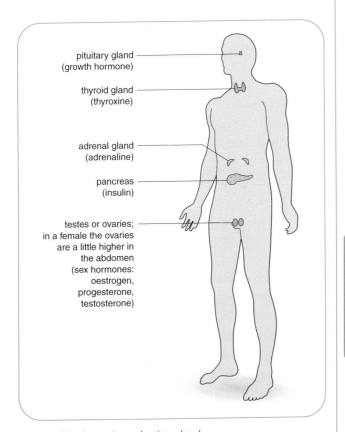

Figure 1f.8 The main endocrine glands.

Hormones are carried around the body dissolved in the blood plasma. Although they arrive at every organ, they only affect certain ones. These are called their **target organs**. Because blood travels more slowly than nerve impulses, the body's reactions to hormones are usually slower than reactions to nerve impulses.

Controlling blood sugar levels

Hormones are responsible for controlling blood sugar levels.

The sugar **glucose** is the **carbohydrate** that is transported in our blood. Cells need glucose as fuel to provide them with energy. They get the energy by breaking down the glucose in respiration. Most cells need a fairly constant supply of glucose. Brain cells are especially sensitive to blood glucose levels.

Two organs work together to control blood glucose levels. These are the **pancreas** and the **liver**.

If blood glucose levels go too high – for example, if you have just eaten food with a lot of sugar in it – the pancreas senses this and secretes the hormone **insulin**. Insulin is carried in the blood to the liver. It causes the liver to take glucose out of the blood and store it.

SAQ

4 What is the target organ for insulin?

Later, if you haven't eaten for a while and your blood glucose levels fall, the pancreas will stop secreting insulin. The liver can break down some of its stores and release glucose back into the blood. In this way, your blood glucose levels stay roughly constant.

When the liver stores glucose, it changes it into a **polysaccharide** called **glycogen**. (Look back at Figure 1b.1 to remind you about polysaccharides.) The glycogen is stored away inside the liver cells.

When more glucose is needed by the body tissues, the glycogen is broken down to glucose again and released into the blood.

Sometimes, this system doesn't work. When the body cannot control blood glucose level, a person has **diabetes**. This can be because the pancreas stops secreting insulin when it should or because the liver cells don't respond to insulin. At the moment, there is no cure for diabetes. Some forms of diabetes can be controlled by eating very carefully, making sure that your food intake never allows blood glucose levels to go too high or too low. Other forms may require regular injections of insulin.

Figure 1f.9 Steve Redgrave has won five Olympic gold medals even though he has diabetes.

Some people with diabetes inject themselves with insulin each day. They have to judge when they need an injection and what dosage they need. This depends on several things, including:
● how much glucose their body is using – which depends on how much exercise they are doing
● how much glucose is getting into their blood from their digestive system – which depends on how much carbohydrate-containing food they have eaten.

If someone has a regular routine each day and has a diet that doesn't change much from day to day, then they probably need the same dose of insulin each day. But many people don't have lives like this. They need to monitor their blood glucose levels so that they can give themselves an appropriate dose of insulin.

The best way of monitoring blood glucose is to take a tiny blood sample – usually done by pricking your finger – and then use an electronic monitor to give you a readout of your blood glucose concentration. This helps you judge whether you have been injecting the right dose of insulin and whether you need another injection soon.

SAQ
5 Make a copy of Figure 1f.7. Alter the words in the boxes and between the arrows so that they apply particularly to the control of blood sugar.

Sex hormones
In both men and women, three hormones called **testosterone**, **oestrogen** and **progesterone** are secreted. In men, the main hormone is testosterone, secreted by the testes. In women, the main hormones are oestrogen and progesterone, both secreted by the ovaries.

These hormones are secreted from birth, but only at very low levels until a person reaches puberty. Then a surge in production causes the **secondary sexual characteristics** to develop.

In both males and females, the sex hormones cause pubic hair to grow and also hair in the armpits, while in males hair growth also happens on the face. Testosterone also causes a boy's voice to break and his body to become more muscular. Sperm production begins in boys. In girls, breasts develop and hips widen. A girl's periods start.

Figure 1f.10 shows how oestrogen and progesterone are involved in the **menstrual cycle**. In this cycle, the lining of a woman's uterus grows thick and full of blood vessels as an egg matures in one of her ovaries. This is timed so that, if the egg is fertilised, the uterus lining will be ready and waiting for the tiny embryo to sink into it and begin its nine months of development.

If the egg is not fertilised, the uterus lining breaks down and is gradually lost through the vagina. This is called a period, and lasts somewhere between 3 and 9 days for most women.

In Figure 1f.10, you can see that the levels of oestrogen and progesterone change as the cycle progresses. The quantity of oestrogen secreted starts to rise just after menstruation ends. It stimulates the repair of the uterus lining and you can see that, as the oestrogen level rises, so does the thickness of the lining. Ovulation happens as oestrogen reaches its peak level.

As the oestrogen level falls, the progesterone level begins to rise. Progesterone helps to maintain the thick uterus lining. Progesterone begins to fall during the last week of the cycle and, when the levels of both hormones are at their lowest, the uterus lining begins to break down and the cycle starts again.

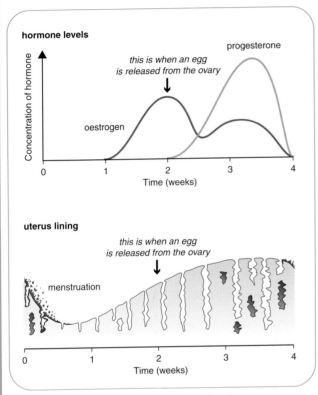

Figure 1f.10 Hormones and the menstrual cycle.

Sex hormones can be used to increase a woman's fertility – that is, to help her to conceive a child. They are also used in the contraceptive pill, which stops her from ovulating so that she cannot conceive.

Fertility treatment

About one in every seven couples has difficulty in conceiving a child. Most of them do eventually succeed but, for some, **fertility treatment** may be needed.

First of all, tests will be done to find out where the problem lies. About 32% of cases are caused by something that is wrong with the man's sperm and 29% with the woman's eggs. (The rest are caused by a combination of factors or else no reason can be found.)

Sometimes, the problem is that the woman is not producing eggs. If so, then she can be given **sex hormones** to stimulate eggs to ripen in her ovaries. If this works then eggs will go from her ovaries into her oviducts, and she can conceive in the normal way.

One hormone that is often used in fertility treatment is called **clomiphene**. Clomiphene stimulates the pituitary gland to secrete two hormones called **LH** (luteinising hormone) and **FSH** (follicle-stimulating hormone). These encourage eggs to ripen in the ovaries and then to be released into the oviduct.

It is difficult to get the dose of the hormones just right. Give the woman too little and she won't be able to conceive. Give her too much and she may produce two or three eggs instead of one. Women who are given this kind of fertility treatment often have twins or triplets.

Contraception

Sex hormones can also be used to *stop* a woman conceiving if she doesn't want to have a child but does want to be sexually active. The hormones that are used are **oestrogen** and **progesterone**. They are taken as pills, so they are called **oral contraceptives**. ('Oral' means that you take them by mouth.) There are many different types but, in most, the woman takes a pill daily for 21 days and then stops for 7 days. During those 7 'no-pill' days, she has a period.

The oestrogen in the pills stops her from producing eggs. Her body behaves as though she was pregnant. However, used on its own, oestrogen can increase the risk of getting some types of cancer. Taking progesterone as well reduces this risk.

Summary

You should be able to:

◆ state that the body keeps temperature, water, oxygen and carbon dioxide at steady levels

◆ explain what is meant by *homeostasis*

Ⓗ ◆ explain how negative feedback mechanisms are involved in homeostasis

◆ describe how body temperature can be measured

◆ describe how the body gains and loses heat, and the dangers of very high or very low body temperatures

◆ explain how sweating reduces body temperature

Ⓗ ◆ explain why it is important to keep the body at 37 °C and how the brain and blood vessels help to control body temperature

◆ name and locate the pancreas, ovaries and testes, and name the hormones they secrete

◆ describe the effects of male and female sex hormones, and how they can be used in the contraceptive pill and as fertility drugs

Ⓗ ◆ describe how progesterone and oestrogen control the menstrual cycle, and how they can be used in contraception and fertility drugs

◆ describe how insulin controls blood sugar levels, and the causes of diabetes

Ⓗ ◆ explain how insulin controls blood sugar levels and how a person with diabetes can judge the dose of insulin they need

Questions

1 Copy this diagram. Then draw lines to link each hormone to the gland that produces it and to its effect.

ovaries	testosterone	causes liver to take sugar out of the blood
pancreas	oestrogen	causes male secondary sexual characteristics
testes	insulin	helps to control the menstrual cycle

continued on next page

Questions - *continued*

2   Figure 1f.11 shows the levels of glucose (sugar) and insulin in the blood after a meal containing glucose.

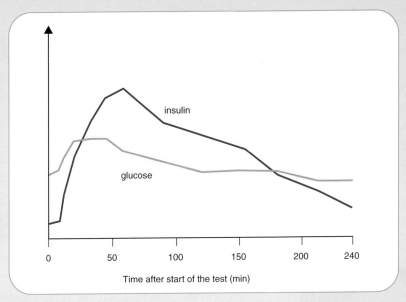

Time after start of the test (min)

Figure 1f.11

a   What happens to the glucose level in the first 20 minutes after eating the meal?

b   What happens to the insulin level in the first 50 minutes after eating the meal?

c   Explain what makes the blood glucose level go up.

d   Explain what makes the insulin level go up.

e   What happens to the blood glucose level between 50 and 240 minutes?

f   Suggest what makes this happen.

3   Patti fell into a river on a cold day. She was pulled out soaked to the skin and shivering.

a   Explain how having wet clothes would increase heat transfer from Patti's body to the environment.

b   Patti was at risk of getting hypothermia. Explain what hypothermia is and why it is dangerous.

c   Suggest two things that should be done as quickly as possible to prevent Patti's body temperature from falling too much and explain how they would work.

Chromosomes and genes

What makes you who you are? At least part of the answer is your **genes**. You have about 30 000 of them, and the chances that your particular collection of 30 000 is the same as anyone else's is practically zero – unless you have an identical twin.

Genes are made of **DNA**. Inside the nucleus of each of your cells, you have 46 long, coiled strands of DNA, called **chromosomes**. A gene is a length of DNA that codes for a particular characteristic – which means that the gene can produce that characteristic. There are many hundreds of genes on each chromosome. This is true for every species of living organism. Different species have different numbers of chromosomes. We have 46, fruit flies have 8 and cauliflower plants have 18.

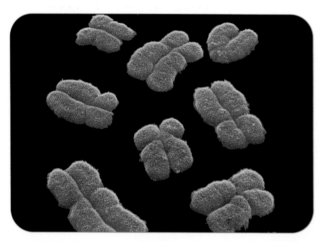

Figure 1g.1 Human chromosomes.

SAQ

1 Arrange these in order of size, smallest first: chromosome, cell, nucleus, gene.

Genetic code

The DNA in your chromosomes carries information in the form of coded instructions, called the **genetic code**. DNA has very long molecules, made up of thousands of **bases** all linked together in a long line. There are four different bases, called A, C, T and G, standing for adenine, cytosine, thymine and guanine (although you don't need to remember those names). The sequence of these bases is the genetic code.

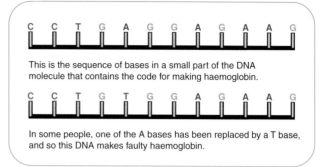

This is the sequence of bases in a small part of the DNA molecule that contains the code for making haemoglobin.

In some people, one of the A bases has been replaced by a T base, and so this DNA makes faulty haemoglobin.

Figure 1g.2 Part of the genetic code.

SAQ

2 Imagine a length of DNA that is made up of just three bases. How many different sequences of bases could there be?

The genetic code provides instructions to the cell about making proteins. For example, you have genes in your cells that tell them how to make keratin (a protein in your hair and skin), haemoglobin (for carrying oxygen in your blood) and amylase (to digest starch in your mouth).

Many of these proteins are enzymes. And because enzymes control the metabolic reactions in your cells, this means that your genes control all your cells' activities. So the base sequence in a gene that instructs the cell about making keratin will determine how that keratin is made, and therefore whether you have straight, wavy or curly hair.

Each of your cells contains every one of your genes. But not every cell needs to *use* all of these genes. For example, a cell in your stomach does not need to use the gene that provides instructions for making keratin. A cell in your skin does not need to use the gene that provides instructions for making amylase. In each cell, a particular set of genes is switched on and used. So, although all the cells in your body contain exactly the same genes, the cells can end up being very different from one another.

Chromosomes and reproduction

When an organism reproduces, it is essential that the new organisms get complete sets of genes. There are two different ways of achieving this.

The Human Genome Project

In 1990, a bold international programme began. It was called the Human Genome Project and its aim was to find out the complete sequence of bases in the DNA in all 23 kinds of human chromosome.

The sequence was published in 2003. It turned out that there were about 3 billion bases altogether in the full human DNA sequence. It's been difficult, though, sorting out which particular length of bases makes up particular genes, and there is still uncertainty about exactly how many genes there are. Of the genes that have been identified, we only know the functions of around half. Chromosome 1, which is the longest chromosome, not surprisingly has the largest number of genes, currently thought to be 2968.

In **asexual reproduction**, just one parent produces offspring by itself. For example, when a spider plant grows new plants on the end of long stems, it makes the new plants by producing lots of exact copies of its own cells. This process is called **mitosis**. Each cell gets a complete set of chromosomes that are identical with the chromosomes of the parent cell, so each of the baby spider plants produced has exactly the same number and kind of chromosomes and genes as the parent. The new plants are **clones**. They are genetically identical to each other and to their parent.

Sexual reproduction is a bit more complicated. The 46 chromosomes in our cells are actually 23 matching *pairs* of chromosomes. So each cell contains two complete sets of 23 chromosomes. Figure 1g.4 shows these arranged in order. The picture was made by taking a photograph of a cell that was just getting ready to divide, then rearranging the pictures of each chromosome in order, from biggest to smallest, and with the members of each pair next to each other.

Figure 1g.3 Spider plant young are clones of the parent.

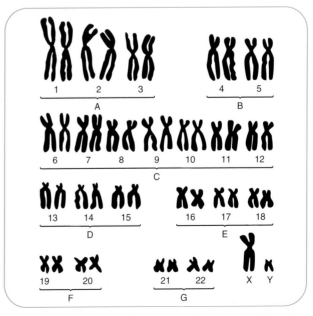

Figure 1g.4 Chromosomes from a normal man, arranged in order.

Sexual reproduction involves sex cells, or **gametes**. These are produced in a person's ovaries or testes. Gametes are made from ordinary body cells, which divide in an unusual way. Instead of making sure that each new cell gets two complete sets of chromosomes, the new gametes get just one set – one of each kind instead of two of each kind. So human eggs and sperm have only 23 chromosomes, not the usual 46 in body cells.

When fertilisation happens, the 23 chromosomes from the sperm get mixed up with the 23 chromosomes from the egg. A new cell is formed, called a **zygote**. It has 46 chromosomes, 23 from the sperm and 23 from the egg.

So the zygote has a mixture of genes from its father and its mother. It will grow into a unique individual, with a mix of genes that is different from either of its parents or any of its brothers or sisters.

SAQ

3 Some animals can produce both eggs and sperm and fertilise themselves. Is this asexual reproduction, or is it sexual reproduction?

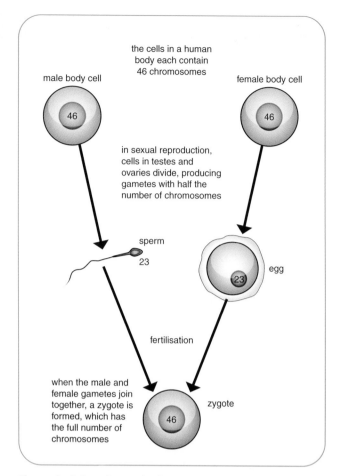

Figure 1g.5 Sexual reproduction.

Down's syndrome

It is not surprising that things sometimes go wrong when a cell is dividing and the 46 chromosomes are being sorted out into the two new cells.

When cells are dividing in an ovary to produce eggs, the pair of chromosome 21s sometimes stick together. Normally, each would head off in a different direction into the two new cells. But if they stick together, one of the new cells gets two chromosome 21s, while the other cell gets none.

The egg with no chromosome 21 dies. The other one behaves normally and it can be fertilised by a sperm. If this happens, then the zygote that is produced gets three chromosome 21s.

The zygote then divides repeatedly by mitosis, just as normal, eventually forming a complete person. But each of their cells contains an extra chromosome 21. This extra chromosome makes some of their cells behave differently and this results in the condition we call Down's syndrome. A

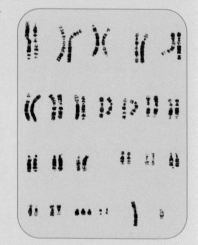

Figure 1g.6 Chromosomes from a person with Down's syndrome.

person with this condition is usually happy and friendly but they may have problems with their heart and other body organs, and they will have learning difficulties, too.

Down's syndrome is much more common in babies born to older mothers. The older the woman, the more likely it is that her cells will make this mistake when they divide.

Summary

You should be able to:

◆ describe what chromosomes are, what they are made of and where they are found

 ◆ explain how DNA controls the function of cells

◆ explain what is meant by the genetic code

◆ state that DNA contains four different bases and that each gene has a different base sequence

H ◆ know that the four bases in DNA are called A, T, C and G

◆ know that humans have 23 matching pairs of chromosomes in each cell and that other species have different numbers of chromosomes

◆ describe the differences between asexual and sexual reproduction

H ◆ explain how fertilisation produces unique individuals

Questions

1 Copy and complete this table, to compare asexual and sexual reproduction.

Asexual reproduction	Sexual reproduction
only one parent	usually two parents

2 Bargles have 32 chromosomes in each of their cells.

a How many chromosomes are there in a sperm cell from a bargle?

b How many chromosomes are there in the zygote of a bargle?

c Explain why every bargle is slightly different from every other bargle.

Figure 1g.7 This is a bargle.

3 Chromosomes are made of DNA. A stretch of DNA that codes for one characteristic is called a gene.

a What can you say about the sequences of bases in the DNA molecules of clones?

b Describe how clones are produced naturally.

c Explain why, even though your cells contain exactly the same DNA as each other, cells in different parts of your body do different things.

Variation

Your genes play a big part in making you who you are. The shape of your earlobes, the colour of your eyes and the shape of your nose are all determined by your genes. So is your gender (sex).

But that isn't the whole of the story. Your environment also affects your characteristics. For example, you may have a scar from an accident or operation. You speak a particular language because that was the language of the people around you when you were learning to talk.

Many of our characteristics are affected by both genes and environment. You have genes that give you the potential to grow to a particular height, but you won't grow to that height unless you had a good diet when you were growing up. Your body mass is partly decided by your genes, but perhaps even more by how much you eat and how much exercise you take. Your intelligence is partly determined by your genes but is also affected by how hard you make your brain work as you grow up.

With characteristics such as intelligence and sporting ability no-one is at all sure how much is determined by our genes and how much by environment. If two parents are top class professional sports people, statistics show that their children are also likely to be very good at sports. It's really difficult, though, to know whether this is because of the genes they have inherited or because they were brought up in a sporty environment and worked hard at sport while they were growing up. Scientists are still researching this topic and there is a big debate about it. Some people think it is wrong to suggest that not every child can grow up to be as intelligent as every other, if only they were given the same opportunities. Others think that maybe genes have a big influence on this.

Mutation

Usually, the characteristics of offspring are very similar to those of their parents – though they may have a mix of features from their father and their

Twin studies

In 1993, St Thomas' Hospital in London began what has become an enormous information-collecting exercise about twins. They launched a media campaign asking for twins to volunteer to answer questionnaires about themselves. Since then, and following several further media campaigns, they have built up a database that records answers from 10 000 twins aged from 18 to 80.

The information in the database is used by researchers trying to find out how much a particular characteristic is influenced by our genes and how much by our environment. Some twins – called dizygotic twins – develop from two different eggs fertilised by two different sperms. They are genetically different from one another. Monozygotic twins, however, develop from one fertilised egg (a zygote) that then

splits into two and develops into two babies. They are genetically identical.

If the twins are brought up in the same family then their environments are probably very similar. By comparing data for monozygotic and dizygotic twins who were brought up together or in different families (for example, if one or both twins were adopted), you can get useful information about the relative influence of genes and environment on a particular characteristic.

For example, a recent study has looked at some aspects of memory. It has found that being able to remember where you put your keys, or what day it is, is strongly influenced by genetic factors. Remembering what clothes you wore yesterday is more influenced by your family environment.

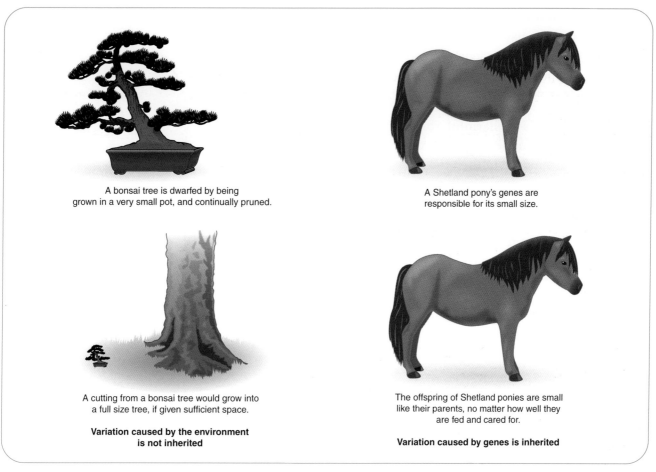

A bonsai tree is dwarfed by being grown in a very small pot, and continually pruned.

A Shetland pony's genes are responsible for its small size.

A cutting from a bonsai tree would grow into a full size tree, if given sufficient space.

The offspring of Shetland ponies are small like their parents, no matter how well they are fed and cared for.

Variation caused by the environment is not inherited

Variation caused by genes is inherited

Figure 1h.1 The inheritance of variation.

mother. But sometimes a new feature suddenly turns up. This is often the result of a **mutation**.

A mutation is a sudden and unpredictable change in an organism's genes. It happens when DNA is being made – the bases get put together into the wrong sequence. So the gene gives different instructions, and the cells behave in a different way. They make different proteins.

A mutation can happen for no obvious reason – it just happens spontaneously. But the chances of it happening are increased by anything that damages DNA. This includes ionising radiation, such as alpha radiation and X-rays, ultraviolet light (in sunshine) and various chemicals. The X-ray dose that people are exposed to is carefully limited. Radiographers, whose work includes giving people X-rays, have to be very careful to keep well away from the X-ray equipment when it is switched on. Chemicals that are used in products like household cleaners, weedkillers and drugs all have to be thoroughly tested before they are allowed to be sold, to check that they don't

increase the risk of mutations happening in someone's cells.

Figure 1h.2 In this albino penguin melanin is not produced because a gene base sequence has changed.

SAQ

1 Radiation therapy may be used to kill cancer cells. Suggest why a woman who could be pregnant will not be given this kind of therapy.

Usually, mutations are harmful. For example, if there is a mutation in the gene that codes for the production of haemoglobin then the haemoglobin that is made may be faulty and not able to carry oxygen as well as it should.

Just occasionally, though, a mutation can be beneficial – the new protein that is made turns out to be even better than the old one. This must have happened as we evolved from our ancestors. Genes that helped with brain development must have mutated, making us more intelligent than the primitive humans that lived tens of thousands of years ago. You can read more about evolution in Item B2f *Staying in balance*.

Mutation	A mistake is made when DNA is being copied. The base sequence gets altered, so the gene may now code for a slightly different protein, or even for no protein at all.
Gamete formation	A cell in an ovary or a testis has two sets of chromosomes and genes – one set from the person's father and one from their mother. When a gamete is made, only half the genes go into it – and they can be every possible mix between the two original sets (Figure 1h.3).
Fertilisation	Any gamete from one partner can fertilise any gamete from the other, so there is an almost infinite variety of different genes that may end up in the zygote.

Table 1h.1 Sources of variation.

Genetics

Genetics is the study of inheritance. Genetics can explain and predict the chances of offspring inheriting a particular characteristic from their parents.

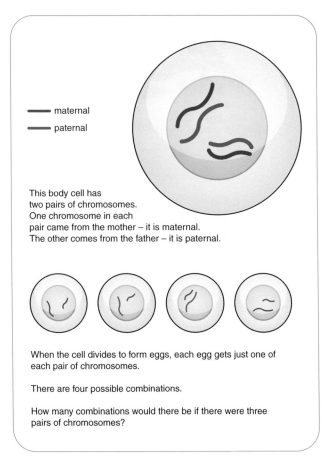

—— maternal
—— paternal

This body cell has two pairs of chromosomes. One chromosome in each pair came from the mother – it is maternal. The other comes from the father – it is paternal.

When the cell divides to form eggs, each egg gets just one of each pair of chromosomes.

There are four possible combinations.

How many combinations would there be if there were three pairs of chromosomes?

Figure 1h.3 Variation in gametes.

Some conditions are caused by faulty genes and therefore can be inherited.

- **Red–green colour blindness.** This occurs mostly in men. They have a faulty gene for one of the three different kinds of colour receptors in the eye, so red and green look almost the same. This is very common – about one in every 15 men has it, and one in every 100 women.
- **Sickle-cell anaemia.** This occurs equally in men and women. It is caused by a faulty gene for making haemoglobin. Less oxygen can be carried in the blood. The red blood cells become curved and stiff when they are short of oxygen and get stuck in blood capillaries. This is very painful and can be dangerous.
- **Cystic fibrosis.** This, too, occurs equally in men and women. Very thick, sticky mucus collects in the lungs and digestive system. As well as making breathing and digestion difficult, the mucus in the lungs increases the likelihood that they will be infected by bacteria.

Figure 1h.4 A person with red–green colour blindness sees the scene like this.

Alleles

You'll remember that we have two copies of each kind of chromosome in our body cells. The two chromosomes of a matching pair are said to be **homologous**. They have the same genes on them (Figure 1h.5).

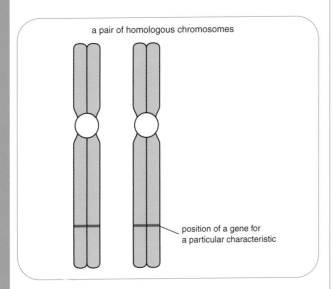

a pair of homologous chromosomes

position of a gene for a particular characteristic

Figure 1h.5 Homologous chromosomes have genes for the same characteristic in the same position.

But not every copy of a gene is the same. Most genes come in more than one variety. The different varieties of a gene are called **alleles**.

For example, there is a gene that codes for a protein needed for lung cells to work properly. The normal allele of this gene helps normal mucus to be produced in the lungs. But there is also a faulty allele that causes very thick mucus to be made. This allele causes the disorder called **cystic fibrosis**.

This gene is found on chromosome 2. Everyone has two copies of chromosome 2 in their cells. So there are three possible combinations of the two alleles for this gene. They are shown in Figure 1h.6.

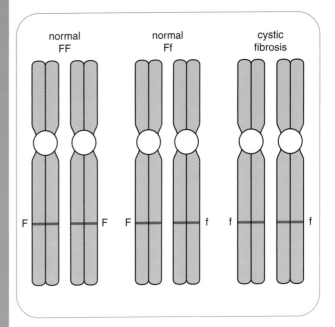

normal
FF

normal
Ff

cystic
fibrosis

Figure 1h.6 Genotypes for the cystic fibrosis gene.

You can see that the alleles have been represented by letters. One letter is chosen to represent the gene. Here, the letter is F. A capital F represents the normal allele of this gene, while small f represents the faulty allele.

The normal allele, F, is **dominant** over the faulty allele f. This means that when the two alleles are found together, it is the dominant one that works. So a person with alleles Ff is fine – they don't have cystic fibrosis. The allele f only works when there isn't an F allele present. It is said to be a **recessive allele**. You only have cystic fibrosis if *both* alleles are f.

The set of alleles that a person has is called their **genotype**. Your genotype could be FF, Ff or ff. Your characteristics are your **phenotype**. In this case, your phenotype is either having cystic fibrosis or not having it.

Someone with two alleles the same is said to be **homozygous**. So a person whose genotype is FF is homozygous, and so is a person who is ff. Someone who has two different alleles of a gene is said to be **heterozygous**. So a person with genotype Ff is heterozygous.

SAQ

2 The dominant allele B codes for brown eyes, and the allele b for blue eyes.

   a What are the three possible combinations of these alleles in someone's cells?

   b Which of these are homozygous, and which are heterozygous?

   c What colour eyes will each of these combinations give?

Inheritance

Let's think about what happens if a man and a woman who are both heterozygous for the cystic fibrosis genes have children.

In the man's testes, cells divide to form sperm. You'll remember that each sperm only gets one of each chromosome. The man has an F allele on one of his chromosome 2s and an f allele on the other. Half of his sperm will get one chromosome of the pair, while the other half will get the other. So half of his sperm will have the F allele and the other half will have the f allele.

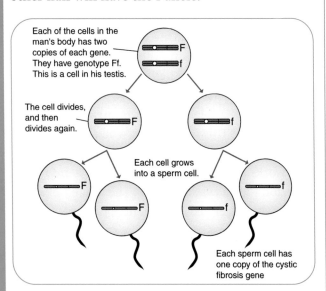

Figure 1h.7 What happens when sperm are formed in a heterozygous man.

The same happens in the woman's ovaries. Half of her eggs will contain the F allele and half will contain the f allele.

When fertilisation occurs, hundreds of sperm cluster round the egg. Only one of them succeeds. Which will it be? And which kind of egg is it that gets fertilised?

The answer is that we have no idea. It is a completely random event. There is a 50:50 chance that the successful sperm will have an F allele or an f allele. And the same holds true for the egg.

We can predict the chances of what happens by drawing a **Punnett square**. This shows the two kinds of eggs and the two kinds of sperm, and the different possibilities that can happen at fertilisation.

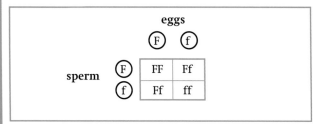

The letters around the edges, with circles around them, are the alleles in the eggs and sperm. The letters inside the square show the possible combinations of alleles in the zygote.

Because there are equal numbers of each kind of sperm and egg, there are equal chances of the zygote having any of these combinations of alleles. Each time the couple have a child, there is a 1 in 4 chance it will be FF, a 2 in 4 chance it will be Ff and a 1 in 4 chance it will be ff.

SAQ

3 Does the Punnett square mean the couple will have four children? Explain your answer.

4 The couple already have one child with cystic fibrosis. What are the chances that their next child will have cystic fibrosis?

Genetic diagrams

A Punnett square is often used as part of a **genetic diagram**. This is not really a diagram at all – it is just a convention for showing the possible offspring from two parents. Here is an example.

Sickle-cell anaemia is caused by a recessive allele of the gene that codes for haemoglobin. The normal, dominant allele is H and the faulty, recessive allele is h.

A man with genotype Hh has a child with a woman who has sickle-cell anaemia – her genotype is hh. What are the chances that their next child will have sickle-cell anaemia?

Parents	man with normal haemoglobin	woman with sickle-cell anaemia
Alleles	Hh	hh
Sperm and eggs	(H) and (h)	all (h)

eggs
(h)

sperm (H) | Hh
(h) | hh

The Punnett square shows that there is a 1 in 2 chance that the child will be Hh and have normal haemoglobin, and a 1 in 2 chance that the child will be hh and have sickle-cell anaemia.

Notice that the Punnett square is simpler than the first one. This is because the woman only produces one kind of egg. You don't always need to have two rows or columns for the eggs or sperm – you only need to put in a row or column for each different kind.

SAQ

5 A man has cystic fibrosis. His partner has two normal alleles in her cells.

   a Using the symbols F for the normal allele and f for the allele that causes cystic fibrosis, write down the alleles that the man must have.

   b What alleles will each of his sperm have?

   c What alleles will each of his partner's eggs have?

   d What alleles will each of their children have? Will any of them have cystic fibrosis?

   e Draw a Punnett square to show the possible genotypes of their children.

6 A woman who is heterozygous for sickle-cell anaemia marries a cousin who is also heterozygous for this illness. Using the symbols H for the normal allele and h for the recessive allele that causes sickle-cell anaemia, draw a complete genetic diagram to show the chances that their next child will have sickle-cell anaemia.

7 Sal and Jo have just had their first child, Amy. Amy has cystic fibrosis. Sal and Jo weren't

expecting this. They try to find out if anyone in their family has ever had this disorder before. Sal finds out that his great-granduncle had it, but Jo can't find anyone with cystic fibrosis in her family.

   a Suggest how it is possible for Amy to have cystic fibrosis, even though the only person in her ancestry known to have the disorder was one of her great-great-granduncles.

   b Although he knows it isn't really his fault, Sal feels guilty about Amy's disorder. Why might he feel guilty? How can Jo explain to him that he should not?

   c Sal and Jo know that their next child might have cystic fibrosis, too. They would like more children but it is a difficult decision to make. Try to imagine how you might feel. What would you decide?

Sex inheritance

Your sex is determined by your genes. A man has one X chromosome and one Y chromosome – you can see these in Figure 1g.4. A woman has two X chromosomes.

A man therefore produces two kinds of sperm. Half contain an X chromosome and half contain a Y chromosome. A woman's eggs all contain an X chromosome.

We can use a genetic diagram to explain why there is an equal chance of a baby being a boy or a girl.

Parents	father	mother
Chromosomes	XY	XX
Sperm and eggs	(X) and (Y)	all (X)

eggs
(X)

sperm (X) | XX
(Y) | XY

The Punnett square shows that there is a 1 in 2 chance that the child will be XX and be female, and a 1 in 2 chance that the child will be XY and be male.

Summary

You should be able to:

♦ give some examples of human characteristics that are determined only by genes and some examples of characteristics that are affected by the environment

♦ describe how sex is determined in mammals

♦ know what causes genetic variation and name some of the factors that can cause mutations

H ♦ explain why mutations change cell activity

♦ use genetic diagrams to explain inheritance, using the correct terminology

♦ give some examples of inherited disorders and know that they are caused by faulty genes

H ♦ explain how inherited diseases are caused by faulty alleles and use genetic diagrams to predict the chances of an inherited disease passing to the next generation

♦ discuss the issues raised by knowledge of inherited diseases in a family

Questions

1 Rob and Jack are identical twins.

Figure 1h.8

a Rob and Jack are clones. What does this mean?

b Give one of the twins' characteristics, which you can see in the picture, that has not been caused by their genes.

2 Copy and complete these sentences.

a A mutation is a change to the .... in a cell. Mutations can be caused by .... or ....

b Men have one .... and one .... chromosome in their cells, whereas women have two .... chromosomes.

continued on next page

Questions - *continued*

H 3 In guinea pigs, hair can be curly or straight. The allele for curly hair, h, is recessive, and the allele for straight hair, H, is dominant.

A breeder has one guinea pig with curly hair and another with straight hair which has come from a family in which all the guinea pigs have always had straight hair.

   a What is the genotype of the curly-haired guinea pig?

   b What is the most likely genotype of the straight-haired guinea pig?

   c Draw a genetic diagram to work out the possible genotypes and phenotypes of the offspring, if these two guinea pigs are bred together.

The breeder wants to find out which of the straight-haired offspring are homozygous and which are heterozygous.

   d Explain what is meant by *homozygous* and *heterozygous*.

   e (This bit is extra hard.) The breeder finds out which are homozygous and which are heterozygous by breeding each of the straight-haired offspring with a curly-haired guinea pig. Use genetic diagrams to explain how this gives her the information she wants to know.

What lives here?

Ecology is the study of how living things interact with their environment. Even the most unlikely environment has at least some living things in it. Your school grounds – even if there's only just a small area of grass, a neglected flower bed or a single tree – probably have many different kinds of plants and animals, not to mention fungi and bacteria, living there.

If you wanted to find and identify every living organism in the school grounds, it could take you years. Imagine trying to do it in a dense wood. So ecologists usually just **sample** the area they are studying. They identify and count the different **species** they find in just a small part of the area. It is important to sample enough of the area to make the results reliable.

Quadrats

A **quadrat** is a square area in which you identify and count living organisms. Quadrats can be different sizes. If you are doing your sampling on a grassy area (for example, a sports field, a lawn, a meadow), a good size to use is one with sides 0.5 m long. You can make a quadrat with wood or wire.

You put the quadrat down on the ground, and then identify what you find inside it. You may need to use a **key** to help you to do this. A key gives you a pair of statements, and you have to decide which one describes your organism. This takes you on to another pair, and so on until you get to its name.

As well as identifying the organisms in the quadrat, you can also estimate their **abundance**.

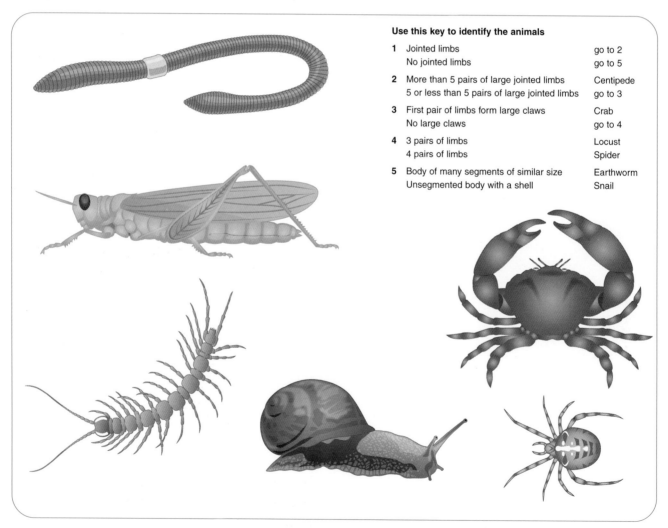

Use this key to identify the animals

1	Jointed limbs	go to 2
	No jointed limbs	go to 5
2	More than 5 pairs of large jointed limbs	Centipede
	5 or less than 5 pairs of large jointed limbs	go to 3
3	First pair of limbs form large claws	Crab
	No large claws	go to 4
4	3 pairs of limbs	Locust
	4 pairs of limbs	Spider
5	Body of many segments of similar size	Earthworm
	Unsegmented body with a shell	Snail

Figure 2a.1 Using a key.

Figure 2a.2 Using a quadrat.

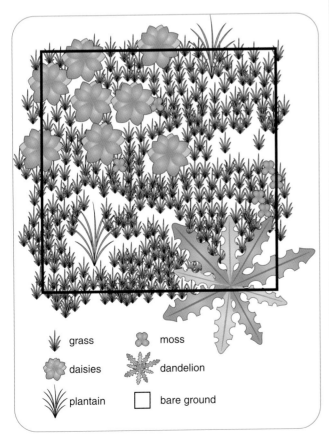

grass

moss

daisies

dandelion

plantain

bare ground

Figure 2a.3 Percentage cover in a quadrat.

Sometimes, you can just count them – for example, you might find 12 daisy plants and 3 dandelions in the quadrat. But counting can be very difficult with some species. For example, you can't really count the number of grass plants in a quadrat. Instead, you could estimate what percentage of the ground they cover. So you might decide that 50% of the ground inside the quadrat is covered by grass, 10% by daisies, 5% by dandelions and so on.

SAQ

1   Estimate the percentage cover of each species of plant in the quadrat in Figure 2a.3. Why is this not easy to do? How could you make it easier?

Placing quadrats

Unless the area you are studying is exactly the same all over, just one quadrat is unlikely to give a good idea of the whole area. One part of a sports field might have very different things growing in it from another area. So you will need to put down many quadrats in different parts of the field.

How do you decide where to put them? One way is to place them **randomly**. This means that you don't *choose* where to put them. The best way to achieve random positions for your quadrats is to use random numbers – some calculators will generate these. You use the numbers as co-ordinates, with two sides of the area you are studying acting like axes on a graph. So, if your first pair of random numbers is 6 and 13, you walk six paces along one side and then thirteen paces out into the field. That's where the first quadrat goes.

You also need to decide how many quadrats to use. Obviously, to get a full picture of what lives in the field, you would need to cover the field completely with quadrats. At the other extreme, if you use only one or two quadrats, you will be missing lots of organisms. The best number will be somewhere in between – a compromise between testing the field really thoroughly, while still setting yourself a task that you can achieve in a short period of time.

SAQ

2 The graph in Figure 2a.4 shows the number of species that were found in a field, plotted against the number of quadrats used.

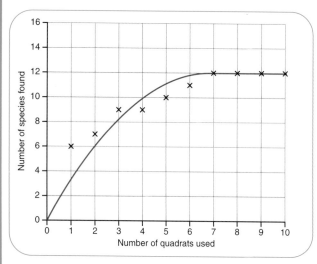

Figure 2a.4 Number of species plotted against number of quadrats.

Use the graph to suggest the best number of quadrats to use when you are sampling the field. Justify your suggestion.

There are occasions when you might not want to place the quadrats randomly. For example, you might want to investigate how the kinds of plants change as you go from a boggy area next to a pond into a grassy field. A good way of doing this is to use a **transect**. This is a line that goes from one area into the other. You can use a piece of string to mark the line. You can then put quadrats down at intervals along the line and record what you find inside them.

Figure 2a.5 Using a transect on a rocky shore.

Counting animals

Quadrats are fine for counting plants but often not so good for finding and counting animals. You will probably need to use a different **sampling technique** to catch and identify them.

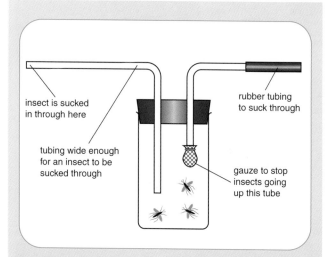

Figure 2a.6 A pooter.

Small insects can be captured using a **pooter**. This is a small tube with a bung in the top, and two tubes leading out. There's a little piece of gauze over the end of one tube. You put the end of the other tube near the insect and then suck quickly through the gauze-covered one. The insect should be sucked into the tube.

Figure 2a.7 A net can be used to catch flying insects.

Nets are useful for capturing organisms in long vegetation, trees, ponds or streams. You need to use a standard 'sweeping' technique with your net to be sure you are sampling in the same way in each place you investigate.

Figure 2a.8 Kick sampling.

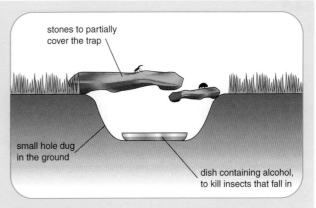
stones to partially cover the trap

small hole dug in the ground

dish containing alcohol, to kill insects that fall in

Figure 2a.9 A pitfall trap.

In a stream, you can use a technique called **kick sampling**, where you stand upstream of the net and shuffle your feet around on the stream bed. This disturbs small animals, which get swept downstream in the current and end up in your net.

Beetles and other small animals that move around on the surface of the ground can be caught in a **pitfall trap**. The animals simply fall in. Pitfall traps are especially useful for collecting nocturnal animals. You will probably need to put some alcohol in the bottom so that the animals are killed. If you don't, you may discover just one very fat carnivorous beetle in the trap the next day, and nothing else.

If you are investigating which animals live in woodland, you may want to search the **leaf litter** for animals. Leaf litter is the layer of dead leaves that collects on the ground. Lots of small animals feed on it – they are **decomposers**, organisms that make a living by feeding on dead plants and animals, or waste materials from them.

The best way of getting the animals out of the leaf litter, so that you can see them clearly and count them, is to use a **Tullgren funnel**. You put the leaf litter over the funnel and shine a bright, hot light onto it. The animals in leaf litter like to be cool, damp and dark, so they burrow down to the bottom of the leaf litter and fall through the tube into a beaker. If the beaker has alcohol in it, it kills them immediately, making it easier for you to look at them closely and identify them.

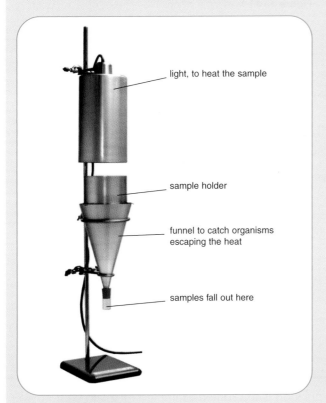
light, to heat the sample

sample holder

funnel to catch organisms escaping the heat

samples fall out here

Figure 2a.10 A Tullgren funnel.

SAQ

3 Suggest two ways to make sweeps with a net standardised.

Making quantitative estimates

All of these sampling methods can be used to help you to make a **quantitative estimate** of the organisms in the habitat you are studying. 'Quantitative' means using numbers. It has to be an 'estimate' because you can never be sure that your sample is really representative of the whole area, nor that you have actually identified and counted all the different organisms in your sample correctly.

The results can be used to estimate the **population** of a particular species of organism. A population is the number of organisms of one species that live together in the same habitat at the same time.

If you have used quadrats then you can 'scale up' your results to make an estimate of the population of an organism in the whole area like this:

number in the whole area

$$= \frac{\text{total number in quadrats} \times \text{total area of habitat}}{\text{total area of quadrats}}$$

Another way of estimating population size is called the **mark–release–recapture** technique. This is useful when you are dealing with small animals that move around, such as woodlice or water beetles.

First, you catch a large sample of the organism (using pooters, for example). You count how many you have caught.

Then you mark the animals, for example with a little spot of paint on their backs. You then let them go in the area they were collected from. They will mix with the unmarked animals.

A couple of days later, you catch another sample of the animals. You count how many you have caught, and how many of them have the mark on them.

Figure 2a.11 Marking a woodlouse.

Worked example 1

A student used quadrats to estimate the number of daisy plants on a lawn.
The total area of the lawn was 40 m^2.
The quadrat had sides that were 0.5 m long. Therefore the area of each quadrat was 0.25 m^2.
He collected results from ten quadrats. These are his results.

Quadrat	1	2	3	4	5	6	7	8	9	10
Number of daisy plants	3	0	7	9	12	12	0	7	2	6

Table 2a.1

Total number of daisy plants found = 58
Total area sampled = 10 × 0.25 m^2 = 2.5 m^2

Total population of daisy plants = $\frac{(58 \times 40)}{2.5}$ = 928

SAQ

4 How reliable do you think the estimate of the daisy population is in Worked example 1? Explain your answer.

Then you can calculate the total population like this:

total population

$$= \frac{\text{total caught 1st time} \times \text{total caught 2nd time}}{\text{number of marked ones caught 2nd time}}$$

Worked example 2

An ecologist caught 65 woodlice, marked them and let them go. He later caught 70 woodlice in the same area. 24 of these were marked.

$$\text{total population of woodlice} = \frac{(65 \times 70)}{24} = 190$$

SAQ

5 Look at Worked Example 2. How reliable do you think the estimate of the woodlouse population is? Explain your answer.

6 Angie caught 21 water boatmen in a pond. She marked them all with yellow paint then let them go. Two days later, she caught 22 water boatmen. Twelve of them were marked with yellow paint.

   a Use Angie's results to estimate the population of water boatmen in the pond.

   b What could Angie have done to make her results more reliable?

Comparing communities

Ecologists often want to compare the organisms that live in two areas. For example, they might want to know the population of orchids in a grass field where cattle are grazing, compared with that in another grass field where sheep are grazing. Then they can decide the best way to conserve the orchids.

Figure 2a.12 The habitat of these orchids is a grass field.

The area where an organism lives is its **habitat**. All the different organisms that live in a habitat make up a **community**. For example, the habitat of the orchids is a grass field. The community is made up of all the different plants and animals that live in the field. The community plus all the

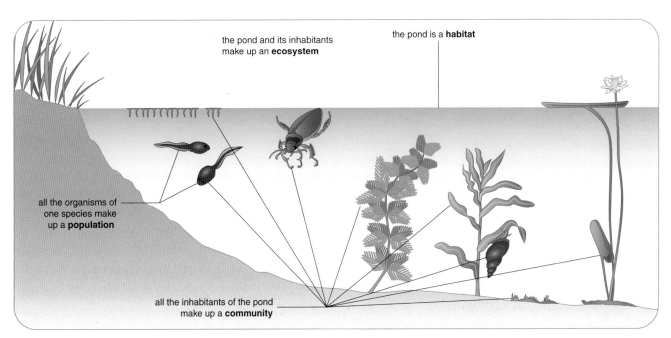

the pond and its inhabitants make up an **ecosystem**

the pond is a **habitat**

all the organisms of one species make up a **population**

all the inhabitants of the pond make up a **community**

Figure 2a.13 A pond and its inhabitants – an example of an ecosystem.

The Census of Marine Life Project

We have absolutely no idea how many different species live on the Earth. So far, about 2 million have been found, named and classified. Biologists have tried to estimate how many more there might be out there that we haven't yet discovered. Some think there might be about 3 million more, but others think it could be as many as 50 million.

You probably won't find any new species on your school playing field, but there are some parts of the world that we have scarcely explored yet. In 2000, a big new project began with the aim of finding new species in the oceans. Called the Census of Marine Life project, it is intended to run for 10 years. The director says: 'We are at the start of a great adventure, like going to the Moon.' They are using several specially equipped ships, unmanned submarines and even aircraft to help them in the hunt.

By 2003, the project had already catalogued 15 300 species of fish, several of which were previously unknown. One of the most exciting areas has been deep in the Arctic Ocean, where many new species of jellyfish and worms have been found.

One animal they would love to find is the 'colossal squid'. No-one has ever seen one alive, but in 2002 one was washed up on a beach in Australia. It was about 5 metres long, but scientists think it was only a baby and that an adult could grow as big as 12 metres. This would make it the largest invertebrate in the world – bigger than most whales.

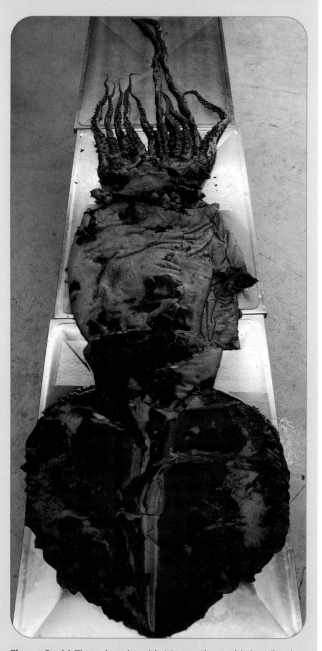

Figure 2a.14 The colossal squid, *Mesonychoteuthis hamiltoni*.

other, non-living things in the field make up an **ecosystem**. The ecosystem in the field includes all the living things that are there and also the air and soil. In an ecosystem, the living organisms and their environment all interact with one another.

Natural and artificial ecosystems

People often use the term 'natural' to mean 'not affected by humans'. There are very few, if any, places in Britain where we can find a really 'natural' ecosystem. Humans have lived here for a very long time and we have altered practically every square centimetre of the landscape around us. Even places that we think of as being 'natural' wouldn't look at all as they do if humans had not affected them.

For example, you may have an area of woodland near your home. Some woodlands have been planted by humans. Others may have grown naturally. But even these have almost certainly been managed by humans in the past. For example, people will have cut down trees to use the timber for building ships or houses. Most trees don't die when they are cut down but simply regrow from their roots and stumps. People will have encouraged some species of trees to grow because they were especially useful to them – oak for making ships, beech for making furniture, hazel for making fences and charcoal.

All the same, an area of woodland is considerably more 'natural' than your sports field. For a lot of the time, the woodland is left alone and allowed just to grow as it will. The sports field is regularly mown, and it may be sprayed with herbicides to kill weeds. If the sports field was abandoned and left to itself for 5 or 10 years, tree seeds would be blown onto the soil and germinate. After about 10 years, you would have a young wood instead of the short grass.

Biodiversity

Some ecosystems have more different species in them than others. They have a higher **species diversity** or **biodiversity**.

In general, the more humans interfere with a habitat, the lower its biodiversity. More natural ecosystems tend to have a higher biodiversity than artificial ones.

The two photographs in Figure 2a.15 show two different fields. One is being used for growing wheat. The farmer wants only wheat to grow in the field, so he has sprayed it with **herbicides** to kill weeds. He has also sprayed **pesticides** to kill insects that would feed on the crop. And he has used **fertilisers** to increase the amount of nitrate, potassium and phosphate in the soil so that the wheat will grow better and give a higher yield.

The other field is a 'set-aside' field. The farmer has not grown a crop in this field for 3 years. He hasn't used herbicides, pesticides or fertilisers. The plants in the field have just grown by themselves, germinating from seeds that were already in the soil, or that have blown in on the wind or been carried in stuck to an animal's coat.

Figure 2a.15 Wheat field and set-aside.

You can see that the set-aside field has many different species of plants in it, whereas the wheat field has only one species. And if you sampled the insects in the two fields, you would find many more different kinds in the set-aside field than in the wheat field. The set-aside field has a much greater biodiversity than the wheat field.

Summary

You should be able to:

- describe how to use different collecting methods, including pooters, nets, pitfall traps and quadrats

- know how to use the data you collect to work out the size of a population

- **H** explain how sample size and collecting methods affect the reliability of your results

- use the terms *habitat*, *population*, *community* and *ecosystem*

- use keys to identify plants and animals

- compare the communities in a relatively natural and an artificial habitat

Questions

1 Copy each of these definitions and then write down the word being defined. Choose your answers from the **bold** words in this chapter.

- the place where an organism lives

- all of the organisms, of all species, that live together in a habitat

- the number of one species living in a habitat

- all of the living organisms, plus the non-living things such as soil and air, that are found in one place

2 These are the results for a transect running from the edge of a pond into a grassy field. A quadrat was put down every 0.5 metres along the transect. The numbers are the estimated percentage cover for each species in the quadrat.

Distance from pond in metres	0	0.5	1.0	1.5	2.0	2.5	3.0
Bare ground	20	10	5	10	15		5
Rush	80	30	5	–	–	–	–
Butterbur	–	30	50	10	–	–	–
Willowherb	–	30	30	30	5	–	–
Grass	–	–	10	30	60	60	60
Clover	–	–	–	10	20	20	15
Buttercup	–	–	–	10	–	20	20

a Name two plants that were found near the edge of the pond but were not found in the rest of the field.

b Suggest why these plants only grew near the pond.

c Which part of the field had the most species growing in it?

Classification

Biologists group organisms to make it easier to study them. It's much simpler to talk about 'mammals' than about 'rats, humans, bears, whales etc.' Grouping living things is called classification.

We classify organisms according to how closely related they are. The living organisms on the Earth today have **evolved** from ones that lived in the past. We are not quite sure when life first appeared on Earth, but we think it was probably about 3.5 billion to 4 billion years ago. The first living things were probably single cells, perhaps something like bacteria. Over time, these simple organisms have gradually given rise to all the species that currently live on Earth.

We can't ever be sure of how closely different organisms are related, because we can't go back in time and check out their ancestry. We have to go by circumstantial evidence and do some detective work.

Most of the time, we look for shared features. If two organisms share a lot of features then we can guess that they might be quite closely related to each other. We therefore classify them in the same group. The more different their features are, the more distantly they are related.

We need to take care with this. Sometimes organisms may have features that look similar but are really very different. For example, insects and birds have wings, but the wings don't have the same basic design. Insects and birds are only very distantly related.

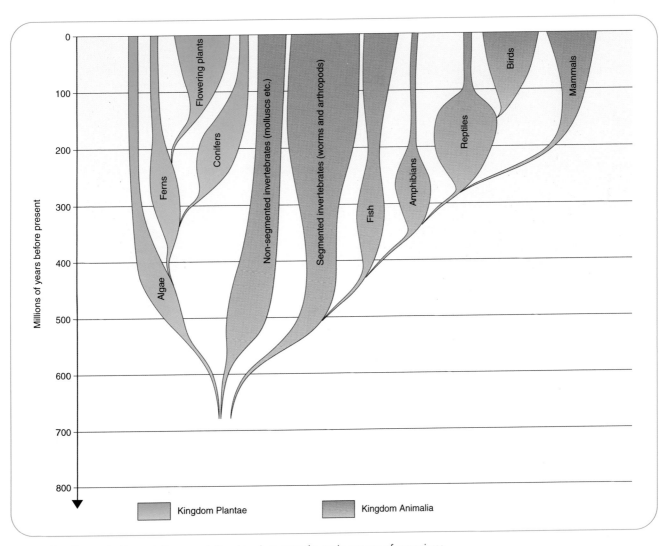

Figure 2b.1 The probable evolutionary relationships between the main groups of organisms.

Plants and animals

Although not all biologists agree about it, one of the standard systems of classification is called the 'Five Kingdom' system. This divides all living organisms into five huge groups called kingdoms. Two of these kingdoms are the **plant kingdom** and the **animal kingdom**.

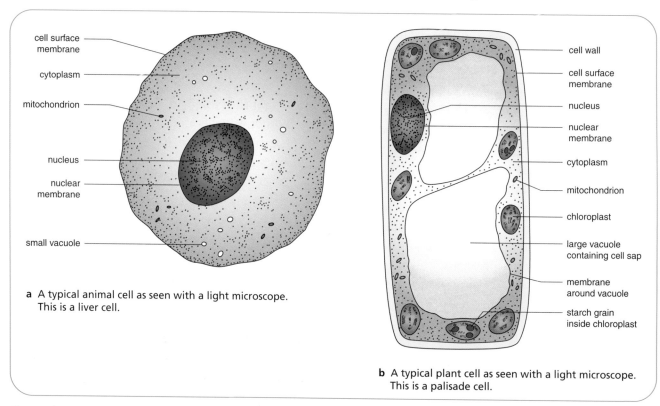

a A typical animal cell as seen with a light microscope. This is a liver cell.

b A typical plant cell as seen with a light microscope. This is a palisade cell.

Figure 2b.2 a An animal cell. **b** A plant cell.

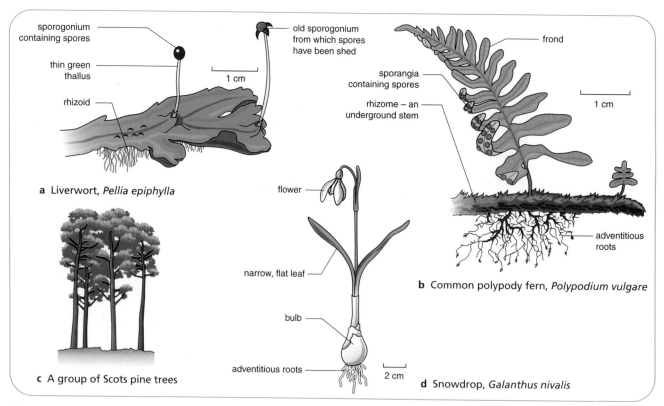

a Liverwort, *Pellia epiphylla*

b Common polypody fern, *Polypodium vulgare*

c A group of Scots pine trees

d Snowdrop, *Galanthus nivalis*

Figure 2b.3 These are all plants.

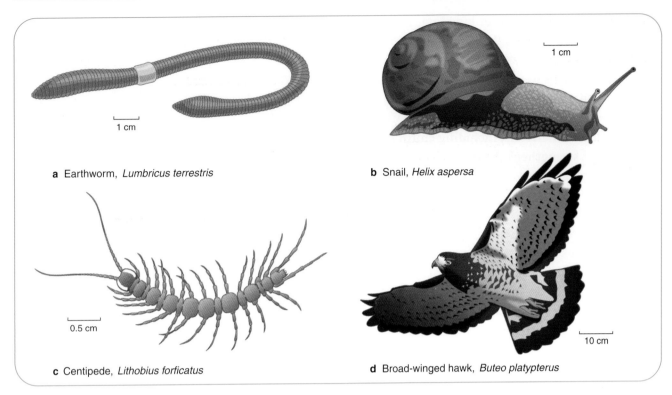

a Earthworm, *Lumbricus terrestris*

b Snail, *Helix aspersa*

c Centipede, *Lithobius forficatus*

d Broad-winged hawk, *Buteo platypterus*

Figure 2b.4 These are all animals.

You probably already know the differences between plant cells and animal cells (Figure 2b.2). Plant cells have a cell wall made of **cellulose**, which is never found in animal cells. Some plant cells have chloroplasts, containing **chlorophyll**, in which photosynthesis takes place. Animal cells never have chloroplasts. Plants are therefore able to make their own food, which animals cannot do.

Plants remain in one place. They don't need to move around – they simply need to hold out their leaves to absorb sunlight and carbon dioxide, and let their roots absorb water and mineral ions from the soil. Most animals move from place to place, searching for food. The body of a plant is therefore often a branching shape, while animals tend to be more compact, as this makes movement easier.

SAQ

1 Copy and complete Table 2b.1, to compare the features of plants and animals.

Plants	Animals
cells are surrounded by a cell wall made of cellulose	cells do not have a cell wall

Table 2b.1

The other three kingdoms

Although plants and animals are the most familiar groups of living things, they are not the most common. These are the **bacteria**, microscopic single-celled organisms that live absolutely everywhere on Earth. They have even been found in rocks hundred of metres beneath the surface. They can live in hot springs, where the temperatures are as high as 80 °C, and in the Arctic, where temperatures rarely rise above freezing. Bacteria have cells unlike those of either animals or plants – they have no nucleus, just a circular molecule of DNA that floats in the cytoplasm.

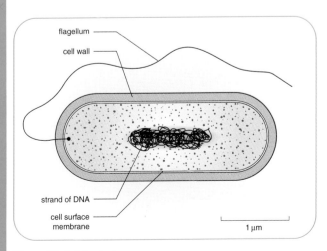

flagellum

cell wall

strand of DNA

cell surface membrane

1 µm

Figure 2b.5 A bacterium.

Fungi belong to yet another kingdom. They have cells that differ from those of plants, animals and bacteria. Like plants and animals, they have a nucleus in which chromosomes are found. They have a cell wall but it is not made of cellulose – fungal cell walls can be made of several different materials, including chitin (the material that strengthens the outer skeleton of insects). Fungi don't have chloroplasts so they can't photosynthesise and make their own food. They live on dead animal or plant material, or waste substances from them such as cow pats. The body of a fungus is made up of long thin threads called **hyphae**, which form a network called a **mycelium**. These grow into and over the material on which they are feeding and secrete enzymes onto it. The enzymes digest the material, forming small, soluble molecules such as sugars that are then absorbed into the hyphae.

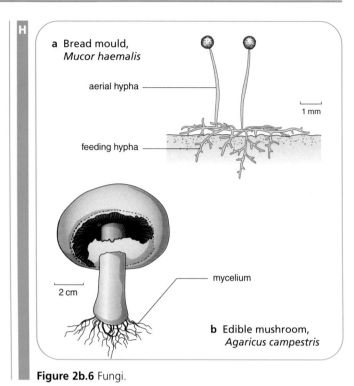

Figure 2b.6 Fungi.

a An amoeba, *Amoeba proteus*

cell surface membrane
cytoplasm
nucleus
0.05 mm
contractile vacuole

b Chlorella, *Chlorella vulgaris*

cytoplasm
cell wall
chloroplast
nucleus
cell surface membrane
0.005 mm

spiral chloroplast
cell surface membrane
cell wall
vacuole
nucleus
strand of cytoplasm
0.1 mm

c Pond weed, *Spirogyra longata*

Figure 2b.7 Some members of the kingdom Protoctista.

H

The fifth kingdom is the **Protoctista**. In some ways, this is a 'dustbin' group – organisms that don't fit into any of the other four kingdoms are classified here. Many of them are single-celled, for example *Amoeba* and *Chlorella*. Others are made up of strings of cells, like the pond weed *Spirogyra*. Many biologists think that the organisms in the kingdom Protoctista aren't really very closely related, and prefer to split them into different kingdoms.

Amoeba is animal-like, and *Chlorella* and *Spirogyra* are plant-like. But many single-celled organisms share features of both animals and plants. *Euglena* is one of these.

SAQ

2 Which feature of *Euglena* is animal like? Which feature is plant like?

H

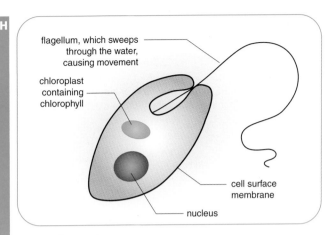

flagellum, which sweeps through the water, causing movement

chloroplast containing chlorophyll

cell surface membrane

nucleus

Figure 2b.8 *Euglena.*

Vertebrates

Vertebrates are animals with backbones. Invertebrates are animals that don't have backbones. They include soft-bodied animals like jellyfish and worms, and also ones with hard outside skeletons like insects and lobsters.

The vertebrates are classified into five groups, called **classes**.

Platypus

Figure 2b.9 Platypus.

In 1799, a preserved specimen of a platypus arrived in England, carried from Australia on a ship. It caused a sensation. Many people thought it was a hoax. The animal had fur on its body but had a beak and webbed feet like a duck, with spurs on its hind legs that contained poison. Its reproductive system looked like that of a bird or a reptile. Later, when people discovered that it laid eggs, the confusion was even greater. Was this strange animal a bird, a reptile or a mammal, or did they need to create a new group to classify it in?

As scientists learned more about the platypus, they decided to classify it as a mammal. Their decision was supported by the discovery that it produced milk and suckled its young. Biologists were certain that only mammals ever did this, so the platypus must share a common ancestor with all the mammals that were already known. The platypus was classified into a group of mammals called monotremes. Now another kind of animal, the echidna or spiny anteater, is also placed in this group.

Class Pisces

The fish all live in water, except for one or two like the mudskipper, which can spend short periods of time breathing air.

Characteristics
vertebrates with scaly skin
have gills
have fins.

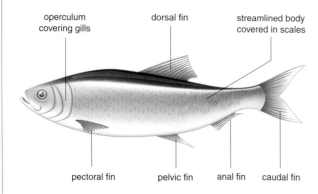

Fish, *Clupea harengus* (herring)

Class Amphibia

Although most adult amphibians live on land, they always go back to the water to breed. Frogs and toads are amphibians.

Characteristics
vertebrates with moist, scale-less skin
eggs laid in water, larva (tadpole) lives in water
but adult often lives on land
larva has gills, adult has lungs.

Frog, *Rana temporaria*

Class Reptilia

These are the crocodiles, lizards, snakes, turtles and tortoises. Reptiles do not need to go back to the water to breed because their eggs have a waterproof shell which stops them drying out.

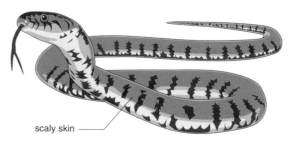

Snake, *Natrix natrix*

Characteristics
vertebrates with scaly skin
lay eggs with rubbery shells.

Class Aves

The birds, like reptiles, lay eggs with waterproof shells.

Characteristics
vertebrates with feathers
forelimbs have become wings
lay eggs with hard shells
homeothermic
have a beak.

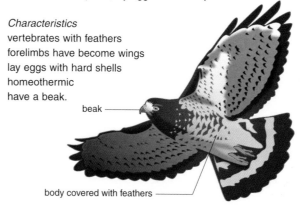

Broad-winged hawk, *Buteo platypterus*.

Class Mammalia

This is the group that humans belong to.

Characteristics
vertebrates with hair, have a placenta, young fed on milk from mammary glands, homeothermic, have a diaphragm, heart has four chambers, have different types of teeth (incisors, canines, premolars annd molars), cerebral hemispheres very well developed.

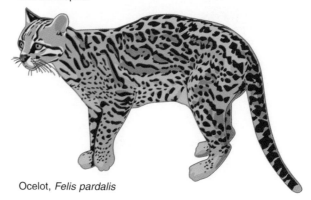

Ocelot, *Felis pardalis*

Figure 2b.10 Classes of vertebrates.

SAQ

3 Match each of these features with a class of vertebrates:

   a have fur and feed their young on milk

   b have wet scales and gills

   c have dry scales and lay eggs

   d have a moist, permeable skin

   e have feathers and a beak.

We know something about the evolutionary history of vertebrates from their **fossils**. Fossil evidence suggests that the first vertebrates to appear on Earth were fish. Amphibians evolved from fish and began to colonise the land. Reptiles evolved from amphibians, and birds and mammals both evolved from reptiles.

If one kind of organism evolved from another kind then we should not be surprised to find 'in-between' organisms. In 1861, a fossilised impression of a feather was found in a rock in Germany. A month later, a small fossilised skeleton was found in the same place, with distinct feather impressions around it. The skeleton looked like a reptile skeleton, but the feathers could only have come from a bird.

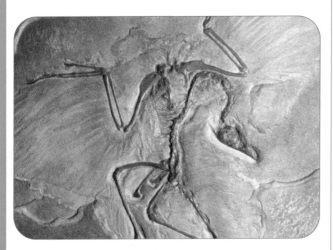

Figure 2b.11 *Archaeopteryx* fossil.

Since then, other fossils have been found of this strange reptile–bird. It is named *Archaeopteryx*, which means 'ancient winged animal'. The fossils are dated at 150 million years old. *Archaeopteryx* had teeth like a reptile and clawed fingers on its wings. If it wasn't for the feathers, biologists would simply have classified it as a reptile.

Archaeopteryx is strong evidence that birds have evolved from reptiles. Some biologists think that reptiles and birds are so similar (apart from feathers) that we ought perhaps to classify birds as reptiles. Maybe birds are just reptiles with feathers.

Species

You belong to the human species. You share a lot of your features with all other humans. Chimpanzees, although they are quite closely related to us, belong to a different species. Organisms of the same species have more features in common with each other than they do with organisms of a different species.

Figure 2b.12 a A human. **b** A chimpanzee.

SAQ

4 What features do you share with a chimpanzee? What features are different?

Binomials

'Binomial' means 'two names'. Every known species has its own two-word Latin name. This is its **binomial**.

The binomial of our species is *Homo sapiens*. The first word is the **genus** that we belong to. A genus is a group of similar species. There are no other living members of our genus on Earth today, but we have found bones and fossils of other human-like species that have been classified in the genus *Homo*, including Neanderthal man, *Homo neanderthalensis*.

The second name of the binomial is the name of the **species**. A species can be defined as a group of living organisms that share the same features and that can breed together to produce fertile offspring. All the humans on Earth belong to the same species.

Notice that the binomial is written in italic. You can't do this when you are writing by hand, so you should underline it instead: Homo sapiens.

Another thing to remember is that the first word in the name (the genus) has a capital letter, and the second word has a small one.

Figure 2b.13 Sorting and naming living things.

This may all seem like a bit of a nuisance. Why not just call us all 'people' or 'humans'? That's probably quite a good argument when it comes to humans, but using a binomial is very helpful when biologists are writing about less well-known organisms or ones that have different English names in different parts of the world. Having one binomial for a particular species, which every biologist in the world uses, avoids confusion.

SAQ

5   These are the binomials of three birds.

chaffinch   *Fringilla coelebs*

greenfinch   *Carduelis chloris*

goldfinch   *Carduelis carduelis*

Which two of these birds are thought to be most closely related? Explain your answer.

Hybrids

Organisms that are in the same species can breed with each other and produce fertile offspring. Usually, organisms in different species can't breed with each other. Dogs and cats don't breed together.

Sometimes, though, two closely related species *can* breed together. Lions and tigers in zoos sometimes breed together to produce ligers or tigons. Horses and donkeys can breed together to produce mules. But these **hybrid** organisms are usually not fertile. They cannot produce young. To produce another mule, you have to cross a horse and a donkey again.

Variation within a species

You've seen that members of the same species share the same features. But that doesn't mean they are all exactly the same. You only have to look around you to see the huge amount of variation there is within the human species.

Humans have been using dogs, which all belong to the species *Canis familiaris*, for thousands of years. People have bred dogs to be all sorts of different shapes and sizes (Figure 2b.15). They are still all dogs and they still belong to the same species. They can all breed with one another (although great danes and dachshunds might have a few difficulties) to produce fertile offspring.

Figure 2b.15 Different breeds of dog.

Similar habitat, similar species

Living things are adapted to live in their habitat. Organisms that live in the same habitat often have similar adaptations, even if they belong to different species.

Take bats and parrots, for example. They both fly and have adaptations to allow them to do this –

Figure 2b.14 a A horse. **b** A donkey. **c** A mule is a hybrid between a horse and a donkey.

they both have wings. But we classify them in different groups: parrots are birds and bats are mammals. If we look closely at their wings, we find that they have completely different structures (Figure 2b.16).

Species that live in different habitats – even if they are closely related – will have different features. For example, the British fox, *Vulpes vulpes*, is the size of a dog and has red fur (Figure 2b.17). The fennec fox, *Vulpes zerda*, lives in deserts in northern Africa. Its small size – smaller than a domestic cat – and large ears help it to lose heat easily in the hot climate. Its sandy-coloured body helps to camouflage it against the sand.

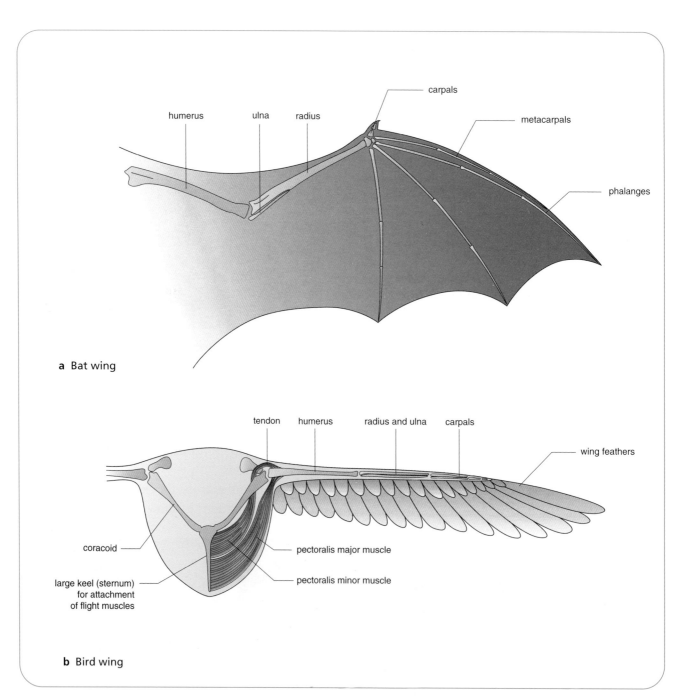

Figure 2b.16 Bat wings and bird wings have completely different structures.

Figure 2b.17 a A British fox. b A fennec fox.

Convergent evolution

Animals that have evolved adaptations that help them to live in a particular environment can look very similar to each other, even if they are only distantly related.

Mammals first evolved from reptiles on land. They were terrestrial animals. However, some groups of mammals then evolved to live in the sea. These aquatic mammals have developed features that they share with other marine animals, such as sharks.

SAQ

6 Describe the similarities between the animals in Figure 2b.18. Why do they have these similarities?

7 Explain why seals and whales are classified as mammals but sharks are classified as fish.

a Seal

b Whale

c Shark

Figure 2b.18 These animals all live in water and have similar appearances, but they are not closely related to each other.

Summary

You should be able to:

◆ describe the characteristics that are used to classify organisms as animals or plants

H ◆ explain why some organisms, such as fungi, are classified as neither plants nor animals

◆ describe the difference between invertebrates and vertebrates

◆ describe the characteristics of fish, amphibians, reptiles, birds and mammals

H ◆ discuss the problems of classifying *Euglena* and *Archaeopteryx*

◆ define the term *species*

◆ use the binomial system for naming species

◆ know that hybrids between species, for example mules, are infertile

◆ explain why distantly related species have similar features if they live in similar environments

◆ explain why closely related species have different features if they live in different environments

Questions

1   A new organism has been discovered on a rock deep under the sea. It is firmly attached to the rock and doesn't move around. It seems to feed by holding out feathery arms into the water and trapping tiny organisms as they float by. Its cells don't have cell walls and they don't contain chloroplasts.

   a   Which features of this organisms are plant-like?

   b   Which features are animal-like?

   c   Is it an animal or a plant? Explain your answer.

2   The photographs (Figures 2b.19 and 2b.20) show an earthworm and a kind of burrowing amphibian that lives in tropical countries, called a caecilian.

   a   Which of these organisms has a backbone?

   b   Describe the kind of skin you would expect the caecilian to have.

   c   Explain why these two animals look so similar, despite being only very distantly related to each other.

continued on next page

Questions - *continued*

Figure 2b.19 An earthworm.

Figure 2b.20 A caecilian.

H **3** Fungi used to be classified as plants but are now put in their own kingdom.

    a Suggest why fungi used to be classified as plants.

    b Explain why they are now put in their own kingdom.

Photosynthesis

Plants make their own food. Unlike us, they don't need to move around to find food or to eat and digest it. All they need to do is stay in one place – everything they need is all around them.

Plants use three things from their environment to make food. These are:

● **carbon dioxide**, which they get from the air
● **water**, which they get from the soil
● **sunlight**.

The carbon dioxide and water are the **raw materials** for **photosynthesis**. Sunlight is the energy source. The energy from the sunlight is captured by the green pigment **chlorophyll**, which is present in the chloroplasts inside some of the cells in the plant's leaves.

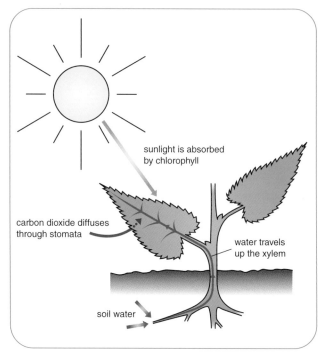

Figure 2c.1 How the materials for photosynthesis get into a leaf.

Rafflesia

The plant with the biggest flowers in the world has scarcely any credentials to be a plant at all. It has no roots, no stems, no leaves – and it doesn't even have any chlorophyll.

Figure 2c.2 *Rafflesia*.

Rafflesia grows in the rainforests of south-east Asia. Its flowers can be more than a metre across. They attract flies to pollinate them and their smell has been described as 'more repulsive than a buffalo carcass in an advanced stage of decomposition'. So, although they are rare, if there is one nearby they are not too difficult to find.

Down on the floor of the rainforest, beneath the dense canopy of leaves of the huge trees above, little light penetrates. It is difficult for any photosynthetic plant to live there. *Rafflesia* has solved the problem by letting other plants do the photosynthesising and then helping itself to the food they have made. It is a parasite, thrusting thread-like strands of its tissue into the body of a host plant, drawing out nutrients and water.

Like most parasites, *Rafflesia* does not take enough from its host to kill it. That would mean its own death, too. Quietly and slowly, it gradually builds up its strength until it can produce its grand performance – the appearance of a huge, stinking, disembodied flower resting amongst the leaf litter on the forest floor.

The food factory 89

In photosynthesis, carbon dioxide and water react to produce **glucose** and **oxygen**. The energy from sunlight drives this reaction. We can summarise the reaction using a word equation like this:

$$\text{carbon dioxide} + \text{water} \xrightarrow[\text{chlorophyll}]{\text{light energy}} \text{glucose} + \text{oxygen}$$

The balanced symbol equation is:

$$6CO_2 + 6H_2O \xrightarrow[\text{chlorophyll}]{\text{light energy}} C_6H_{12}O_6 + 6O_2$$

SAQ

1 Most plants have a branching shape with broad, flat leaves. How does this help them to get more of the raw materials that they need for photosynthesis?

2 Name the reactants and the products of photosynthesis.

Products of photosynthesis

The two products of photosynthesis are glucose and oxygen. The oxygen diffuses out of the plant's leaves and into the air. The glucose is used by the plant to provide its cells with energy, and also to make all the different chemicals that it needs.

Because photosynthesis only happens in the leaves, glucose needs to be transported to other parts of the plant. It is changed into another sugar, **sucrose**, when it is to be transported. Sucrose is carried inside long tubes called **phloem tubes**, which run all the way up and down the plant, carrying the sucrose to wherever it is needed. If necessary, it can be changed back into glucose when it arrives at its destination.

To provide energy, glucose is reacted with oxygen. This is **respiration**. Respiration happens in all the plant's cells all the time, because they need energy all the time.

SAQ

3 Write down the word equation for respiration (see if you can do it without having to look it up). How does it differ from the photosynthesis equation?

Glucose is a **carbohydrate**. It is a simple sugar. This means that it has small molecules – look at Figure 1b.3 (page 10) if you have forgotten this.

Glucose molecules can be linked together to produce **starch** molecules (starch is also a carbohydrate). This is the way in which plants store carbohydrates. The starch may be stored inside the chloroplasts in the leaf cells or in parts of the root or in underground tubers (like potatoes). When the plant gets short of glucose – perhaps because it is winter and there hasn't been enough light for photosynthesis – it can break the starch down into glucose again.

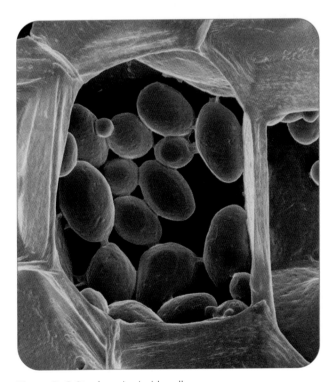

Figure 2c.3 Starch grains inside cells.

If the glucose molecules are linked together in a different way, they make **cellulose**. This is the material from which the plant's cell walls are made.

The glucose molecules can be changed into **fats** and **oils**. These, like starch, are used for storing energy.

H

Starch is a good way of storing energy. Starch molecules are big but very compact, because the chain of glucose molecules is curled into a spiral (see Figure 1b.3). So you can pack a lot of it into a small space inside a cell.

Starch, fats and oils are all **insoluble** in water, so they don't dissolve in the cytoplasm. This means that they don't affect the concentration of the cytoplasm. If the plant stored glucose instead, the glucose would dissolve and could affect the metabolic reactions taking place inside the cell. The dissolved glucose would make the concentration of the solution inside the cell greater than the concentration outside. This would cause water to move into the cell by a process called **osmosis**. This could upset the normal activities that take place inside the cell.

SAQ

4 Explain why sucrose, and not starch, is the carbohydrate that is transported around in a plant.

With the addition of some extra materials obtained from the soil, the plant can turn some of the glucose into **proteins**. To do this, it needs **nitrates** from the soil. (Proteins contain carbon, hydrogen, oxygen and nitrogen, whereas glucose only contains carbon, hydrogen and oxygen.) The proteins are used for growth and repair.

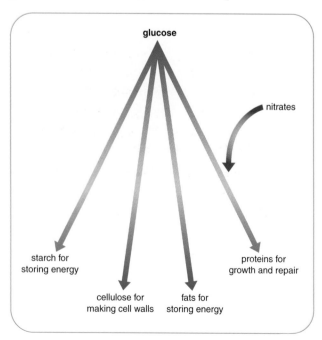

Figure 2c.4 Products of photosynthesis.

Speeding up photosynthesis

We've seen that plants need carbon dioxide, water and light for photosynthesis. If we give them more of these things, we might expect that they will photosynthesise faster.

A good way of measuring the rate of photosynthesis is to use a water plant. We can watch the bubbles of oxygen that it produces and count how many are made in one minute. The more oxygen bubbles, the faster the rate of photosynthesis.

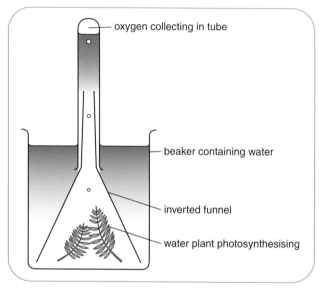

Figure 2c.5 Measuring the rate of photosynthesis.

We could give the water plant extra light by putting a lamp close to it – the closer the lamp, the more light the plant gets. We could give it extra carbon dioxide by adding some sodium hydrogencarbonate to the water, which will break down and release carbon dioxide. (We obviously can't give it any extra water.) Both of these will speed up its rate of photosynthesis.

SAQ

5 a Using the graph in Figure 2c.6a, describe what happens to the rate of photosynthesis as light intensity is increased.

b Explain why the curve rises between A and B.

Another way of speeding up photosynthesis is to increase the temperature. The warmer it is, the faster a plant will photosynthesise. However, if it

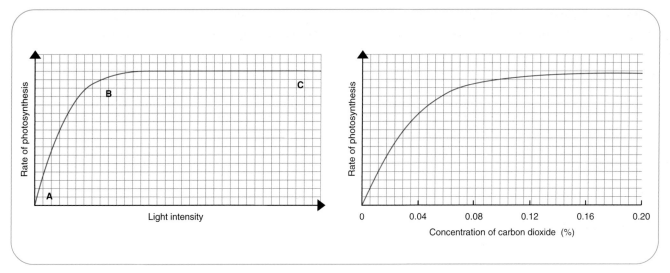

Figure 2c.6 Some factors affecting the rate of photosynthesis. **a** Light intensity. **b** Carbon dioxide concentration.

gets too hot then photosynthesis will actually slow down, because the high temperatures will denature enzymes in the plant's cells.

SAQ

6 Explain why plants usually photosynthesise faster in the summer than in the winter. (You should be able to think of two reasons.)

7 The enzymes in plant cells usually have optimum temperatures around 25 °C, whereas human enzymes have optimum temperatures around 38 °C. Suggest why this is.

8 Sketch a graph, like the ones in Figure 2c.6, to show how temperature affects the rate of photosynthesis. Remember that photosynthesis is actually *slowed down* when the temperature gets too high.

Limiting factors

Look back at Figure 2c.6a. You can see that light intensity increases the rate of photosynthesis, but only up to a point. At B, if we continue to increase the light falling on the plant, photosynthesis just carries on at the same rate.

This means that the plant must be getting more light than it can use. Something must be stopping it from photosynthesising any faster, no matter how much extra light it gets.

The thing that is stopping it is called a **limiting factor**. In this case, it is probably the amount of carbon dioxide it is getting. Carbon dioxide is often the factor that limits the rate of photosynthesis, because there isn't very much of it around. Only about 0.04% of the air is carbon dioxide. Temperature can also be a limiting factor, if the temperature is below the optimum for the plant's enzymes.

A good way of finding out what is limiting the rate of photosynthesis is to give the plant more of something. If that makes it photosynthesise faster, then the 'something' must have been limiting the rate up until then.

SAQ

9 a Look at Figure 2c.6a. These results were obtained when the carbon dioxide concentration and the temperature were kept constant. What is limiting the rate of photosynthesis just above point A on the graph? How can you tell?

b Look at Figure 2c.6b. What is limiting the rate of photosynthesis when the concentration of carbon dioxide is 0.02%? How can you tell?

c Suggest *two* factors that could be limiting the rate of photosynthesis when the concentration of carbon dioxide is 0.16%. How could you find out if you were right?

Plant respiration

Respiration is a metabolic reaction that provides energy for cells. In aerobic respiration, glucose reacts with oxygen:

glucose + oxygen → carbon dioxide + water

This releases energy from the glucose, in a form that the cells can use.

SAQ

10 Where did the energy in the glucose originally come from?

All cells respire, all of the time. This is just as true for plant cells as for animal cells, though plants do tend to respire more slowly than animals. This is because plants don't need so much energy, because they don't move around.

At night, when there is no light, plants cannot photosynthesise. But they carry on respiring. At night, therefore, plants take in oxygen and give out carbon dioxide.

In daylight, they photosynthesise as well as respiring. Usually, they photosynthesise faster than they respire. So they use up carbon dioxide (in photosynthesis) faster than they give it out (in respiration). Overall, they take in carbon dioxide during daylight.

SAQ

11 What gas do plants give out during daylight? Why?

Summary

You should be able to:

◆ state the word equation for photosynthesis

◆ state the balanced symbol equation for photosynthesis

◆ know the other substances that glucose can be converted to, and what these substances are used for

◆ explain why starch, fats and oils are good for storing energy

◆ describe how the rate of photosynthesis can be increased by providing more light or carbon dioxide, or a higher temperature

◆ explain what *limiting factors* are

◆ explain why plants respire all the time

◆ explain, in terms of *respiration* and *photosynthesis*, why plants give out and take in different gases during daylight and at night

Questions

1 Copy and complete these sentences about the products of photosynthesis.

   a Photosynthesis produces glucose and the gas ....

   b Glucose molecules can be strung together to make .... for storage and

      .... for cell walls.

   c With the addition of .... , glucose can be made into proteins for

      .... and ....

2 A student investigated the effect of light intensity on the rate of photosynthesis of a water plant, using the apparatus in Figure 2c.5. He started off with a lamp close to the plant (giving a high light intensity) and then moved it away. These are his results.

Distance of lamp from plant in cm		10	20	30	40	50
Number of bubbles per minute	1st try	34	35	29	16	8
	2nd try	38	37	25	18	5
	3rd try	36	39	27	17	5

Table 2c.1 Results of investigation.

   a Calculate the mean number of bubbles per minute for each distance of the lamp. Use these to draw a line graph of the results.

   b Explain why your graph doesn't look like Figure 2c.6a.

   c Explain why it was a good idea to take three readings at each distance of the lamp.

   d The student was surprised with the results for 10 cm and 20 cm. Explain why he got these results.

   e One important variable was not controlled in this experiment, because the lamp gave out heat as well as light. Explain how this could have affected the results.

The struggle for existence

Life isn't easy. If you are an animal or a plant living in the wild, your chances of surviving long enough to have offspring aren't good. A pair of wild rabbits might have three litters a year, each with six or seven young. If the rabbit population is staying about the same year on year, this must mean that, on average, only two of all the young that one pair of rabbits have in their lifetimes will survive. If it was more, the population would go up.

Figure 2d.1 Wild rabbits.

So what is it that kills so many young rabbits? It could be that they get killed by predators, or perhaps they get a disease and die. Perhaps there isn't enough space for them to make a burrow, and there isn't anywhere suitable nearby that they can move away to, so they have nowhere to hide from predators. Perhaps there isn't enough food to go around.

All of these factors affect how many rabbits there are – their **population size** or **abundance**. They also affect where they live – their **distribution**.

Competition

In the wild, there often aren't enough resources to go around. When two organisms both need the same resource and it is in short supply then they **compete** for it.

Animals can compete for a number of different resources, especially:
- food
- water
- shelter and breeding sites.

Figure 2d.2 Competition for water, for example between these wildebeeste, can be fierce in the African dry season.

Plants also compete with each other. The resources they compete for include:
- light
- water
- minerals.

Figure 2d.3 Desert plants, such as these Joshua trees, compete for water. Their roots spread in a circle around them, so other Joshua trees cannot grow too close

'Competition' conjures up visions of two animals fighting tooth and claw over whatever it is that they both want. That *can* happen, but it isn't always the case. It might be that one animal finds a place to make a nest and another one doesn't, because there aren't enough nesting sites to go around. They probably don't fight over the nesting site – it is just a case of first come, first served.

SAQ

1 How might competition for water affect the distribution of wildebeeste?

2 How might competition for water affect the population size of the Joshua trees?

Competition within the same species

Usually, all the animals or plants within a particular species need the same resources. Imagine a population of frog tadpoles in a small pond. They all need the same sort of food. If the pond is very crowded, there may not be enough food to go around. Some tadpoles will get enough to eat but others won't, and they will die.

This kind of competition affects the size of the population of a species. If the population size gets too high then there aren't enough resources to go around, and some of the organisms die. This reduces the size of the population, so now there are fewer organisms competing for the resource. If the population goes up again, the same thing will happen. Competition between members of a species affects how many there are of them – their population size.

SAQ

3 a Describe how the death rate of the fruit-fly larvae shown in Figure 2d.4 is affected by the amount of food they are given.

b Explain why this happens.

Competition between different species

Plants or animals of different species often compete for a particular resource that is in short supply. For example, trees in a forest all need light. Some species of trees grow tall, increasing the amount of light that falls onto their leaves. This means that less light reaches the ground beneath the trees. Other, shorter plants can't compete with the tall trees, so they can't grow on the forest floor. Competition for light between the trees and the shorter-growing plants means that the smaller plants can't grow in the forest.

Figure 2d.5 There is not enough light on the forest floor for other plants to grow.

The same kind of thing happens between different species of animals. Red squirrels used to be common in English woodlands. They eat a range of different foods but, in the winter, they rely mostly on hazelnuts and pine seeds. Grey squirrels have been introduced into Britain from

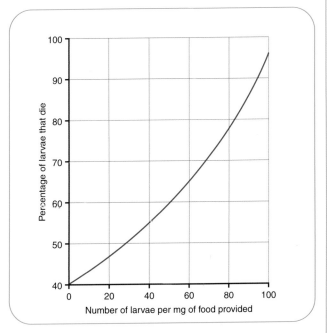

Figure 2d.4 Effect of competition for food on fruit-fly larvae.

Figure 2d.6 a Red squirrel. **b** Grey squirrel.

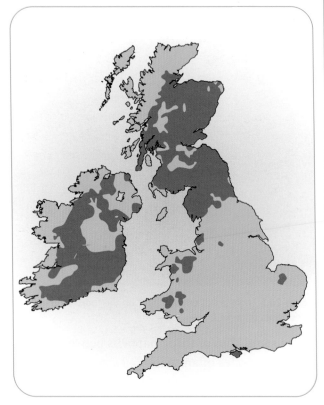

Figure 2d.7 Map of red squirrel distribution. The purple areas show where they are still found.

America. Grey squirrels have a very similar lifestyle to red squirrels but they are a bit larger and eat a wider variety of foods. Although red and grey squirrels don't fight, they do compete with each other for food, especially in the winter when food is in short supply.

Gradually, red squirrels have disappeared from most of our woodlands, although you can still find them in a few areas, such as parts of northern England. The reds have been outcompeted by the greys. The larger greys put on more fat to see them through the winter, and seem to be able to find more food supplies. As a result of this competition, the abundance of reds has fallen and their distribution has been severely reduced.

The lifestyle of a species of organism – the way the species fits into an ecosystem and what it needs to survive – is called its ecological **niche**. The more similar the niches of two species, the more likely they are to be in competition with one another. Red squirrels and grey squirrels have niches that are very similar to each other. They both live in woodland, they both eat nuts and they both need to fatten up before the winter to see them through the months when food is scarce. This is why competition between them is so fierce. The two species of squirrel both need the same resources.

Predator-prey relationships

A **predator** is an animal that kills other animals and eats them. The **prey** is the animal that is killed.

If a predator tends to eat mainly one particular species of prey then the populations of the predator and prey may be closely related to each other. Figure 2d.8 shows an example. In Finland, voles (small brown rodents) are the main prey species of weasels. The graph shows how the populations of voles and weasels go up and down – **oscillate** – in a similar way.

You can see that the ups and downs for the predator and prey populations don't happen at the same time. Start at 1950 and follow through what is happening. First, the vole population goes up, reaching a peak in 1952. As the numbers of voles increase, the weasels have more food, so

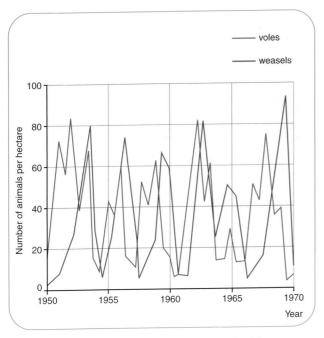

Figure 2d.8 Weasel and vole populations in Finland between 1950 and 1970.

more of them survive and their numbers go up, too. They reach their peak 2 years later, in 1954.

By 1954, the vole population has fallen right down again. Maybe there wasn't enough food to go around, or maybe the big population of weasels killed and ate more of them. So the weasel population also falls, because there aren't enough voles to eat. So now the voles increase again, because not so many of them are being eaten by voles … and so on. The predator and prey populations go up and down in **cycles**, always just out of phase with each other.

Parasites

A **parasite** is an organism that lives in close association with its **host**. It feeds on the host and does it harm.

SAQ

4 How does a parasite differ from a predator?

Some parasites live on the skin of their host. For example, cats and dogs may get fleas that live amongst their fur. Wild rabbits are almost always infested with fleas.

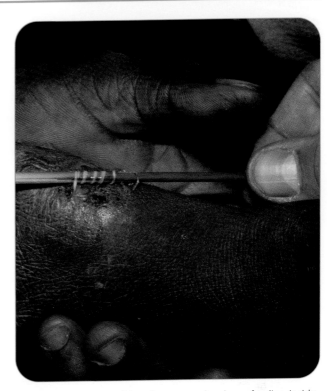

Figure 2d.9 This parasitic guinea worm has been feeding inside the person's body. Now it is being carefully removed from the leg of its human host.

Other parasites live inside their host's body. For example, cats and dogs may get tapeworms inside their digestive system.

Parasites tend to be very specialised. They can only live on one particular host. Cat fleas can't live on people – they might bite you a couple of times but they won't stay. They can only live and reproduce on cats.

Obviously, the distribution and numbers of a parasite depend on the distribution and numbers of its host. Rabbit fleas can only live where there are rabbits. The more rabbits there are, the more fleas there will be.

Sometimes, the distribution and numbers of the host are affected by its parasites. For example, rabbit fleas pass on a virus that causes a deadly disease of rabbits called myxomatosis. When rabbit populations get large, there are more fleas and it becomes easier for the fleas to pass the disease around. More rabbits get myxomatosis, so more of them die.

5 The graph shows what happened to a rabbit population after the introduction of the myxomatosis virus.

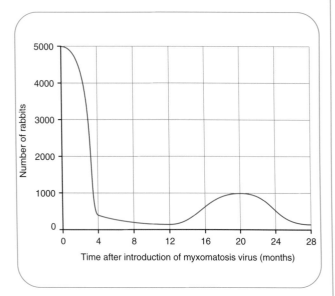

Figure 2d.10 The effect of myxomatosis on a rabbit population.

a How long did it take for the virus to take effect?

b Suggest two reasons why not all of the rabbits were killed.

c Describe what happened in the second year after the virus was introduced. Suggest reasons for this.

Mutualism

Interactions between organisms aren't always harmful. There are many cases where two different species help each other to survive. If they live closely together, and both get benefit from the relationship, it is called **mutualism**.

Cleaners

There are many different examples of one animal getting food by cleaning parasites from the skin of another. Ox-peckers do this, sitting on the backs and heads of buffalo or other animals and pecking inside their ears or in folds of skin where fleas and ticks are attached. The ox-pecker gets food and the buffalo has its parasites removed.

Fish do this, too. Every coral reef has its population of cleaner fish. Larger fish go to them

Figure 2d.11 An ox-pecker helps to remove a buffalo's parasites.

Figure 2d.12 A cleaner fish working on a predatory fish.

to have parasites removed from their skins. They lift up their gill covers and open their mouths so the cleaners can reach inside. Even predators that normally eat smaller fish do this. The cleaners are quite safe.

There are, though, a few species of fish that just *pretend* to be cleaners. They look really similar to real cleaners. But instead of cleaning parasites from the unsuspecting fish that thinks it is being cleaned, they bite lumps out of their flesh.

6 There is some disagreement about whether the pretend cleaner fish are predators or parasites. What do you think? Explain your decision.

Probiotics

Your digestive system is full of bacteria. Most of them are good for you. For example, a bacterium called *Lactobacillus* and another called *Bifidobacterium* help to stop harmful bacteria growing. There's also some evidence that they might help the immune system to work well.

This is an example of mutualism. We need the bacteria and they need us.

In the past few years, a big market has grown up for foods and food supplements containing 'probiotics'. Probiotics are live cultures of these beneficial bacteria. We shouldn't really need them if our own intestines already have healthy populations of these bacteria. But some people think that they do seem to be helpful in some circumstances. For example, if you've had a bad bout of diarrhoea, you may have 'flushed out' a lot of your gut bacteria. Taking probiotics *might* help you to recover a bit more quickly.

Figure 2d.13 Probiotic foods.

Nitrogen-fixing bacteria

We've seen that plants need nitrogen so that they can make proteins. They first make glucose in photosynthesis and then use the glucose to make amino acids, adding nitrogen that they get from nitrates in the soil. The amino acids are then strung together to make proteins.

The Earth's atmosphere is almost 80% nitrogen, so you'd think there would be no problem for a plant in getting enough nitrogen. But nitrogen gas, N_2, is very unreactive. There's no way a plant could make nitrogen gas react with glucose. They need nitrogen that has already combined with something else – that is, **fixed** nitrogen. Normally, they use nitrate, NO_3^-, which they get from the soil.

Often, nitrates are in short supply in the soil, so plants compete for them. But some – especially plants of the pea family, known as **leguminous plants** – have a mutual arrangement with a particular kind of bacteria that helps them out. The bacteria, called **nitrogen-fixing bacteria**, live inside the plant's roots, in little swellings called **root nodules**. The bacteria take nitrogen gas from the air spaces in the soil, and use it to make ammonium ions, NH_4^+. This is called **nitrogen fixation**. The plant can use these ions to make amino acids. The bacterium also benefits from this arrangement. It gets sugars that the plant has made in photosynthesis.

Figure 2d.14 Root nodules.

SAQ

7 Farmers and gardeners sometimes grow beans
and let the roots stay in the soil after harvest.
This helps a crop, such as wheat, that grows in
the field the next year to give a higher yield.
Suggest how this happens.

Summary

You should be able to:

◆ explain what *competition* is and know some of the resources that plants and animals compete for

◆ explain how competition between species can influence their distribution, and give some examples

◆ explain how competition within species can influence their population size (abundance) and give some examples

H ◆ explain that species with similar ecological niches are more likely to compete with each other

◆ recognise organisms as predators or prey

◆ explain how the population sizes of predators and prey can influence each other

H ◆ explain why predator and prey populations sometimes show cyclic fluctuations

◆ explain what a *parasite* is and give some examples of parasites and their hosts

◆ explain what *mutualism* is and give some examples

H ◆ describe the relationship between nitrogen-fixing bacteria and leguminous plants, and explain why this is beneficial to both species

Questions

1 Copy and complete these sentences. Use some of these words.

prey   light   space   predator   parasite   good   food   water   harm   host   minerals
carbon dioxide   air   oxygen   nitrogen

a A .... is an animal that kills and eats its ....

b A .... lives in or on its host, feeds on it and does it ....

c Plants compete for .... , ... and ....

d Animals compete for .... , ... and ....

continued on next page

Questions - *continued*

2 A group of goats was introduced onto an island. They ate grass. There were no predators on the island. The graph in Figure 2d.15 shows what happened to the goat population over the next few years.

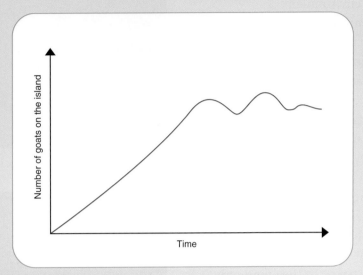

Figure 2d.15 Goat population.

a Suggest why the goat population didn't keep going up for ever. Use the word *competition* in your answer.

b A few years later, sheep were introduced to the island. Using the words *niche* and *competition*, explain why this might cause the population of goats to fall.

Adaptation

An **adaptation** is a feature that helps an organism to survive. Every living organism has adaptations that help it to live in its environment.

We have seen that plants and animals have to compete to survive. Only those that are well adapted to their environment are likely to live long enough to become adult and reproduce. This limits particular species of animals and plants to living in particular habitats.

For example, polar bears have adaptations that allow them to live in very cold conditions. Grizzly bears are less well adapted for those conditions. In the Arctic, grizzly bears cannot compete well with polar bears, so grizzlies are not found in the really cold, icy parts of the Arctic. Polar bears can live in warmer places but they can't compete successfully with grizzly bears, so they tend only to live in the Arctic.

SAQ

1 Suggest what resources polar bears and grizzly bears might compete for.

Figure 2e.1 A polar bear on pack ice.

Figure 2e.2 A grizzly bear.

Global warming threatens polar bears

Polar bears travel long distances by walking across sea ice. The extent of the sea ice limits the southern edge of their range.

As global warming takes place, the sea ice is retreating. On average, the air temperature in the Arctic has increased by 5 °C in the past 100 years. As a result, sea ice is melting and breaking up earlier in spring. Less sea ice is forming in winter. Polar bears catch most of their food on the ice, so this is limiting their ability to find food. One study has found that for every week earlier the ice breaks up, the polar bears are about 10 kg lighter.

Female polar bears spend the winter in a den that they dig out of the snow. Here they give birth to their cubs. If it rains rather than snows in late winter, the rain can make the den collapse before the cubs are ready to leave. This can kill the young bears and perhaps their mother, too.

Some scientists think that, if global warming continues, there will be no sea ice at all in the Arctic summers by 2040. There will be nowhere for the polar bears to live. Might there come a time when the only polar bears in the world are in zoos?

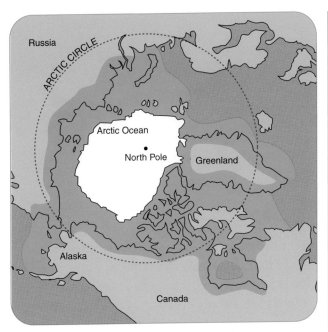

Figure 2e.3 Polar bear distribution (dark green).

Predators and prey

Some adaptations are fairly obvious, and we can see them just by looking at the organism. For example, a hawk is a predator, and it has a sharp beak and talons that help it to kill its prey. Sometimes we can't immediately see the adaptations. They might be to do with the organism's behaviour rather than its structure. For example, the hawk knows how to swoop down and catch its prey.

A tiger is a predator. Figure 2e.4 shows some of its adaptations.

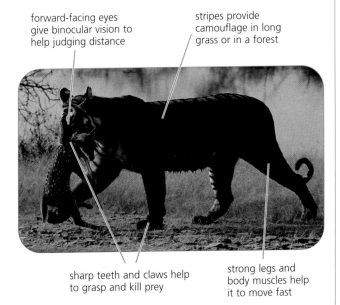

forward-facing eyes give binocular vision to help judging distance

stripes provide camouflage in long grass or in a forest

sharp teeth and claws help to grasp and kill prey

strong legs and body muscles help it to move fast

Figure 2e.4 Adaptations of a tiger.

2 A scorpion is a predator. It eats insects.

Figure 2e.5 A scorpion.

How is a scorpion adapted to be a successful predator?

An antelope is a herbivore – it eats grass. It is hunted by big cats such as lions and leopards. Figure 2e.6 shows some of the adaptations an antelope has to help it to survive.

ears detect sounds made by predator

body a similar colour to the ground, for camouflage

live in groups to help with spotting predators before they get too close

eyes at side of head for all-round vision

hooves and long legs help it to move fast over the ground

good sense of smell helps to detect predators

Figure 2e.6 Adaptations of an antelope.

Frogs are prey for snakes. Poison arrow frogs live in tropical rainforests. They have poisonous slime on their skin. The poisons are used by native people in the Amazon rainforest to put onto the tips of arrows. Poison arrow frogs don't need to be camouflaged. Their bright colours are a warning that tells potential predators that they are poisonous.

Figure 2e.7 A poison arrow frog.

SAQ

3 Hedgehogs are prey for badgers.

Figure 2e.8 A hedgehog.

How are hedgehogs adapted to help them to survive?

Different environments

Living things can be found almost everywhere on Earth. They live in water, on land and in the air. Being successful in these different environments requires different adaptations.

Figure 2e.9 A fish.

Fish live in water. Most fish have a streamlined body that cuts easily through the water. The body is covered with smooth scales, which often have a slimy covering of mucus, which reduces friction and helps them to slip through the water. They have fins and a tail to help to propel them through the water, and to balance them. They have gills for gaseous exchange.

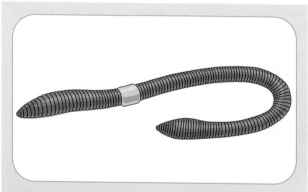

Figure 2e.10 An earthworm.

Earthworms live in the soil. Earthworms have a long, thin body. This helps them to move easily through burrows under the soil. They have tiny bristles on their underside, which can grip against the sides of the burrow to help them to push themselves forwards. They don't have eyes, because these would not be much use underground, and they usually only come out of their burrows at night. However, they are sensitive to light and tend to move away from it. An earthworm's body is covered with slippery mucus, which helps it to slide through its burrow. Earthworms feed by eating dead leaves and soil.

Figure 2e.11 Swifts.

Swifts live in the air. Like all birds, their front pair of limbs is modified to form wings. Swifts have especially long, swept-back wings, which help them to fly very efficiently for long periods of time. They spend almost all their lives in the air. Their bodies are covered with feathers, which are light and provide heat insulation. They have air spaces in their bones, which cuts down their weight. Swifts feed on insects, which they catch while flying.

Deserts

A desert is a very hostile place to live. There is very little water. All plants and animals need water to survive so, if they live in a desert, they must have ways of not letting their body dry out and making the very best use of the little water that there is.

Figure 2e.12 Some plants have adaptations that allow them to live in a desert.

Deserts can often be extremely hot in the day time. This makes the water problem even worse, because water evaporates quickly at high temperatures. But at night it can get very cold.

Many different kinds of plants have adaptations that help them to live in deserts. The ones everyone knows about are **cacti** (singular: cactus).

A surprising number of animals live in deserts, but most of them are small. One of the largest animals that can live in a hot, dry desert is the **camel**. There are few wild camels now because they are nearly all kept by people, who use them for transport. A camel can last for many days without water. And they can allow their body

swollen stem stores water

stem is covered with a thick, waterproof, waxy cuticle to prevent water loss

stem is green and photosynthesises (instead of leaves)

leaves are reduced to spines to cut down water loss by transpiration (evaporation from the leaf surface); the spines also stop animals from eating the plant

rounded shape gives a low surface area to volume ratio, so there is less surface from which water can evaporate and more volume in which water can be stored

roots spread wide and deep, so they have a good chance of reaching water

Figure 2e.13 Adaptations of a cactus.

Figure 2e.14 Camels can live in the desert.

temperature to rise up much higher than usual – as high as 40 °C – without causing any long-term harm to their cells. This means that they don't need to sweat to keep their temperature down – so more water is conserved inside the body.

bushy eyebrows and hair-lined nostrils help to keep wind-blown sand out of eyes and nose; the nostrils can be closed completely

hump stores fat, so there is no fat elsewhere, allowing heat to be lost easily from the body

large feet help to spread the load, so the camel does not sink into soft sand

Figure 2e.15 Adaptations of a camel.

Snow, ice and cold

Near the North and South Poles, temperatures are so low that everything is permanently covered with snow. The sea freezes. This is a really difficult place for plants and animals to live.

thick layer of fat beneath the skin provides insulation to reduce heat loss

large body gives a small surface area to volume ratio, so there is proportionally less surface through which heat can be lost

thick white fur provides camouflage against the snow, and insulation to cut down on heat loss from the body

strong legs for running on snow and ice, and for swimming

small ears keep surface area small, so less heat is lost

Figure 2e.16 Adaptations of a polar bear.

Polar bears are superbly adapted for these conditions. Polar bears are only found in the Arctic. Their main food source is seals. The bears walk across the sea ice and wait for seals to come up for air. Then they have a brief chance to catch one.

large feet spread the load on the snow and ice, so the bear does not sink into the snow or break through the ice

fur on the underside of the paws provides insulation and grip

sharp claws hold and kill prey

Figure 2e.17 Adaptations of polar bear feet.

SAQ

4 Fennec foxes are adapted to live in deserts. Arctic foxes are adapted to live in the Arctic.

a

b

Figure 2e.18 a A fennec fox. b An arctic fox.

Use what you can see in the pictures to explain how the differences between a fennec fox and an arctic fox are related to the places where they live.

H Pollination

Many plants reproduce using flowers. This is sexual reproduction.

The male gametes of a plant are inside its **pollen grains**. The female gametes are inside its **ovules**.

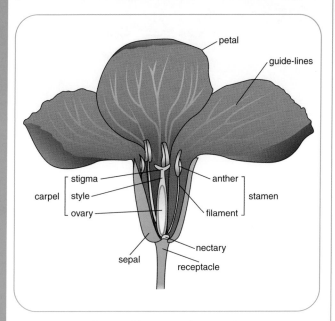

Figure 2e.19 A flower with one petal removed.

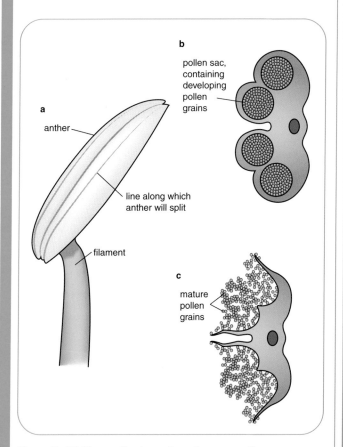

Figure 2e.20 How pollen is made. **a** A young anther.
b Transverse section through a young anther. **c** Transverse section through a mature anther.

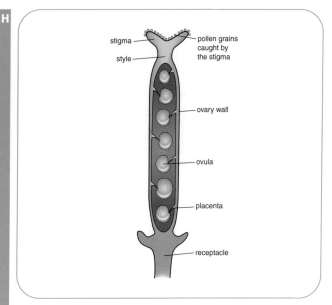

Figure 2e.21 Section through the ovary, style and stigma of the flower in Figure 2e.19.

Before fertilisation can happen, the pollen grains have to be moved from the anther where they were made, onto a stigma. This is called **pollination**.

Pollen can't move itself around. So plants have become adapted to making use of their environment to help pollen to get from one flower to another.

Some plants are pollinated by insects. The flower in Figure 2e.19 is an **insect-pollinated** flower. Insects don't pollinate flowers for nothing – they do it because they get something out of it. Insect-pollinated flowers contain sweet-tasting nectar that insects collect. The nectar is at the base of the petals, so the insect has to push past the anthers to reach it. Some of the pollen rubs off the anthers and sticks onto the insect's body.

Figure 2e.22 The yellow structures in the middle of the flower are its anthers, covered with yellow pollen. You can't see the stigma.

The insect then flies off to another flower. As it pushes down into the flower to get more nectar, it brushes against the stigma, and pollen from its body rubs off onto the stigma.

Insect-pollinated flowers usually have brightly coloured petals that attract insects. They often have a smell, too, which insects can detect from some distance away. Many flowers smell nice to us, but some smell just like rotting meat. These flowers attract flies to pollinate them. The flies – whose larvae feed on rotting meat – are fooled into thinking there is a dead body there, onto which they could lay their eggs.

Figure 2e.23 Pollen from an insect-pollinated flower is usually sticky or spiky, so it sticks to the insect's body. These pollen grains have been magnified about 1000 times.

Other flowers use the wind to transfer their pollen. These **wind-pollinated** flowers don't need colours, smell or nectar to attract insects, so their flowers are usually small and insignificant.

The flowers are often held high above the ground on long stems, so it is easy for the wind to catch them. The anthers are long and dangle out of the flower, so pollen easily blows away. The stigmas also dangle outside the flower. They are feathery, which makes it easier for them to catch pollen blowing past on the wind. The anthers produce huge quantities of very light pollen, which blows away easily. They need to make a lot because it blows everywhere, and only a very small proportion will land on a stigma of the same kind of plant. Insect-pollinated flowers get a better-directed delivery service, so they can get away with making less pollen.

Figure 2e.24 Each grass stem carries many small flowers. You can see the yellow anthers dangling out of each of the colourless flowers. These flowers are wind-pollinated.

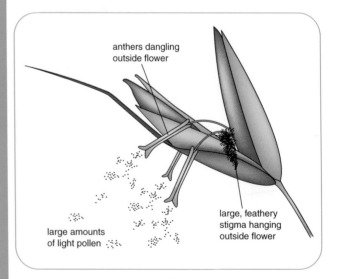

anthers dangling outside flower

large, feathery stigma hanging outside flower

large amounts of light pollen

Figure 2e.25 A wind-pollinated flower.

SAQ

5 Construct a table that compares the adaptations of insect-pollinated flowers and wind-pollinated flowers.

Summary

You should be able to:

◆ describe how the adaptations of an organism determine where it can live

◆ explain how organisms compete for limited resources and how their success is affected by their adaptations

◆ describe some of the adaptations of predators and prey

◆ describe how fish, birds and earthworms are adapted to survive in their habitats

◆ describe some of the adaptations of camels to living in deserts

Ⓗ ◆ explain how its adaptations help a cactus to survive

◆ describe how polar bears are adapted to living in the Arctic

Ⓗ ◆ explain how the structures of wind-pollinated and insect-pollinated flowers help them to reproduce

Questions

1 Make a table with two columns headed 'Predator' and 'Prey'. Write each of these features into the correct column. You may need to write some of the features in both columns.

● eyes at the front of head for binocular vision

● eyes at the side of the head for all-round vision

● sharp teeth and claws

● high speed

● camouflage

● poisonous sting for defence

● live in groups to increase chance of spotting danger

2 Polar bears are found in the Arctic. Imagine that someone wants to introduce polar bears to the Antarctic.

a Could polar bears live in the Antarctic?

b How might they affect other animals that live there, such as penguins?

3 How is the seal in Figure 2e.26 adapted for its way of life?

Figure 2e.26 A seal.

Fossils

On 28 June 1924, the mountaineer and geologist Noel Odell was climbing on the slopes of Everest when he spotted something amazing. At a height of almost 8000 m above sea level, there were **fossils** of sea creatures in the exposed rock.

Fossils are the remains of long-dead organisms, embedded in rocks. The fossils that Odell found demonstrated that once, long ago, the rocks that are now at the highest point on Earth were underneath the sea. These fossils have helped us to understand how the seemingly solid Earth is always changing, always moving. Fossils give us an insight into times so long ago that no-one can comprehend the vast times that have elapsed as the Earth has developed.

How fossils form

There are many different ways in which a fossil can form, but they all happen when an organism dies in a place where new rocks are forming. For example, a sea-living animal might die and fall to the bottom, where sediment is slowly building up.

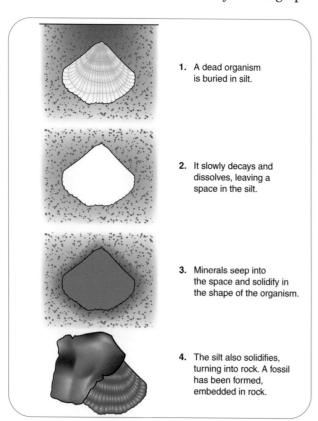

1. A dead organism is buried in silt.

2. It slowly decays and dissolves, leaving a space in the silt.

3. Minerals seep into the space and solidify in the shape of the organism.

4. The silt also solidifies, turning into rock. A fossil has been formed, embedded in rock.

Figure 2f.1 One way in which a fossil may be formed.

Over hundreds of thousands of years, the sediment gets squashed and hardened, eventually becoming a rock with the animal remains inside it.

Figure 2f.1 shows one way in which a fossil might form. The hard body parts of the animal dissolve and are replaced by minerals that gradually seep into the space the animal occupied. These minerals take up the shape of the animal, so we get an animal-shaped piece of rock. On other occasions, all we might get is an impression of where the animal lay in the rock.

SAQ

1   Fossils of shelled animals such as ammonites are very common. It is much rarer to find fossils of animals such as jellyfish or slugs. Suggest why.

Sometimes, the animal itself may not be fossilised but traces of its existence do get preserved in the rocks. For example, we can find fossilised worm burrows or fossilised dinosaur footprints.

Figure 2f.2 An insect in amber.

Some of the most beautifully preserved animal and plant remains are found inside amber. Many trees produce a sticky sap that oozes from their trunks if they are wounded. It is easy for an insect to get trapped in this. As the sap dries and hardens, it turns into amber, with the insect inside.

On a bigger scale, animals and plants have been preserved inside tar pits. In some parts of the world, such as Rancho Las Brea in California,

Figure 2f.3 A tar pit in Trinidad.

Figure 2f.4 This is Tollund Man. His body was buried in around 350 BC in Denmark. He was preserved by the low temperature, lack of oxygen and the chemicals in the peat, so his body hadn't rotted despite being over 2300 years old.

thick asphalt (tar) bubbles up from under the surface of the ground. It is very sticky. An animal that accidentally comes into contact with it can get trapped. Carnivores may then try to catch the trapped animals and get trapped themselves. This is happening now, but there are also old tar pits that were trapping animals in the Ice Age, over 10 000 years ago. Bones of Ice Age animals are perfectly preserved in them.

Peat bogs are another place where we can find preserved animals and plants from long ago. Peat forms where plants have been growing on waterlogged soil. When the plants die, they don't rot completely, because the bacteria that would normally bring about their decay can't get enough oxygen to survive. Over time, the partly decayed plants get buried and squashed under more and more layers, forming peat. Some large pieces of plants, such as pieces of tree trunks, might hardly decay at all. Animals can also be preserved. For example, several extremely old human bodies have been found in peat bogs.

Some animals and plants may be preserved in ice. Mammoths have been found frozen into solid ice in northern Europe and Russia. They died many thousands of years ago.

Fossil sequences

Sedimentary rocks are formed in layers. The deepest layers are the oldest, with newer rocks on top.

So fossils found in deep layers of rock are older than fossils found in layers above them. We can use this to work out a time line of what kinds of animals and plants lived at different times in the past.

For some animals and plants, we have enough fossils to get at least some idea of how a particular kind of organism has changed over time. Horses are a good example. Their fossils show us how horses have grown larger and how their hooves have developed from five-toed feet over the past 54 million years.

However, for most organisms, fossils only give us a few patchy glimpses of their past histories. This is not surprising. Most animals and plants will die in places where their bodies will rot, rather than turning into fossils. And, of course, there must be millions of fossils that we have not yet discovered.

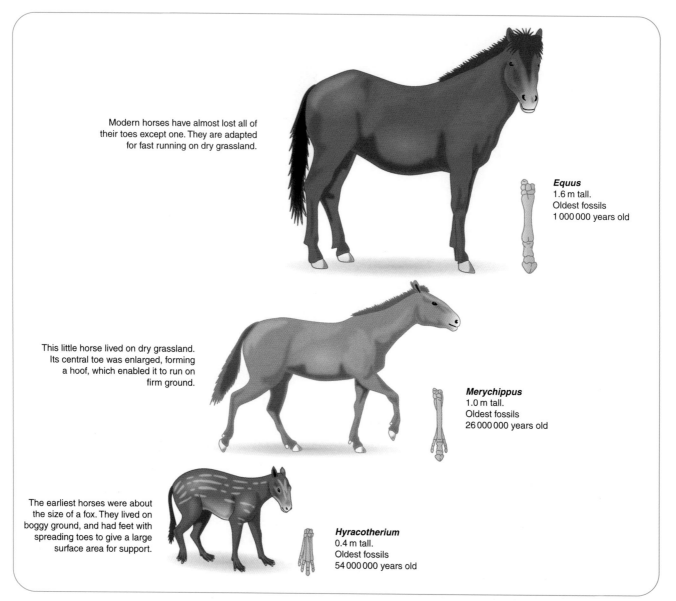

Modern horses have almost lost all of their toes except one. They are adapted for fast running on dry grassland.

Equus
1.6 m tall.
Oldest fossils
1 000 000 years old

This little horse lived on dry grassland. Its central toe was enlarged, forming a hoof, which enabled it to run on firm ground.

Merychippus
1.0 m tall.
Oldest fossils
26 000 000 years old

The earliest horses were about the size of a fox. They lived on boggy ground, and had feet with spreading toes to give a large surface area for support.

Hyracotherium
0.4 m tall.
Oldest fossils
54 000 000 years old

Figure 2f.5 Evolution of the horse. Many fossils of horses and their ancestors have been found in North America. The fossil sequence shows changes in the number of toes, structure of teeth and size of these animals. Only three examples are shown here, but there have probably been hundreds of species of horse. Each was well adapted to live in the conditions which existed at that time, in that place. As the climate changed, different features were favoured by natural selection.

Interpreting the fossil record

Biologists interpret fossils as evidence that life on Earth has changed over huge periods of time. We now know that our planet first formed about 4.5 billion years ago. The oldest fossils that have so far been found date back to around 3.5 billion years ago (although some people think they have found traces of bacteria that are even older than this). The fossil records suggest that the earliest life on Earth was simple, single-celled organisms. We can trace a progression through the rocks – albeit a patchy, disconnected one – showing us that more complex organisms have appeared over time.

Today, most scientists believe that the changes in the life forms that we see as fossils happened through natural selection and evolution. But there are still some people who don't believe in evolution. They believe that the stories told in the Bible and other religious books, in which God created each species separately, are true. They suggest that we find fossils on Mount Everest because there was once a huge flood that carried them up there, not because rocks that were once at the bottom of the sea have been pushed upwards.

Figure 2f.6 The ages of some of the earliest fossils of some types of organism.

Natural selection

We've seen how organisms have to compete with each other when a resource is in short supply. In a wild population of rabbits, for example, most young rabbits will die before they are old enough to reproduce. There might not be enough food to go around, or they may be killed by foxes, or they may get myxomatosis and die.

What determines which ones die and which ones survive? It could be just luck. Most of the time, though, some rabbits will have slight advantages over the others. They are not all exactly the same.

Perhaps some of them are better at finding food than others, at seeing predators early, at running from them or at surviving the effects of the myxomatosis virus.

Some of the variations between the rabbits are caused by their genes. Imagine, for example, that some rabbits inherit genes that make them resistant to myxomatosis. If myxomatosis strikes the population, only the resistant rabbits survive and breed. They pass on their genes to the next generation. After a while, all the rabbits in the population have the resistance gene. All the

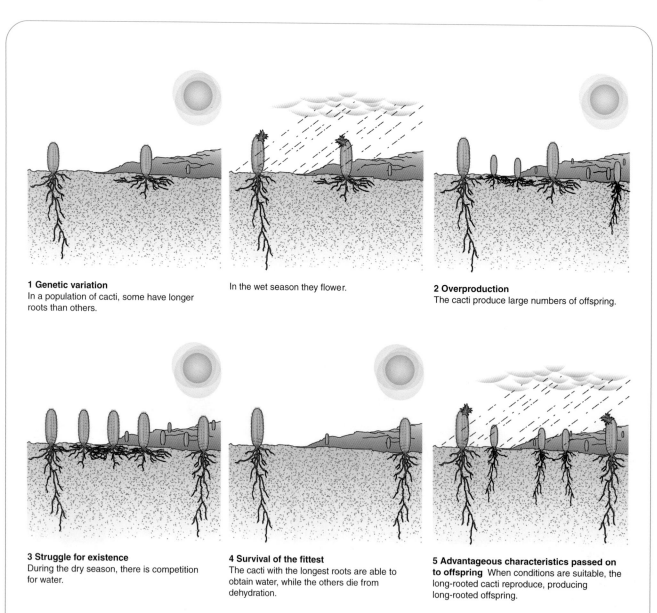

1 Genetic variation
In a population of cacti, some have longer roots than others.

In the wet season they flower.

2 Overproduction
The cacti produce large numbers of offspring.

3 Struggle for existence
During the dry season, there is competition for water.

4 Survival of the fittest
The cacti with the longest roots are able to obtain water, while the others die from dehydration.

5 Advantageous characteristics passed on to offspring When conditions are suitable, the long-rooted cacti reproduce, producing long-rooted offspring.

Figure 2f.7 An example of how natural selection might occur.

others die before they have a chance to reproduce and pass on their genes.

This is **natural selection**. The organisms with variations that help them to survive are more likely to survive, reproduce and pass on the genes that give them their advantages to the next generation.

Figure 2f.7 shows another example, explaining how some of the adaptations that cacti have for living in deserts may have evolved. And we have already seen how antibiotic resistance can evolve in bacteria (see Item B1c).

Peppered moths

The peppered moth, *Biston betularia*, lives in most parts of Britain. It flies by night and spends the daytime resting on tree trunks. It has speckled wings, which camouflage it beautifully on lichen-covered tree trunks.

In 1849, an unusual peppered moth was spotted near Manchester. It was almost black. By 1900, 98% of the peppered moths living near Manchester were black.

The black colour is caused by a gene. There have probably always been a few black moths around but, because they stood out on pale tree trunks, they were easily seen and eaten by birds. They didn't survive. However, in the 19th century, big coal-burning industries grew up in Manchester and other parts of Britain. The air became polluted, so lichens were killed. The tree trunks became blackened by soot. Now it was the black moths that had the advantage and that survived and reproduced, passing on their gene for the black colour to their offspring.

SAQ

2 The prevailing winds in Britain come from the south-west. In the 19th century, the peppered moths in Cornwall were still all speckled, while the black ones were mostly found in the east of the country. What is the connection between these two statements?

Today, most peppered moths in Britain are speckled. As we have cleaned up air pollution, lichen has come back to the trees and the speckled moths are the better survivors.

Figure 2f.8 a Lichen-covered bark hides a speckled moth perfectly. **b** Dark moths are better camouflaged on lichen-free trees.

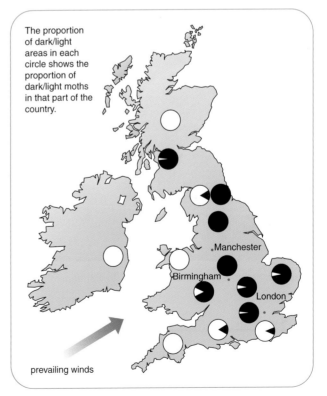

The proportion of dark/light areas in each circle shows the proportion of dark/light moths in that part of the country.

Manchester

Birmingham

London

prevailing winds

Figure 2f.9 The distribution of pale and dark forms of the peppered moth in 1958.

Resistant rats

Rats can be pests. They can eat stored foods and even chew through electricity cables. Their urine and faeces can carry pathogenic microbes. If they are causing problems, rats can be killed using poisons.

One rat poison is called **warfarin**. Warfarin stops the blood from clotting normally. People who have diseases of the circulatory system are sometimes prescribed warfarin, to make it less likely a blood clot will form in their arteries or veins. Rats readily eat warfarin and most of them are killed by it.

However, a few rats have a gene that makes them resistant to warfarin. This same gene also makes them need more vitamin K. As there is not usually very much vitamin K in the food that they eat, rats with this gene are usually at a disadvantage compared with all the other rats, so not many of them survive long enough to pass on the gene to the next generation. But when warfarin is being used, these rats have a big advantage. They are not killed, while all the non-resistant rats are. Nearly all the rats in the next generation will therefore have the warfarin-resistance gene. The population of rats can no longer be killed by this rat poison.

Figure 2f.10 Some rats are now resistant to warfarin.

SAQ

3 Can you suggest anything that could now be done to make warfarin useful as a rat poison once more? Or are we stuck with warfarin-resistant rats for ever?

Stability and change

Most of the examples of natural selection we have looked at – peppered moths, antibiotic-resistant bacteria and rats resistant to warfarin – involve some kind of change. The environment has changed so that the organisms that used to be best adapted are now at a disadvantage.

Another reason for change happening in a species could be that a new gene appears. This is called **mutation**. Mutations happen a lot but most of them don't produce lasting change because they are usually disadvantageous. The organism with the mutation dies before it can reproduce and pass the mutation on.

But just occasionally, a mutation happens and produces a new version of a gene that is actually better than the normal version. For example, some butterflies have eye patterns on their wings. The eyes fool a predator into grabbing the wing rather than the body, giving the butterfly a chance to escape. A gene that determines an eye pattern on the wing of a butterfly might mutate so that the colours form an even more realistic-looking eye. A butterfly with this gene could have a better chance of survival. Over many generations, that gene could become so common that most of the butterflies have this new eye pattern.

Figure 2f.11 Eye patterns on its wings help to protect this moth from predators.

Most of the time, though, natural selection keeps things the same. If organisms are already well adapted to their environment and the environment stays the same, then natural selection will keep the characteristics of the organisms the same, too. You only get change

Bighorn sheep

Bighorn sheep live in the Canadian Rocky Mountains. They are much prized by hunters, who shoot them and keep the horns as trophies.

Figure 2f.12 A bighorn ram.

Since 1973, 57 rams (male sheep) have been shot on Ram Mountain in Alberta, Canada. A study of the size of the horns of these rams shows that they have gradually been getting smaller and smaller. In 1972, the mean horn length was about 65 cm, but by 2004 it was only 50 cm.

What has been happening? Each year, hunters have shot about 10% of all the rams on Ram Mountain. They have gone for the rams with the biggest horns. So these rams have been at a disadvantage. Rams with smaller horns are more likely to survive. Over time, the genes for big horns have been passed on less often than genes for smaller horns. Each generation, the mean length of horns has become smaller. The hunters have caused the population of bighorn sheep on Ram Mountain to evolve in exactly the opposite way than they would have hoped.

when the environment changes or when a new form of a gene is produced by mutation.

Sometimes, the changes in the environment can be so huge that some species can't survive at all. They become **extinct**. We think this is why there are no longer any dinosaurs on Earth. Fossils tell us that the last dinosaurs lived on Earth about 65 million years ago. Then they just disappear from the fossil record. There is evidence to suggest that a giant asteroid – about 10 km in diameter – hit the Earth round about that time. It made a crater in the Yucatan peninsula (part of what is now Mexico) almost 180 km wide. It would have sent a huge cloud of vapour and dust into the atmosphere, blocking light and heat from the Sun. The big dinosaurs could not survive in this cold, dark world during the many years it took for the atmosphere to recover. They became extinct. (See Figure 2g.3 in Item P2g on page 351.)

Charles Darwin

Charles Darwin was born in 1809. His family was well-off and he had an excellent education. For a while, he trained to be a doctor, but he was unable to cope with the horror of watching blood-soaked

operations carried out without anaesthetic. When he was offered the chance to sail to South America on the *Beagle*, he jumped at it.

In the mid-19th century, ideas about the history of the Earth were beginning to change. Most people still believed in the stories about creation in the Bible. They thought the Earth and all the species on it had been created at the same time. Some people still believe this today. Despite all the evidence we have around us, they think that the Earth was created somewhere between 6000 and 8000 years ago.

Figure 2f.13 Darwin's voyage on the *Beagle* helped to spark his ideas about evolution.

H But then scientists began to realise that the Earth was much, much older than this. They began to understand how rocks were formed and how mountains could be lifted up high and then worn down by rain. When Darwin sailed on the *Beagle*, he knew about these geological theories and he looked for evidence to support them on his voyage.

Some biologists were also realising that species of living things might change over time, too. Darwin saw many different animals and plants on his voyage, and when he came back to England he began work on his most important book, *On the Origin of Species*. Darwin not only suggested that new species might have evolved from old ones but he also explained how this might happen. His reasoning went like this:

- Most species produce many more young than ever survive to adulthood.
- Individuals within a species are not all the same – there is variation.
- Living things compete with one another for limited resources.
- Only the individuals with the best adaptations – those who are 'fittest' – will survive and reproduce.
- The young born to these individuals will inherit the advantageous characteristics from their parents.

Darwin knew that his theory would be met with fury and disbelief. He did not dare to publish his book until 1859. He was absolutely right to expect it to be met with horror. Perhaps the thing that people found hardest to accept was that we are descended from the same common ancestors as monkeys and apes. This went against everything they had ever believed, and for some it seemed to make mankind less important – just another animal, instead of something special that God had created to be the pinnacle of life.

But Darwin's arguments were so logical, and the scientific evidence so strong, that today almost all biologists accept the theory of evolution by natural selection.

H ## Lamarck

Jean-Baptiste Lamarck was born in France in 1744, 65 years before Darwin. He, too, had put forward a theory of evolution. Like Darwin, he believed that one species could change into another over time.

Figure 2f.14 Jean-Baptiste Lamarck.

But Lamarck thought that a characteristic that an organism had acquired in its lifetime could be passed on to its offspring. For example, a giraffe might repeatedly stretch up to reach leaves above those that it could normally reach. This could make its neck grow longer. Its offspring would also have longer necks than usual.

Lamarck was suggesting that **acquired characteristics** could be passed on. This wasn't such a crazy idea. At that time, no-one knew anything about genetics or what exactly was passed from parents to their children.

Darwin's idea, though, proved to be the better one. We can now explain Darwin's theory in terms of genes, passed on from generation to generation. Characteristics that an organism acquires in its lifetime don't alter its genes. And it is only its genes that are passed on to its offspring.

SAQ

4 How can Darwin's theory of evolution by natural selection explain the evolution of giraffes with long necks?

Speciation

If the theory of evolution is correct, it must mean that one species can change into another species. This is called **speciation**.

Darwin called his book *On the Origin of Species*, but he did not actually explain how he thought a new species might arise. We've defined a species as a group of living organisms that share the same features and that can breed together to produce fertile offspring (see Item B2b, page 82). So if one species turns into another species, the new group must be unable to breed with the old one.

It is difficult for scientists to do experiments on speciation, because we think that it probably takes a long time. There are various theories about how speciation might happen and it is likely that there are many different ways, some that happen over a very long time and maybe others that happen quickly.

One way in which speciation might happen is by **geographical isolation**. Imagine there is a population of birds called Long-legged Nitwits living in a country near the sea. They have long legs that help them to paddle around in mud and find food.

Figure 2f.15 In different environments, different leg lengths may be best for survival.

In a storm, a few Long-legged Nitwits get blown onto an isolated island. There isn't any mud there – it is all rough and rocky – so it isn't helpful to have long legs. In fact, it is better to have short legs, so that they don't get hurt when walking round on the rocks. Any Long-legged Nitwits that have slightly shorter legs have a better chance of surviving and reproducing than ones with long legs. So, in each generation, genes that produce short legs are more likely to be passed on than genes that produce long legs.

Over many generations, the legs of the isolated island-dwelling Nitwits become gradually shorter. After hundreds of years, they have become Short-legged Nitwits. If you put a Long-legged Nitwit and a Short-legged Nitwit together, they are so different that they are not attracted by each other and don't breed together.

Summary

You should be able to:

◆ explain how fossils provide evidence for evolution

◆ outline some of the ways in which fossils may be formed

◆ describe how the fossil record can be interpreted in different ways

◆ explain why the fossil record is incomplete

◆ understand how natural selection works

◆ describe examples of natural selection: peppered moths, antibiotic resistance in bacteria, rats resistant to warfarin

◆ outline Darwin's theory of natural selection and explain why it was at first given a hostile reception

◆ outline Lamarck's theory of the inheritance of acquired characteristics, and explain why this is now discredited

◆ explain how natural selection can cause a new species to be formed

Questions

1 Copy these sentences, and use some of the words to complete them.

  fossils   natural selection   evolution   adapted   resistant

  a Animals and plants that are better .... to their environment are most

  likely to survive. This is called ....

  b .... provide evidence that animals and plants have changed over long

  periods of time.

2 The graph in Figure 2f.16 shows the number of reports of infections by MRSA between 1990 and 2004.

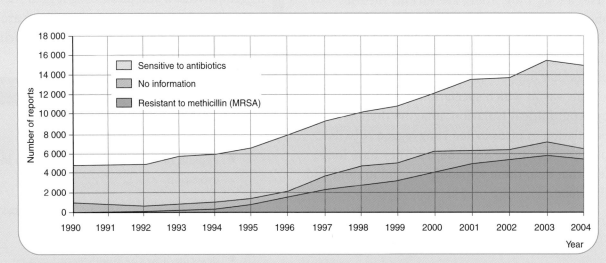

Figure 2f.16 Number of reports of infection by *Staphylococcus aureus*, including MRSA, in England and Wales, 1990–2004.

  a How many reports were there of MRSA in 1990?

  b When were the first reports of MRSA made?

  c How many were there in 2004?

  d Explain, using what you know about natural selection, why this increase has occurred.

3 Discuss arguments for and against the idea that natural selection no longer acts on humans.

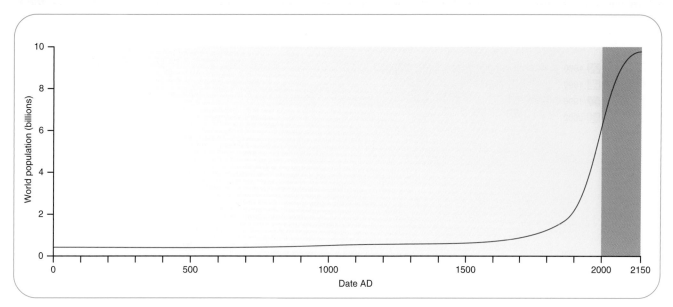

Figure 2g.1 Global human population growth (estimated, yellow; predicted, blue). Data from the UN Population Division.

More and more people

The Earth's human population is growing. Figure 2g.1 shows what has happened to our population in the past 2000 years, and what may happen by 2150.

Since about 1800, the number of people on Earth has been doubling again and again. This is called **exponential growth**. There seems to have been nothing to stop this increase. In 1980, there were estimated to be 4.5 billion (4 500 000 000) people. In 1990, there were 5.3 billion. In 2000, there were 6.1 billion. It is estimated that the population will be 6.8 billion by 2010.

There are some hopeful signs that this increase is slowing down. You can see in Figure 2g.1 that it is predicted that the graph will flatten off in the next 150 years or so.

Using resources

The more people there are, the more resources we use. For example, we use more fossil fuels to provide energy. We use more minerals, such as iron ore to make steel. Eventually these resources will run out.

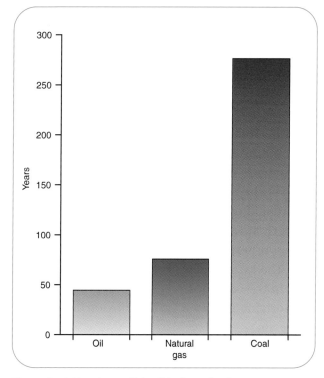

Figure 2g.2 How long will fossil fuel supplies last?

SAQ

1 The graph in Figure 2g.3 shows the number of motor vehicle sales in nine countries between 1980 and 2002.

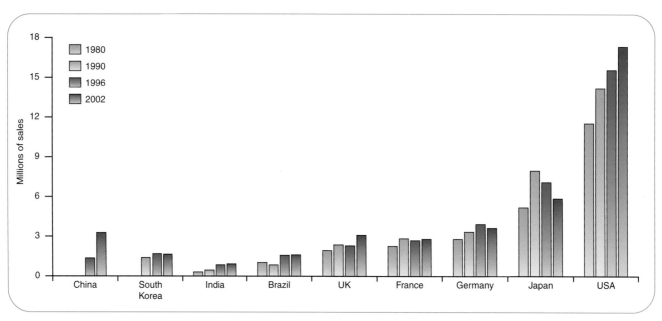

Figure 2g.3 Motor vehicle sales.

    a Describe the general trend in motor vehicle sales between 1980 and 2002. Suggest why this has happened.

    b Which countries do not show this trend? Suggest reasons for this.

    c What do these figures suggest may have happened to fossil fuel consumption between 1980 and 2002?

Carbon dioxide pollution

Carbon dioxide is found in the Earth's atmosphere. There isn't very much of it – only 0.04% of the air around us is carbon dioxide.

Carbon dioxide is very important for life on Earth. It is the gas that plants need for photosynthesis and so it is the source of all of our food. The carbon atoms in all the food that you eat – in the carbohydrates, fats and proteins – were originally part of carbon dioxide molecules in the air.

Carbon dioxide also helps to keep the Earth warm. It prevents long-wavelength radiation (infra-red or heat) from the ground escaping from the Earth and into space. Without it, the Earth would be so cold that it is unlikely any life could exist.

However, too much carbon dioxide is not a good thing. The more carbon dioxide there is in the atmosphere, the less heat escapes and the warmer the Earth becomes. This is happening now. We are experiencing **global warming**.

It's proving difficult to predict how fast global warming will happen or what its effect might be. Most scientists agree that it will cause a rise in sea levels, as the water in the oceans expands (because it gets warmer) and as icecaps melt. This could flood low-lying cities and even whole countries such as the Maldives. Global warming will probably also cause more extreme weather events, such as hurricanes and floods. It will affect the distribution of animals and plants. Organisms that have adaptations for living in cool places may have to move northwards or southwards to escape rising temperatures.

The percentage of carbon dioxide in the atmosphere is increasing. This is probably partly caused by the increase in use of fossil fuels. When fossil fuels are burnt, they release carbon dioxide into the atmosphere.

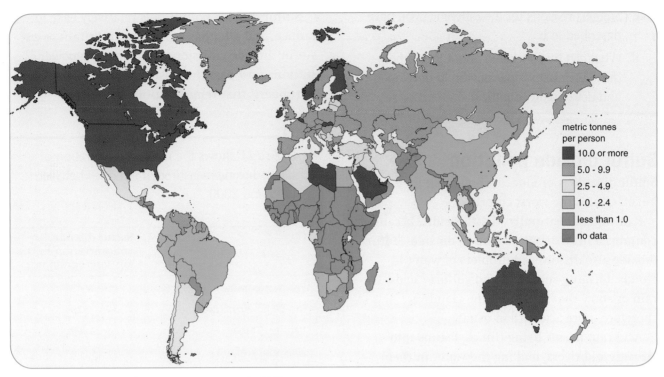

Figure 2g.4 Carbon dioxide emissions per person in 2000.

SAQ

2 The graph in Figure 2g.5 shows carbon dioxide emissions between 1980 and 2000.

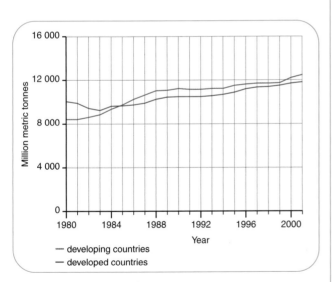

Figure 2g.5 Carbon dioxide emissions.

a Describe the overall trend in carbon dioxide emissions between 1980 and 2000.

b Where does this carbon dioxide come from?

c Compare the carbon dioxide emissions for developed countries and developing countries.

H 3 The graph in Figure 2g.6 shows the carbon dioxide emissions per person for nine countries in 2002.

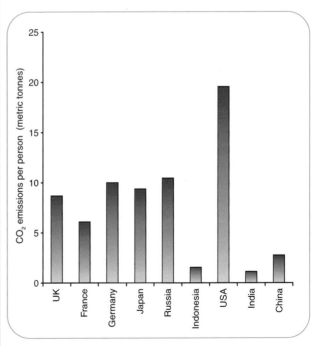

Figure 2g.6 Carbon dioxide emissions per person for selected countries. (Source: International Energy Agency.)

a List these countries in order of the quantity of carbon dioxide emitted per person.

b How many times greater are the values for the USA compared with India?

H

c Suggest reasons for the difference you have described in **b**.

d What can be concluded about the carbon dioxide emissions per person in developed and developing countries?

Sulfur dioxide pollution

Sulfur dioxide is produced when fossil fuels are burnt. Coal is the worst culprit.

Rainwater is naturally slightly acidic, because it contains a little dissolved carbon dioxide. Sulfur dioxide also dissolves in droplets of water in clouds, forming sulfurous and sulfuric acids. The rain or snow that falls from the clouds is therefore more acidic than usual.

Acid rain harms living things. It runs into streams and rivers, making the water in them acidic. Few fish can live there. It damages leaves on trees and other plants. It soaks into the soil and makes it more difficult for plants to absorb the minerals that they need. It leaches metals such as aluminium out of the soil so that they are washed into streams and lakes, where they are toxic to fish and other animals.

Figure 2g.7 Acid rain damages trees.

Sulfur dioxide in the air also affects people's health, especially those who already have breathing problems such as asthma, bronchitis or emphysema.

Most fossil fuels are burned in or near cities, but the sulfur dioxide they produce can travel very large distances in the air. So pollution produced by one country can affect another.

Sulfur dioxide pollution is relatively easy to reduce, and a lot has been done in Britain to cut sulfur dioxide emissions. Coal-burning power stations now pass the waste gases through 'scrubbers' that remove sulfur dioxide.

SAQ

4 Table 2g.1 shows the mean concentration of sulfur dioxide measured in ten cities between 1995 and 2000.

City	Sulfur dioxide concentration in the air (microgram/m^3)
Perth, Australia	5
Rio de Janeiro, Brazil	129
Beijing, China	90
Helsinki, Finland	4
Delhi, India	24
Dublin, Ireland	20
Kuala Lumpur, Malaysia	24
Moscow, Russian Federation	109
Istanbul, Turkey	120
London, UK	25

Table 2g.1 Mean concentration of sulfur dioxide measured in ten cities between 1995 and 2000.

a The World Health Organization's air quality guidelines say that there should be no more than 50 micrograms of sulfur dioxide per cubic metre. Which cities exceeded these limits?

b i Suggest reasons for the high figures in Rio de Janeiro or Moscow. (There are many possible answers you might think of.)

ii How could you find out if your suggestions are correct?

iii What could these cities do to improve the air quality?

Ozone depletion

Ozone is a gas. It has the formula O_3. Ozone tends to collect in a layer high up in the atmosphere, between 20 km and 35 km above your head.

The **ozone layer** absorbs many of the ultraviolet rays that reach the Earth from the Sun. These rays can damage DNA in living cells. Without the ozone layer, more ultraviolet light would reach the Earth's surface. This could damage plants, and increase rates of skin cancer in humans.

In the second half of the 20th century, we released a lot of pollutants into the air that damaged the ozone layer. These included gases called CFCs, which were used as refrigerants and in aerosols. The CFCs diffused up to the high levels of the atmosphere, where they reacted with the ozone, breaking it down into oxygen. This happened especially over the poles. The places where there is less ozone are said to have an 'ozone hole'.

The ozone concentration is measured in Dobson units, named after G. M. B. Dobson, one of the first scientists to investigate atmospheric ozone. CFCs and most other chemicals that can harm the ozone layer are no longer used in most developed countries, but they are still manufactured and used elsewhere. All the same, despite the increase in the Earth's population, we have managed to cut down the release of ozone-damaging substances into the atmosphere. But CFCs take many years to break down, so it will be some time before we see the ozone layer recovering completely.

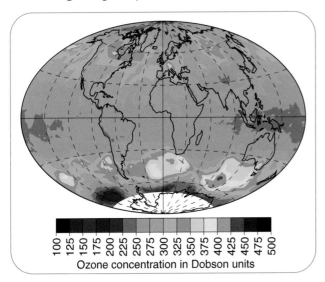

Figure 2g.8 Ozone concentration around the world on 15 August 2005.

Other sources of pollution

We have seen how an increasing population has increased air pollution by releasing more carbon dioxide, sulfur dioxide and CFCs. There are other kinds of pollution that have also increased.

People produce rubbish. We throw away huge quantities of household waste each year. A lot of this is packaging – boxes, jars, tins and bottles. There are also old newspapers, garden rubbish and uneaten food. All of this has to be disposed of. In the UK, increasing quantities of household waste are being recycled. But there is still a lot of rubbish that goes into landfill sites or for incineration. The more of us there are and the more money we spend, the more rubbish we produce. It is expensive to get rid of it. Even if it is recycled, the fuel used in transporting it adds to carbon dioxide emissions.

We also produce waste from our bodies. Urine and faeces flushed down the toilet go into the sewage system, along with water that runs off streets and other places when it rains. If the sewage went straight into rivers, it would cause serious pollution.

Raw (untreated) sewage contains a lot of things that bacteria and algae (small floating plants) can use as nutrients. When raw sewage gets into a river, the population of bacteria increases because they have extra food. Most of them respire aerobically, so they use up oxygen from the water. This makes it difficult for fish and many invertebrates to live there. They die or move away.

In the UK, almost all sewage is now treated before the water from it runs into streams, rivers or the sea. Sewage treatment gets rid of any harmful bacteria or viruses that might cause disease. It also gets rid of the nutrients, so that it won't cause increases in populations of bacteria or algae in the water. It is really only untreated sewage that does harm.

Incident in the Thames

On 4 August 2004, more than 100 000 dead fish were found floating in the river Thames in London. It was a real setback for the Environment Agency and Thames Water, who between them have made huge improvements to the quality of the water in the river.

In the 1970s, the river Thames was in a poor state. The water was polluted with sewage and industrial waste. Since then, the level of pollution has been greatly decreased. Gradually, more and more species of fish have been seen in the river. Salmon have returned. In June 2004, there was even a seahorse discovered in the Thames estuary. Seahorses are very sensitive to pollution and would not be able to live in the kind of water that was flowing out to sea in the 1970s.

So what happened in August 2004? There was a sudden, torrential downpour of rain. So much rain fell that the sewage system could not cope with all the extra water flowing through it. The rain flushed thousands of tonnes of untreated sewage into the Thames. Thames Water did their best to stop the disaster happening. They put boats out onto the river that bubbled oxygen into it. But they were powerless in the face of the huge quantities of untreated sewage that had entered the water.

Over the next few days, the sewage gradually broke down. Oxygen levels were restored. There don't seem to have been any serious, lasting problems. Thames Water and the government are looking at ways of improving the sewage system to stop this happening again.

Figure 2g.9 Thousands of fish were killed in the Thames on one August day in 2004.

What happens next?

What might happen if the world's population keeps on increasing? Just like any population, we are limited by the resources that we need. But probably it won't be fossil fuels or minerals that will run out first. It may well be water that causes the biggest problems. There are already water shortages in many parts of the world.

Food supply is also likely to become even more problematic than it is now. We have seen how food shortages can cause serious illness and death (see Item B1b, which describes kwashiorkor). Even when there is enough food to go around, it may not get to the right people, perhaps because of transport difficulties, incompetence or fighting.

Infectious diseases are also a major threat. The more of us there are and the more closely we live together, the easier it is for pathogens to spread. It is probable that a new infectious disease will emerge some time in the future and spread rapidly around the world. If we are lucky, we will be able to bring it under control, as happened with a new illness called SARS which appeared in 2003. If we are unlucky, it will cause a pandemic – in which many people all over the world get the disease – and possibly huge numbers of deaths.

But none of this is inevitable. Birth rates have decreased greatly in developed countries in the past 50 years, and it looks as though this trend is gradually spreading into developing countries, too. As people have more security in their lives and don't have to worry about who will look after them in their old age, they don't feel quite such a need to have a lot of children.

Years	World population growth (percentage increase per year)
1950–2000	1.76
2000–2005	1.22
2045–2050 (predicted)	0.33

Table 2g.2 Percentage increase in population.

So, there are many reasons to think that the population won't continue to rise as it is doing now. Table 2g.2 shows how population growth rates have changed since 1950 and are predicted to change in the future.

Monitoring pollution

Everyone is now very aware of the dangers of air and water pollution, and regular monitoring takes place in the UK. Sometimes, this is done by directly measuring the concentration of a pollutant – for example, the concentration of sulfur dioxide in the air or the pH of river water. We can also get a good idea of the level of water pollution by measuring the concentration of oxygen in it, because pollution by substances that bacteria can feed on means that they use up more oxygen.

Indicator species

A very good way of getting an idea of whether water or air is polluted is to find out what is living there. Many species of organisms can only live in certain conditions. For example, mayfly nymphs need a lot of oxygen in the water in order to survive, so you don't find them in polluted water. Rat-tailed maggots are especially adapted to live in water that contains very little dissolved oxygen, so if you find them you can assume that the water is polluted. Figure 2g.10 shows some organisms that live in streams and can indicate how clean the water is.

SAQ

5 Suggest how rat-tailed maggots get their oxygen.

Lichens are a good indicator of air pollution. They are very unusual organisms, made up of a very close association between an alga and a fungus. They are able to live in very inhospitable places, such as on the surfaces of rocks or tree bark.

Most species of lichen cannot live where there is sulfur dioxide in the air. Doing a lichen survey can therefore give you information about the level of pollution in the air. The more lichen species there are, the less polluted the air.

Indicator species are sometimes allocated a number called their biotic index. The highest possible biotic index is 10, and this number is given to animals that can only live in an unpolluted environment. At the other end of the scale, with a biotic index of 1, are animals that can live in a very badly polluted environment.

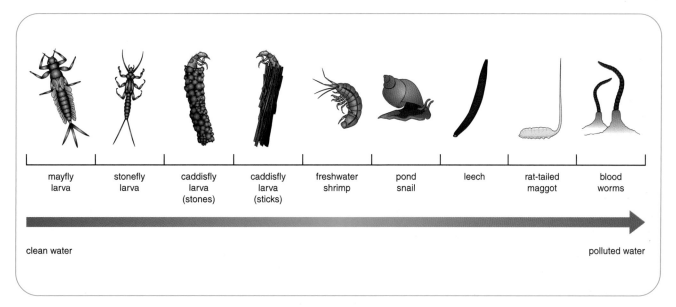

| mayfly larva | stonefly larva | caddisfly larva (stones) | caddisfly larva (sticks) | freshwater shrimp | pond snail | leech | rat-tailed maggot | blood worms |

clean water   polluted water

Figure 2g.10 Indicator species for water pollution.

SAQ

6 Table 2g.3 shows the results of a survey done in two rivers. The same sampling techniques were used in each river, and they were done at the same time.

Animal	Biotic index	Number found in River A	Number found in River B
stonefly larvae	10	0	15
mayfly larvae	10	0	6
caddis larvae with cases made of stones	10	0	1
caddis larvae with cases made of sticks or leaves	7	0	2
freshwater shrimps	6	0	30
leeches	3	3	12
snails	3	31	59
rat-tailed maggots	2	92	0
bloodworms	1	103	0

Table 2g.3 Indicator species in two rivers.

a Which river is the more polluted? Explain your answer.

b Do you need all of the information in the table in order to decide which is the more polluted river? Explain how you could cut down the time that must have been taken to do this survey, and still get a reliable answer.

Summary

You should be able to:

◆ explain that the human population has been increasing exponentially, but that this rate of increase is now slowing

◆ describe how the increasing population means we use more of the Earth's finite resources

◆ describe how the increasing population means that we produce more pollutants, including carbon dioxide, sulfur dioxide, CFCs (which damage the ozone layer), household waste and sewage

◆ explain that the environmental impact per person is greater in developed countries than in developing countries

◆ discuss the possible consequences of a continuing increase in global population

◆ explain what is meant by an *indicator species* and describe how they can be used to monitor pollution

◆ interpret data about indicator species

Questions

1 Copy the diagram in Figure 2g.11. Draw a link from each box in the first column to the correct box in the second column and then the third column.

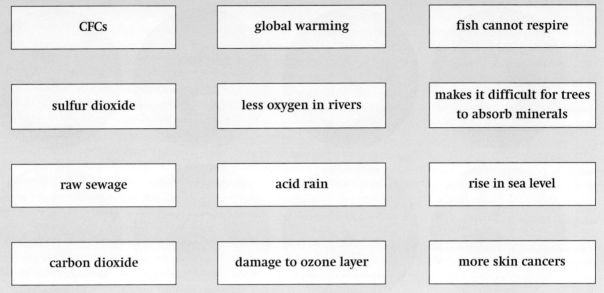

CFCs	global warming	fish cannot respire
sulfur dioxide	less oxygen in rivers	makes it difficult for trees to absorb minerals
raw sewage	acid rain	rise in sea level
carbon dioxide	damage to ozone layer	more skin cancers

Figure 2g.11

2 The graph in Figure 2g.12 shows the percentage of different fuels used in developed countries between 1980 and 2000.

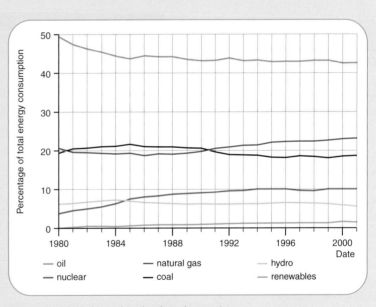

Figure 2g.12 Fuels used in developed countries.

a Which of the fuels shown in the graph are fossil fuels?

b Suggest one source of energy that would be included in 'renewables'.

c Describe how the proportion of energy derived from non-fossil fuels changed between 1980 and 2000.

d Calculate the percentage of energy that was obtained from fossil fuels in the year 2000.

continued on next page

Questions - *continued*

3 These images show the changes in the ozone layer over the South Pole between 1970 and 1997.

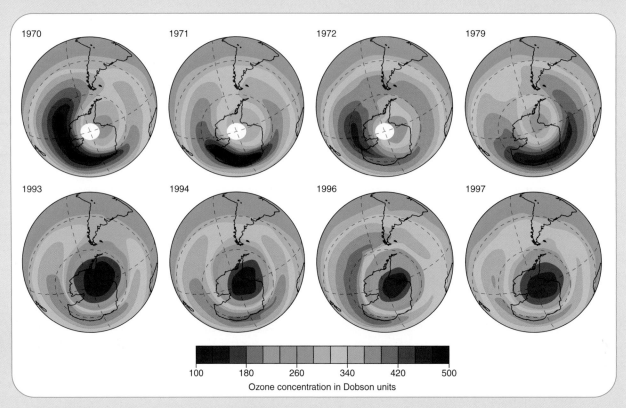

Figure 2g.13 Changes in ozone over the South Pole between 1970 and 1997.

a Describe what happened to the ozone concentration around the South Pole between 1979 and 1993.

b Explain why these changes happened.

c Explain why people are worried about these changes to the ozone layer.

d What is being done to try to prevent further changes to the ozone layer?

e Compare the results for 1997 shown above with the image in Figure 2g.8. Do you think there was there any improvement between 1997 and 2005? Suggest reasons for this.

Endangered species

Many species of plants and animals that once lived on Earth are no longer alive. They have become **extinct**.

Some of these extinctions happened a long time ago. We know about them because we find fossils of plants and animals that are not found today. For example, all of the dinosaurs appear to have become extinct before about 65 million years ago. Palaeontologists (people who study fossils) have found fossilised bones of around 700 species of dinosaur. It is estimated that this is likely to be less than 1% of all the dinosaur species that ever existed. So the dinosaurs alone account for 70 000 species of extinct animal.

Other extinctions have happened more recently. In 1662, the last living dodo was seen on a small island off Mauritius, in the Indian Ocean. Dodos were giant, flightless pigeons. They did not need to be able to fly, because there were no predators on the islands where they lived. So when humans arrived in ships in 1581, the dodo was completely defenceless. It was easy to kill them for food – people could just walk up to a dodo and club it to death. Rats, which ran off the ships and onto the land, ate dodo eggs. The dodo had lived on Mauritius for hundreds of thousands of years but was wiped out in less than a century.

Figure 2h.1 Dodos used to live in Mauritius.

SAQ

1 Many native birds that live in New Zealand cannot fly and are in danger of extinction.

   a Suggest why they are endangered.

   b What could be done to save them?

Reasons for past extinctions

We know exactly why the dodo became extinct, because it happened at a time when people wrote down or painted things that happened. Extinctions that happened longer ago are more difficult to explain.

Scientists have come up with several different possible reasons for the extinction of the dinosaurs, but at the moment most of the evidence points to the impact of a huge asteroid which altered the Earth's climate so drastically that probably around 60% of all species of living things were destroyed. **Climate change** (see Item B2g *Population out of control?*) may also explain the extinction of mammoths. These huge, hairy relatives of elephants lived in northern America and Europe up to around 10 000 years ago. They were well adapted for the icy conditions of the last Ice Age, and may not have been able to adapt to the warming of the climate since then.

Figure 2h.2 Woolly mammoths became extinct soon after the last Ice Age.

Sabre-toothed cats, whose remains are quite common in the Rancho Las Brea tar pits (see Item B2f, pages 110–111), became extinct about 1000 years before that. We don't know if they, too, were victims of climate change. Perhaps either they or their prey were no longer suitably adapted for the warmer conditions that were developing around then. If the prey animals became extinct because of climate change then, unless the predator found alternative food supplies, it too would die out.

Figure 2h.3 A sabre-toothed cat.

Humans may also have had something to do with the extinction of the mammoth. Certainly, prehistoric humans hunted mammoths. The extinction of the North American mammoths does seem to coincide with the first arrival of humans there. Or perhaps it was due to a combination of both factors together – climate change and hunting.

Species can also become extinct because of **competition** with another, better-adapted species. This can happen when a species moves into an area where it has not lived before. We have seen how competition between the introduced grey squirrel and our native red squirrel has almost wiped out the red squirrels, because they do not compete well with the grey squirrels for food and other resources (see Item B2d, page 95). This must have happened to millions of species in the past, as newly evolved species with better adaptations out-competed older species.

SAQ

2 Look back to Figure 2f.5 (page 112) about the evolution of the horse. We think that 54 million years ago, the climate was wet and much of the ground was boggy. By 26 million years ago, the ground was becoming drier. Suggest why *Hyracotherium* and *Merychippus* became extinct.

Recent extinctions

There have been several times in the past when huge numbers of extinctions have happened all at once – for example, the extinction of the dinosaurs. This is called a **mass extinction**. Many scientists think that we are in the middle of another mass extinction event now. And this time the cause is not something from outer space. It is us.

It is estimated that at least 150 species of vertebrates have become extinct in the last 300 years. No-one has any idea how many invertebrates and plants have become extinct during this time. There may be hundreds of species of insects, for example, that are becoming extinct before we even know they exist.

The International Union for Conservation of Nature and Natural Resources (IUCN) publishes a list of all the species we are aware of that are in danger of extinction. They are called **endangered species**. The list is called the **Red List**. In 2004, the Red List contained 15 589 species. And these are just the ones that we know about.

We've seen how **hunting** caused the extinction of the dodo and may well have caused the extinction of the mammoth. Hunting is also threatening many species today. For example, the scimitar-horned oryx was hunted almost to extinction in the semi-deserts of northern Africa. Local people had always killed a few for meat and skins, but once firearms were developed the numbers of oryx that were killed increased hugely. Things became even worse when oil workers moved into the area, because they shot oryx for sport.

Gorillas and other animals living in the rainforests of central Africa are shot for food, and many people are worried that gorillas will soon become extinct in the wild.

Figure 2h.4 Scimitar-horned oryx are at risk of extinction because of hunting.

Figure 2h.5 In some parts of Africa, people kill and eat almost any animal for food, including endangered species.

However, hunting is the not the main way in which we can cause species to become extinct. The biggest danger is **habitat destruction**. As our population increases, we use up more and more land for farming and for building towns and roads. We drain wetlands. We cut down forests so that animals adapted for forest living have no suitable habitat where they can survive. Giant pandas, which live in forests in China and feed on bamboo, have become rarer as their habitat has been destroyed.

Pollution can also endanger species. We have seen how pollution by untreated sewage can decrease the oxygen content of rivers and streams, making it impossible for fish and many kinds of invertebrates to live there. And the extra carbon dioxide that we are adding to the atmosphere is bringing about climate change – perhaps not as drastic as the change that happened when the asteroid hit the Earth

65 million years ago, but still probably the single greatest threat to living things on Earth today. For example, the IUCN Red List contains 122 amphibians (frogs and toads) that are classified as Critically Endangered. It is possible that at least 113 of these are already extinct. The cause is thought to be pollution by carbon dioxide affecting global warming, which is increasing the rate of infections of frogs by a deadly fungus that grows on their moist skins.

Some success stories

We can do a lot to protect endangered species. We have many success stories here on our doorstep.

The **red kite** is a huge and impressive bird of prey. With a wingspan of 1.5 metres and a forked tail, it is difficult to mistake it for anything else if you see it flying above you.

Red kites used to be common in Britain. But they were shot and poisoned until only a very few of them were left, living in a small area of wild hills and valleys in Wales. In 1989, some red kites were imported from Spain and released into the Chilterns, close to the M40 just north of High Wycombe. The kites have survived and reproduced, and now you can easily see them soaring high above the hills and roads, looking down for dead animals or any other edible things they can find. They will readily fly into people's gardens to find food. There are now thought to be more than 200 breeding pairs in the Chilterns, and kites have also been successfully reintroduced into several other areas of England and Scotland.

Figure 2h.6 Red kites are now a common sight in the Chiltern Hills.

Another striking bird that has returned to live in the Britain is the **osprey**. Ospreys feed on fish, which they catch on the wing and carry off in their impressive talons. Once again, these used to live all over Britain but had been shot to extinction. However, ospreys continued to live in other parts of the world, including mainland Europe, and in 1955 a pair returned to Scotland, at Loch Garten, and nested. Egg thieves took their first eggs.

The Royal Society for the Protection of Birds (RSPB) was undecided whether to try to keep the nest site secret or whether to make it public and give people the chance to watch these beautiful birds. They decided to get the public involved and, in 1959, a hide was opened for watching the ospreys at Loch Garten. More than 14 000 people visited the hide in the first six weeks. The ospreys bred successfully for several years running, and their offspring returned to the area. Now there are about 100 pairs around that area. Ospreys are also being reintroduced near Rutland Water in England.

Figure 2h.7 Ospreys now breed in several places in Britain.

Efforts are also being made to conserve **red squirrels** in Britain. Red squirrels are endangered because of competition with the introduced grey squirrels. But they still live in some parts of Scotland, Wales (including the island of Anglesey) and a few spots in England, including Brownsea Island off the south coast and Formby Dunes on Merseyside.

Red squirrel conservation involves making sure they have a suitable habitat and keeping grey squirrels away. In general, red squirrels compete better with grey squirrels in conifer woodland than in woodland containing broad-leaved trees (such as oak and ash). However, if no grey squirrels are present, they do well in broad-leaved woodlands, too.

Figure 2h.8 This red squirrel is being fed peanut butter as part of a conservation project, because of a lack of natural food.

SAQ

3 Wolves used to live in Britain. Some people would like to reintroduce them to parts of Scotland. Discuss the benefits and problems that this could produce.

How to help endangered species

The successful rescue of red kites, ospreys and squirrels in Britain shows how we can save species if we are prepared to try. But protecting endangered species isn't always as easy or successful. Each endangered species has its own problems and we need to work out what these problems are before we can design a suitable rescue plan.

Protecting habitats

For the red kites and ospreys, there were plenty of suitable habitats in which they could live. With red squirrels, we need to make sure that we preserve habitats where they can successfully out-compete the grey squirrels.

In England, Scotland and Wales, a number of Sites of Special Scientific Interest – SSSIs for short – have been designated by the government. They include a very wide variety of habitats, such as heathland, wetlands, woodlands or meadows. Many charities are also involved in habitat conservation, especially the Wildlife Trusts. They rely on donations and voluntary work from their members and others who want to support nature conservation.

Figure 2h.9 Volunteers help to preserve habitats.

Conservation conflicts

Large predators such as lions and tigers could become extinct unless we make efforts to conserve them. But sometimes conserving predators can cause major problems for people who live near them.

Tanzania has the biggest population of lions in Africa. Over the past 15 years, lions have killed at least 563 people in Tanzania and seriously injured 308. The number of lion attacks seems to be increasing. Not surprisingly, many people who live in the countryside in Tanzania are not whole-heartedly behind projects to conserve lions.

A study into the reasons for these attacks suggests that lions are most likely to attack people where their normal prey (mostly antelope) is scarce and where there are a lot of bush-pigs. Bush-pigs damage crops at night, so farmers often spend the night out near their fields so that they can chase off the pigs. Unfortunately, hungry lions may also be chasing the pigs. Perhaps conservation of lions in Tanzania should involve making sure that they have plenty of their normal prey, and that there are not too many bush-pigs around.

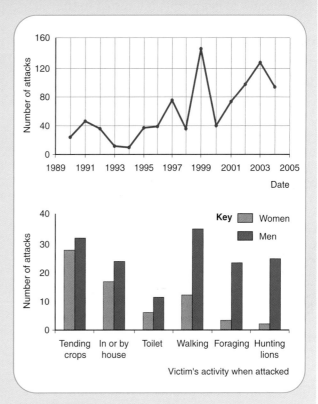

Figure 2h.10 Attacks by lions in Tanzania.

Legal protection

In Britain, endangered species are protected by law.

It is illegal to disturb ponds in which great crested newts live. Sometimes builders get permission to build on land that includes great crested newt habitats. They have to make new ponds elsewhere and carefully transport the newts to them.

It is also illegal to disturb bats. Bats sometimes roost in the roof spaces of houses. You can't move them without special permission. The bats are protected by law.

Education programmes

If people don't know about endangered species, then they can't help to protect them. The RSPB was amazed at the response they got when they publicised the return of the ospreys to Loch Garten.

Many zoos and botanic gardens run education programmes to get people interested in the species and habitats they are trying to conserve. The more people can get involved, the more likely they are to provide support, either by giving donations or by volunteering their time.

Figure 2h.11 Dogs can disturb nesting birds.

For example, a wildlife charity may be trying to conserve an area of woodland that people from a local village people use for walking their dogs. The charity is concerned because the dogs running loose are disturbing species of birds that nest on the ground. Just putting up a notice saying 'Dogs must be on leads' has no effect. So the charity sends out leaflets to everyone in the village explaining what they are doing in the woodland and why. They ask people if they would like to volunteer to help. They give a talk in the village school. They have a meeting one evening in the village hall, where people can ask them questions. As a result, the people living in the village feel part of the conservation project and get involved in it themselves. They not only keep their own dogs on leads but tell other dog walkers to do the same.

Captive breeding programmes

Sometimes just protecting an endangered species in the wild is not enough. We need to try to build up their numbers in some other way.

The scimitar-horned oryx had become almost extinct by 1960. If nothing had been done, they would all have been shot within the next few years. So some of the few remaining wild oryx were captured and taken to zoos – some in Russia, some in the USA and some in Europe. In England, Marwell Zoo took several oryxes.

The oryxes bred readily in the zoos. Now there are over 2000 of them in captivity. The **captive breeding** programme saved them from extinction.

In the 1970s, the zoos got together to try to reintroduce the oryxes into their natural environment. They were able to work with the Tunisian government to designate some large areas of suitable habitat as reserves. The areas had to be fenced to keep out sheep and goats, and left for several years to let the vegetation recover. The local people were involved in the project, and some of them were given jobs looking after the reserve. Oryxes from Marwell and another zoo were introduced into the reserves, and now there are good-sized populations doing well in the wild.

Figure 2h.12 Scimitar-horned oryx bred in captivity are now being returned to the wild.

SAQ

4 Apart from captive breeding programmes, how else can zoos help with conserving endangered species?

Creating artificial ecosystems

An 'artificial ecosystem' is one that has been produced by human activity. Some artificial ecosystems, such as a field of wheat, are not good for wildlife (see Item B2a, page 73). But others can provide important habitats for wildlife that could not otherwise live there. For example, a lake in a disused gravel pit can become an important habitat for amphibians, birds and aquatic plants.

Artificial ecosystems can be specially created for the purpose of conserving endangered species. This can be on a large scale – for example, creating artificial coral reefs around tropical islands – or on a small one. For example, you might be able to create a small pond in your garden. The pond will provide a habitat for frogs, toads, newts, dragonflies and many other animals.

Gardens are very important wildlife havens. They contain more different species of animals than you find in land that is being farmed. Bumblebees, hedgehogs and a wide variety of birds use gardens. Many gardeners now garden with wildlife in mind, taking care to leave some areas of rough grass and not using pesticides such as slug pellets, which could harm birds and animals that eat the poisoned slugs.

Why should we conserve species?

Conserving endangered species is not easy. It can cost a lot of money and it takes up a lot of someone's time. So why should we try to do it?

For many people, it is simply obvious that we have a responsibility to ensure that our activities don't make species extinct. We share the Earth with all of them and we should do our best to make sure that they can survive. Wildlife charities get donations of money from people because they simply feel they want to do this. Unfortunately, not all species get the same degree of support. People may be very willing to give money to support conservation of giant pandas or snow leopards, but not for an endangered species of blood-sucking leech.

There are more practical reasons why species should be conserved. For example, many of the medical drugs that we use, including aspirin, have been derived from chemicals that plants make. Only a tiny proportion of plant species have been investigated to see if they might contain more chemicals that could be used as life-saving drugs. If we let them become extinct, then we will never know how they might have helped us.

Figure 2h.13 The Madagascan periwinkle provides a drug called vincristine, which helps to cure some types of cancer. It grows only in Madagascar.

Conserving whales

Whales are air-breathing mammals that live in the sea. Not all whales are huge. Dwarf sperm whales only grow to about 2.5 m long. At the other extreme, the blue whale is the largest mammal that has ever existed. They can grow to 32 m long, with a mass of 146 tonnes.

There are many different species of whale, and many of them are endangered. This is largely due to hunting. Although they are so large, we know little about the lives of many kinds of whales, and this makes their conservation more difficult.

Figure 2h.14 a Sperm whales. **b** Humpback whale. **c** Southern right whale. **d** Fin whale.

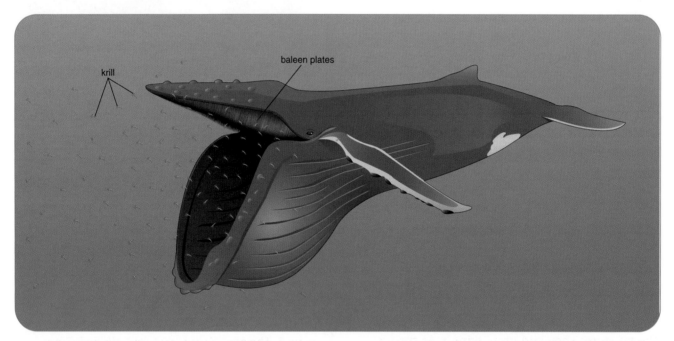

Figure 2h.15 How a humpback whale feeds using baleen plates. When the whale closes its mouth, the tiny krill are filtered by the baleen hanging from the whale's top jaw.

Blue whales, right whales and humpback whales are thought to be the most at risk. Grey whales, sperm whales and fin whales are also threatened.

How whales feed

There are two major groups of whales. The first is the toothed whales – whales that have teeth and use them to kill and eat other marine animals.

These include killer whales and sperm whales. The second group is the baleen whales. Instead of teeth, baleen whales have huge strainers made of fringed plates running from top to bottom of their jaws, called baleen. They swim through the water with their mouths open and strain out plankton from the water.

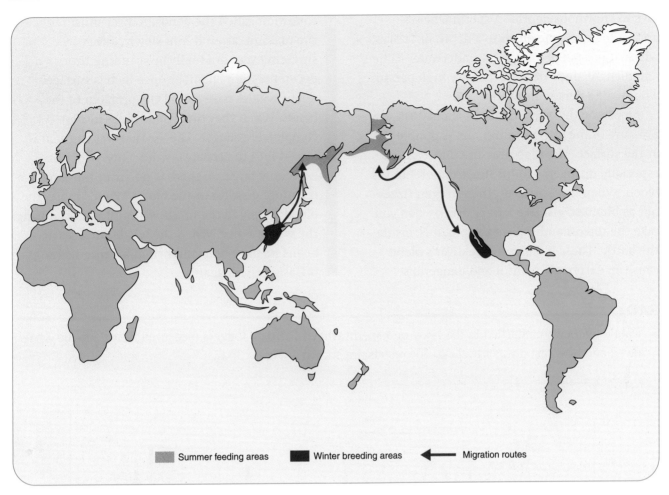

Figure 2h.16 Migration pattern of the grey whale.

Whales move around the oceans to the best places to find food. This varies at different times of year and they may make huge migrations from one place to another. For example, blue whales spend the summer in the cold waters near the poles, where there is a lot of plankton. As it becomes colder, they move towards the equator. There is little food for them there, so they rely on the fat stores they laid up in winter. As spring approaches, they move back to the polar waters again. Grey whales follow a similar pattern. They cover around 18 000 km in each circuit.

Killer whales, on the other hand, don't follow regular migrations like this. They feed on penguins and seals as well as fish and squid. They simply go wherever they can find food.

How whales live

There is a huge amount that we don't know about whales. Whales tend to do things very slowly, so it isn't always easy to work out what they are doing. Even experienced whale researchers can sometimes be unsure whether they are watching a whale playing, hunting or displaying to another whale.

Whales communicate with one another by sound. The sounds are very low pitched. Low-pitched sounds travel further in water than high-pitched ones. Large whales also make very loud sounds, which can be heard by other whales as much as 200 miles away. The sounds are often rhythmical and repeated, like a slowed-down bird song. We still don't understand exactly why they make these sounds and what they mean.

Whales are difficult to follow – they have the vast oceans at their disposal and some species can dive to enormous depths. A sperm whale is known to have gone down to a depth of 1134 m, where it got tangled up in an underwater cable and died. At these depths, the very high pressures would kill a person instantly.

If a person dives to a depth of about 10 m, the pressure of the water around him is double that at the surface. The high pressure makes gases – especially nitrogen – in his lungs dissolve in his blood. When he resurfaces, the nitrogen fizzes out as bubbles, just like a fizzy drink when you take the top off and reduce the pressure inside the bottle. These bubbles in the diver's blood cause an extremely painful and dangerous condition called the bends. A diver must reduce the pressure around him slowly, either by surfacing very gradually or spending time in a decompression chamber once he has surfaced.

But whales have solved the problem of the bends. When they dive, their lungs almost completely collapse. The air that was in them is forced into the trachea (wind-pipe), which is especially large. So there is no air in the lungs, so it cannot dissolve in the blood. What's more, the blood supply to the lungs is almost cut off during the whale's dive. When the whale surfaces, it opens its blow-hole and brings air into its lungs, inflating them again.

SAQ

5 Data loggers were attached to the body of a sperm whale. The data loggers measured how deep the whale dived and how fast it was travelling. The results are shown in Figure 2h.17.

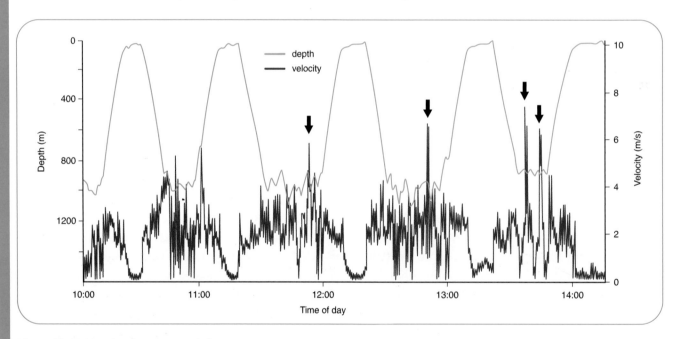

Figure 2h.17 Dive data for a sperm whale.

a What is the maximum depth to which the whale dived?

b For how long did it stay at the surface after each dive?

c How long did it stay under water during each dive?

d The researchers thought that the times marked with arrows might be when the whale was chasing prey. What is the evidence for this hypothesis?

e What patterns can you see that relate the activity of the whale and its depth below the water?

Making use of whales

Ever since people found ways of killing them, whales have provided a whole range of substances that people can use. Whale hunting increased greatly in the 17th century, and the whaling industry provided jobs for thousands of people.

Almost every bit of a whale can become something useful. Their meat is prized as food by people in some countries, such as Japan and Iceland. It can also be made into animal feed. Oil extracted from their bodies has been used for lamps and in industry, and for making cosmetics. Whalebone – actually not bone at all, but the baleen plates from the jaws of baleen whales – was used for making corsets, horsewhips and umbrellas.

Today, most of us don't need any of these things. We have much better materials that we can use now. We can get oils from many different kinds of plants, as well as from oil deposits underground. We can make springy steel rods instead of using whalebone. We have many other things to eat and don't need whale meat. However, some Inuit societies in northern Canada and Greenland do genuinely depend on whales for their existence, and probably would not be able to live where they do if they could not exploit whales.

Figure 2h.18 Whalebone was once used for making corsets.

And there is now another way of making money out of whales. People will pay to see them. Whale-watching expeditions take people out on boats into areas where they can get close to whales. You don't need to go abroad to do this – you can see whales around the coasts of England, Wales, Scotland and Ireland. Sometimes, you can even see them from the shore.

Figure 2h.19 A whale-watching expedition.

SAQ

6 Suggest how the development of whale-watching facilities can help to conserve whales.

Whales in captivity

Most whales are much too big to keep in zoos. But killer whales (orcas) are relatively small, and they are active whales that entertain zoo visitors.

Can we justify keeping whales in captivity? It must be hard for them to be confined in a small space, when they are used to having the whole ocean to move around in. On the other hand, they do get a regular supply of food and veterinary attention to keep them healthy. Some people have suggested that they may even enjoy the stimulation and company of their human keepers.

There are several arguments supporting the keeping of a few captive whales. People who have enjoyed being close to a whale may better understand why we should conserve them and contribute to conservation programmes. Research done on captive whales could help us to understand their biology, which could in turn help us to understand how to conserve them in the wild.

H **Whaling**

Conserving whales has to be an international effort. Whales travel huge distances, often far out to sea. It is no good one country deciding not to kill whales if other countries continue to do so.

The International Whaling Commission was formed in 1946. Its purpose was to provide for the conservation of whales, so that they were not hunted to extinction. It did not intend to ban whaling completely but rather to regulate it so that no-one caught so many whales of a particular species that they became extinct. Any country with an interest in whaling can join the IWC.

In 1986, the IWC imposed a complete ban on commercial whaling. But by the early 21st century, some countries wanted to partially lift this ban so that they could hunt and kill whales to sell products from them. Japan was especially determined to get the ban lifted. They would like to be able to hunt minke whales in what is currently a 'sanctuary' for whales in the Southern Ocean. They argue that there are plenty of whales there and that, even if some are caught, there will be enough to breed and keep the whale population reasonably high.

It is really difficult to get complete international agreement on whaling. Each time the IWC meets, fierce disagreements are argued out. And even when rules are agreed, it is difficult to enforce them. For example, Japan has been allowed to kill a certain number of minke whales 'for research purposes'. But no-one can actually be sure of how many they have killed, or what the whale bodies are actually used for.

Sustainable resources and development

If we had gone on killing whales at the rate that was happening in the 18th century, there might not be any whales left now. The whale hunting was not sustainable. A **sustainable resource** is one that we can go on and on using, without using it up.

Sustainable fishing

Fish is an excellent protein-rich food. People have always caught fish to eat. But as our population

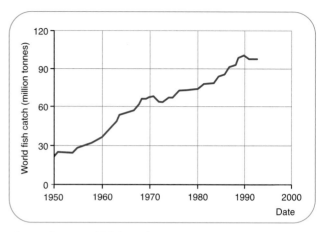

Figure 2h.20 World fish catches since 1950.

goes up and our fishing technology gets better, the number of fish caught has increased tremendously.

Our fishing activities have greatly reduced the populations of several species of fish. In the North Sea, cod populations have fallen to such low levels that some people think they may not be able to recover. The level of fishing activity has not been sustainable.

Since 1982, the European Union has had a Common Fisheries Policy. This regulates the fishing activities of European countries. Scientists have monitored the size of the fish populations and have made recommendations to the EU about how many fish we can catch without endangering the fish populations. The EU has then decided on fishing rules that each country should follow.

There are several ways in which fishing can be limited. These include:

● imposing fish quotas – that is, limiting the quantity of fish of particular species that each country can catch

● limiting the number of fishing boats that can be licensed and the length of time that they can fish each year

● imposing lower limits for the mesh size of nets, so that little fish can escape and have time to grow and reproduce.

But it has proved very difficult to make and enforce regulations that actually work. Each country wants to make sure that they get their fair share. For example, if fishermen in Scotland are only allowed to catch a few fish or are persuaded to give up fishing completely and have their boats destroyed, they are not very pleased if they then see Spanish fishing boats catching what might have been 'their' fish.

Figure 2h.21 Modern fishing vessels have technology that helps them to find fish shoals.

SAQ

7 Suggest the effects that the imposition of fish quotas has on:

a employment in fishing ports

b the price of fish in the shops.

Sustainable harvesting of wood

Trees are a very valuable resource, and have been as long as humans have been on the Earth. We use their wood for building and making furniture, and for fuel.

If you cut down a broad-leaved tree, it doesn't die. It just grows back up again from the base. In Britain, people have used this fact to get a sustainable harvest of wood from woodlands and forests. In many old woodlands some of the trees have been cut down like this many, many times. Each year, just a part of the woodland would be harvested. The other parts were left to regrow. This is called **coppicing**.

Coppicing is a sustainable way of using woodlands. You can go on doing it for ever. It does not harm the animals that live in the woodland. Because only part of the wood is harvested each year, they still have places to live.

Another way of harvesting wood is **clear-felling**. This means cutting all the trees down at once. This is much more harmful to animals, because

Figure 2h.22 Coppiced woodland provides a sustainable harvest.

their habitats are destroyed. It can also damage the soil, because now it is all exposed to rain and can be more easily washed away, especially if it is on a steep hillside. Sometimes the soil is so badly damaged that the trees cannot regrow.

Conifer woodlands are often harvested by clear-felling. Unlike broad-leaved trees, conifers don't regrow if you cut them down. If you want to keep the woodland there so that you can use it again in the future, you must replant the area with new trees.

Sustainable development

Humans always want to improve things. We build new houses, new roads, new factories, new hospitals. People want things like new and better cars, computers, washing machines and mobile phones.

All of these things put pressure on our environment. The UK is densely populated and almost everything that we do affects the animals and plants that share our country with us.

The UK government has a Sustainable Development Strategy. These are guidelines that they hope to follow, to make sure that we do not damage our environment too much. They define sustainable development as 'development which meets the needs of the present without compromising the ability of future generations to meet their own needs.'

We have seen how, as the human population grows, we produce more waste products and use more energy (Item B2g *Population out of control?*). We eat more food, so we need more land for farming, more factories processing food and more shops to buy it in. If we are to leave our environment in a fit state for future generations then careful planning will be needed.

At a local level, **local authorities** have strict planning rules that people must follow. You cannot just build a new house or a new supermarket when and where you want to. The local authority must give permission. So, for example, they will probably refuse permission if the new house cannot be connected up to the local sewage treatment plant or if there isn't already good access to it. They won't allow a new housing development if this would mean that the water supply in the area would not be big enough.

The local authorities themselves have to follow rules laid down by the government at a national level. For example, they may decide on a policy that encourages people to use less energy in their homes or to recycle more of their waste. They may then set targets that local authorities are to achieve, asking them to provide recycling facilities for people to use and trying to make sure that a certain percentage of waste is recycled.

And governments themselves make agreements with others. **International meetings** thrash out treaties that commit countries to making sure their development is sustainable. For example, many countries have signed a treaty in which they pledge to keep their carbon dioxide emissions under control. This took place in Japan and is called the Kyoto Agreement. Unfortunately, there is not really any way these treaties can be enforced, so targets are often not met. And not all countries sign the treaties. The USA, for example, has not yet signed up to targets for any emissions. By contrast, many other countries have agreed to meet fairly strict targets within the next few years. The USA is by far the biggest producer in the world of greenhouse gases.

SAQ

8 Think of *three* examples of how sustainable development could help to protect an endangered species.

Summary

You should be able to:

- give examples of extinct and endangered animals, and describe the reasons why some animals become extinct

- explain why it is important to try to conserve endangered species and habitats

- describe some of the ways in which we can protect endangered species

- describe how the distribution patterns of whale species are related to their feeding habits

- describe some of the aspects of whale biology that we still don't fully understand

- give some examples of the uses of products from whales

- discuss whether whales should be kept in captivity

- discuss the problems of agreeing and enforcing international agreements to reduce whaling

- explain what is meant by a *sustainable resource* and *sustainable development*

continued on next page

Summary – *continued*

◆ explain how fish stocks and woodland can be sustainably harvested

H ◆ explain how our increasing population makes sustainable development more difficult, and explain how local, national and international agreements can help

◆ describe how sustainable development can help to protect endangered species

Questions

1 Explain what is meant by each of the following terms:

a sustainability

b endangered species

c captive breeding programme

d fishing quotas.

2 Table 2h.1 shows the numbers of critically endangered and endangered species in some different groups listed in the Red List between 1996 and 2004.

Group	1998	2000	2002	2003	2004
mammals	484	520	520	521	514
amphibians	49	63	67	67	1142
insects	160	163	164	164	167
plants	2106	2280	2337	2910	3729

Table 2h.1

The percentage increase in the number of endangered species of mammals between 1998 and 2004 can be calculated like this:

$$\frac{(514 - 484)}{484} \times 100 = 6.2\%$$

a Calculate the percentage increase in the number of endangered species for the other three groups.

b Display your results as a bar chart.

c Which group has had the greatest percentage increase in endangered species?

continued on next page

Questions – *continued*

    d There are more insects on Earth than any other group of animals. Suggest why the numbers of endangered species of insects in the Red List are so low.

    e Suggest the reasons why many plant species are in danger of extinction.

H **3** Write a paragraph explaining how our increasing population is threatening other species.

    Write a second paragraph explaining how planning and cooperation at a national level can help to conserve endangered species.

Cooking food

You may be surprised to find that the chemistry section of this book starts off with a section on food and cooking. But actually, every time you cook food, what happens is an excellent example of a **chemical reaction**. The raw uncooked food is the 'starting material'. In chemistry, we call the starting materials **reactants**. When you heat these starting materials, they react and change into the final cooked food. In chemistry, we call the things that are made **products**. In every chemical reaction, reactants are changed into products.

Figure 1a.1 Reactants and products.

SAQ

1 Look at the six photos in Figure 1a.1. Group them into three pairs. Each pair consists of some cooked food and the raw ingredients it would be made from. Decide which drawing in each pair shows the reactants and which shows the products. Copy and complete Table 1a.1.

Pair	Reactants	Products
1		
2		
3		

Table 1a.1 Reactants and products.

Changing the raw ingredients into the cooked food is a chemical reaction because it is a permanent change. Once you've hard-boiled an egg, you can't change it back into runny white and runny yolk. Most chemical reactions are permanent changes.

Ways of cooking food

We eat some foods uncooked or **raw**. For example apples, strawberries and most other fruit are

Method	What this means
Frying	Hot oil in a pan or a deep fryer is used to heat the food.
Grilling	When we cook food under a grill we are heating it with direct heat radiation.
Boiling	Vegetables are often cooked by boiling them in water. Unfortunately the vegetables can lose a lot of vitamins into the water.
Steaming	If vegetables are cooked by surrounding them in hot steam, only small amounts of vitamins are lost.
Roasting	Roasting food means putting it into a hot oven where the hot air in the oven will heat the food.
Microwaving	A microwave oven heats the food very quickly by using microwaves. This way of cooking has become popular since the 1980s.

Table 1a.2 Cooking methods.

often eaten raw. Some people like sushi – that's raw fish. However most of our food is cooked. Meat is usually cooked before we eat it and bread is made by cooking dough, which is mainly wheat flour. During cooking, the food can be heated in several different ways.

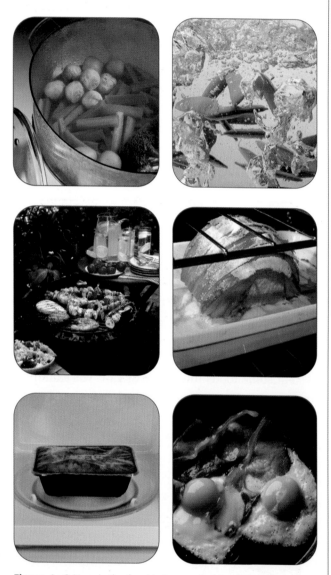

Figure 1a.2 How is the food being cooked in each picture?

Why we cook food

We cook food for several different reasons.

- Some foods are much easier to **digest** if they are cooked properly.
- Some food is very tough and chewy if eaten raw. Imagine eating raw dried pasta! Cooking it makes it easier to chew up and swallow, so cooking pasta improves its **texture**.

- Some food can contain **micro-organisms** such as bacteria, and cooking the food kills them and makes the food safe to eat.
- Cooking improves the **flavour** of many foods. What tastes better – a raw potato or a plate of chips?

Figure 1a.3 All these problems might have been avoided if they had cooked their food properly.

SAQ

2 What problem is each person in Figure 1a.3 having? Use the four words in **bold** type from the list above the cartoons in your answer.

Cooking changes food

You would never mistake a piece of raw meat for a piece of cooked meat, or a raw egg in a bowl for a bowl of scrambled egg. They look different – cooking has changed their appearance. Raw meat is red, while cooked meat is brown. Raw egg white is colourless and clear – you can see through it. When it is cooked, the egg white actually does become white and you can't see through it any more. We say it is **opaque**.

The 'feel' of the meat and of the egg change too when they are cooked. Raw meat can be very tough but cooking makes it tender. Raw egg is a liquid but cooking turns it into a solid. When you cook an egg or some meat, the **texture** of the food changes. The texture of potato changes too when it is cooked: it becomes much softer and easy to chew up and swallow. Cooking also changes the taste of potato – raw potato is unpleasant. Potatoes contain carbohydrates which you can use as an important energy source. It is much easier for your body to get the carbohydrate from the potato if it has been cooked.

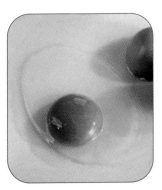

Figure 1a.4 A raw egg and a hard-boiled egg.

SAQ

3   Look at the photographs of the two eggs and list all the differences between the raw and cooked eggs. Make it clear which of the changes are changes of appearance and which are changes of texture.

Cooking – a chemical reaction

When meat, egg or potato are cooked, they change into something different. Although we still call them meat, egg and potato, they have changed into **new substances**. Because these changes cannot be reversed we say they are **irreversible** or **permanent changes**. These changes also involve an energy change. Heat is a form of energy and heat goes into the food during cooking. Changes like this

1   which form new substances
2   which are permanent
3   which involve an energy change

are called **chemical changes**. Cooking is a chemical change. A change like melting ice or boiling water is a **physical change**.

SAQ

4   When food is cooked it changes. Give three reasons why this change is a chemical change.

Heat changes food

Everything is made of tiny particles called **atoms**. These atoms join together to form groups called **molecules**. Atoms and molecules are too small to be seen even with good microscopes, but it is true – everything is made of them. The paper and ink you are looking at and the eyes you are using to look with are all made of atoms joined into groups called molecules.

Because the word 'everything' includes food, this means our food is made of atoms grouped into groups called molecules. These molecules have particular, exact shapes. In some substances, for example compounds made of both metal elements and non-metal elements, the molecules are known as **formula units**.

Meat and egg white are made of a substance called protein. When they are cooked the protein molecules change shape and then they stay in their new shapes. This is why the chemical change that happens when food is cooked is permanent.

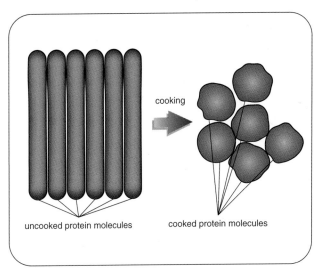

Figure 1a.5 Cooking protein.

SAQ

5   Look at Figure 1a.5 above.

   a   A lamb chop is made of molecules. What is a 'molecule'?

b Describe what has happened to the protein molecules in a lamb chop when the chop is cooked.

c What differences do these changes make to you when you eat the chop?

H Changes in protein and cellulose

If you look at Figure 1a.5, you will see why the texture of protein changes when it is cooked – the protein molecules change shape and pack together differently. This happens to the proteins in egg and meat when they are cooked. This permanent changing of the shape of the protein molecules is called **denaturing**. If you would like to know more about the size of protein molecules, why they have particular shapes and why they

H denature during cooking, you could do some independent research.

Cooking potatoes also changes the molecules they are made of. Like all living things, potatoes are made of millions of cells. (Each cell is made of billions of molecules.) Because potatoes are plants, these cells have a tough cell wall made of a substance called **cellulose**. The carbohydrate you can use as an energy source is inside this cell wall.

You can't digest cellulose, so if you eat raw potato you can't get at the carbohydrate, because it stays locked in the cells. Cellulose is not a protein but the molecules it is made of are also changed by heat during cooking. This makes the cellulose soften and become weaker so that you can now gain access to the carbohydrate inside the cells. Now you can digest the carbohydrate.

A new way of cooking

In 1921, Albert Hull invented a device called the magnetron. The magnetron produced waves that were like radio waves but had a much shorter wavelength. Researchers found out some interesting things about these waves. The waves produced by the magnetron would go straight through clouds. If they hit something solid, like an aircraft, they would bounce off it. Some of the reflected waves would arrive back to where they had originally come from. By detecting these reflected waves, it was possible to 'see' the original aircraft, even in thick cloud and even at night.

This discovery led to the development of radar, one of the inventions that possibly saved the UK from defeat in 1941. The radar stations were able to tell the RAF when German aircraft were attacking. The RAF fighters were able to take off and be ready to meet them in the air.

The next part of the story is the sort of piece of luck that seems to crop up all over the history of science and technology. In 1946, Percy Spencer was trying to learn more about the possibilities of the magnetron when he noticed that the chocolate bar in his pocket had melted. He tried putting other foods near the magnetron and

found that it would cook popcorn. When he put an egg near it, the egg exploded. While wiping very hot egg yolk of his face, Percy Spencer realised that he had stumbled across a very fast way to cook food.

The waves that were originally used to detect German bombers are called microwaves, and the ovens that use Percy Spencer's discovery are called microwave ovens. They are very efficient, although they can't do everything. A baked potato done in a microwave comes out well cooked but with a very disappointing 'jacket'. Microwave ovens work by selectively heating the water in the food. The microwaves make the water molecules vibrate. This vibration is one form of heat, so the microwaves make the water molecules in the food heat up.

There was recently a scare caused by microwave ovens with doors that didn't seal properly. It was possible that microwaves were leaking out and partly cooking anyone standing nearby! Percy Spencer and his colleagues must have been constantly exposed to microwaves, but as he lived to be 76 it seems he wasn't too badly affected by them.

Baking powder

Figure 1a.6 Types of flour.

Look at Figure 1a.6. You can buy several different types of flour, but the ones that are used most often are plain flour and self-raising flour. Plain flour is basically ground-up wheat. Self-raising flour is plain flour with **baking powder** added to it. Self-raising flour is used to make scones and cakes. If you want to make a cake but you have run out of self-raising flour, you can make some by adding baking powder to plain flour.

When the cake mixture is cooking in the oven the baking powder gives off carbon dioxide gas. This carbon dioxide gas makes the cake rise, so when the cake is cooked it is light and fluffy, not heavy and puddingy.

Figure 1a.7 Different types of cake.

SAQ

6 a Which of the cakes in the photographs in Figure 1a.7 was made with self-raising flour? How can you tell?

b Which of the cakes in the photographs was made with plain flour? How can you tell?

c What is added to plain flour to make self-raising flour?

You can see what happens when you heat baking powder, in a school laboratory. If you heat some baking powder in a test tube, it will give off carbon dioxide. If the carbon dioxide is passed into lime water, the lime water will turn cloudy. This is called the **chemical test** for carbon dioxide. When the lime water turns cloudy this proves the gas is carbon dioxide, because no other gas will do this.

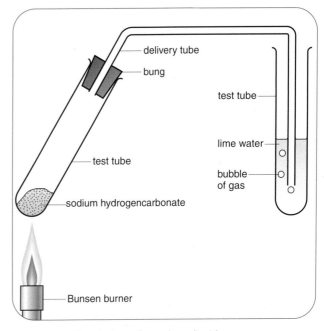

Figure 1a.8 Chemical test for carbon dioxide.

Baking powder decomposes

The chemical name for baking powder is **sodium hydrogencarbonate**. Its formula is $NaHCO_3$. This formula tells us several things.

1 The formula includes the symbols for four elements, Na, H, C and O. A substance made from more than one element is a **compound**, so sodium hydrogencarbonate is a compound.

2 The formula tells you exactly which atoms are joined together in one 'formula unit' of sodium hydrogencarbonate. These atoms are one Na atom, one H atom, one C atom and three O atoms. Every sodium hydrogencarbonate formula unit is the same, made from exactly the same atoms joined together in a group.

The meaning of the term 'formula unit' is very similar to the meaning of the term 'molecule'.

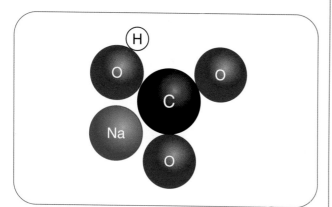

Figure 1a.9 How a sodium hydrogencarbonate formula unit might look if we could see it. (We've added the colours and the symbols!)

SAQ

7 a How many different elements are there in sodium hydrogencarbonate?

b How many atoms are there in one formula unit of sodium hydrogencarbonate?

c Use the Periodic Table on page 393 to find the full names of the atoms represented by Na, H, C and O.

d The formula of water is H_2O. What are the atoms in one molecule of water?

e The formula of carbon dioxide is CO_2. What are the atoms in one molecule of carbon dioxide?

Word equations

When sodium hydrogencarbonate is heated, it **decomposes**. This means it breaks down into simpler substances. The simpler substances that form are sodium carbonate, water and carbon dioxide. These two sentences can be summarised by writing a **word equation**. A word equation has the reactants on the left, followed by an arrow which means 'changes into' and the products on the right.

The word equation for this decomposition reaction is:

$$\text{sodium hydrogencarbonate} \rightarrow \text{sodium carbonate} + \text{water} + \text{carbon dioxide}$$

SAQ

8 a What happens to sodium hydrogencarbonate when it is heated?

b Which substances are the **products** of this reaction?

c Which substance is the **reactant** in this reaction?

d How could you prove that carbon dioxide is the gas produced when sodium hydrogencarbonate is heated?

Balanced symbol equations

The formula of sodium hydrogencarbonate is $NaHCO_3$. The formula of sodium carbonate is Na_2CO_3. The formula of water is H_2O. The formula of carbon dioxide is CO_2. We can rewrite the word equation using symbols like this:

$$NaHCO_3 \rightarrow Na_2CO_3 + H_2O + CO_2$$

This is not completely correct, however. Symbol equations tell us the **formulae** of each substance, but they must also be **balanced**. The symbol equation written above includes the correct formula for each substance, but it is not balanced.

Being 'balanced' means that all the atoms in the reactants also appear in the products. Being 'balanced' means that all the atoms in the products also appear in the reactants. Atoms are not lost or made in a chemical reaction. The symbol equation above shows more atoms in the products than there are in the reactants. This cannot be true.

SAQ

9 a How many atoms of each of sodium, hydrogen, carbon and oxygen are there on the left of the arrow above?

b How many atoms of sodium, hydrogen, carbon and oxygen are there on the right of the arrow above?

c Why can't the equation above be the whole truth?

H In order to make one formula unit of sodium carbonate and one molecule each of water and carbon dioxide, two formula units of sodium hydrogencarbonate must decompose. This is shown in a completely correct **balanced symbol equation** like this:

$$2NaHCO_3 \rightarrow Na_2CO_3 + H_2O + CO_2$$

H **SAQ**

10 Look at Figure 1a.9. Use blue circles for sodium atoms, white circles for hydrogen atoms, black circles for carbon atoms and red circles for oxygen atoms to show this balanced symbol equation as an 'illustrated equation'.

Summary

You should be able to:

◆ state six ways of cooking food

◆ describe the desirable changes that take place when food is cooked

◆ describe the cooking of food as a chemical change

◆ describe the changes that take place when protein is cooked

◆ describe the changes that take place when potatoes are cooked

H ◆ explain the changes that take place when protein and potatoes are cooked

◆ describe what happens when baking powder is heated

◆ describe a chemical test to identify carbon dioxide

H ◆ write a balanced symbol equation for the decomposition of sodium hydrogencarbonate

Questions

1 a List the changes that happen when food is cooked.

b Why do we call the change caused by cooking a chemical change?

c How can the chemical changes caused by cooking make the food safer to eat and more pleasant to eat?

2 a What is the chemical name for baking powder?

b What does baking powder do when it is heated?

c What products are made when baking powder is heated?

d Write a word equation for the reaction that takes place when baking powder is heated.

e Explain why baking powder mixed into the flour makes scones rise in the oven.

continued on next page

Questions - *continued*

3   A young scientist is told that chalk gives off carbon dioxide gas when it is heated strongly. She
    wants to do an experiment to find out if this is true.

   a   Draw a fully labelled diagram of the apparatus she should use to do her experiment. Label any
    chemicals she would use.

   b   What safety precautions should she take when doing the experiment?

   c   What will she see during the experiment if the chalk does give off carbon dioxide gas when it
    is heated?

H 4   **a**   Write a balanced symbol equation for the decomposition of sodium hydrogencarbonate.

   b   Explain what the word *balanced* means here.

   c   Explain what each number and symbol in the balanced symbol equation means.

5   **a**   Name a food that is a good source of carbohydrate.

   b   Name a food that is a good source of protein.

   c   Everything is made of tiny particles called atoms which form into groups. What are these
    groups called?

   d   Use diagrams to show what happens to the shape of the groups of atoms in a piece of protein
    as the protein is cooked.

H 6   **a**   Explain the meaning of the term *denaturing* and its relevance to the cooking of meat.

   b   Explain why cooked potato is easier to digest than raw potato.

7   A chef keeps his plain flour and his self-raising flour in glass jars because paper bags go soggy in
    the kitchen. One day he finds the labels have fallen off the jars onto the floor. He can't remember
    which flour is which. How can he find out?

Chemicals and additives

Vegetables are made from substances that the plant has absorbed from the air through its leaves and from the soil through its roots. Meat is made from the substances in the food the animal ate. All these substances are **chemicals**. The plants and animals change these chemicals into the different ones that they themselves are made of. All foods are a mixture of different chemicals. However, some foods we eat have extra chemicals added to them while they are being blended or packaged. These extra chemicals are called **additives**.

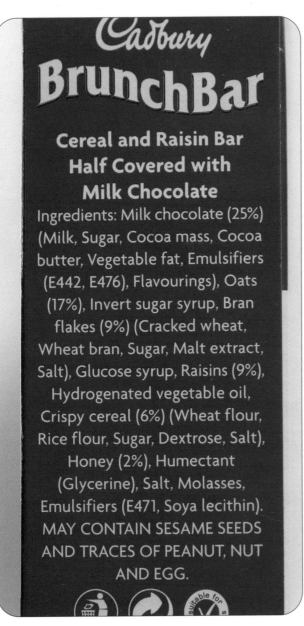

Cadbury
BrunchBar

Cereal and Raisin Bar Half Covered with Milk Chocolate

Ingredients: Milk chocolate (25%) (Milk, Sugar, Cocoa mass, Cocoa butter, Vegetable fat, Emulsifiers (E442, E476), Flavourings), Oats (17%), Invert sugar syrup, Bran flakes (9%) (Cracked wheat, Wheat bran, Sugar, Malt extract, Salt), Glucose syrup, Raisins (9%), Hydrogenated vegetable oil, Crispy cereal (6%) (Wheat flour, Rice flour, Sugar, Dextrose, Salt), Honey (2%), Humectant (Glycerine), Salt, Molasses, Emulsifiers (E471, Soya lecithin). MAY CONTAIN SESAME SEEDS AND TRACES OF PEANUT, NUT AND EGG.

Figure 1b.1 The ingredients of a cereal bar.

Look at the ingredients label on the cereal bar packet. The list includes the foods that the maker has used, like oats and bran flakes. The list is in order of amounts – the bars contain more milk chocolate than anything else, so milk chocolate comes first. It also includes additives like E471, soya lecithin. Many additives are identifiable by their 'E numbers' (the 'E' stands for European). The E number of soya lecithin is E471.

Types of food additive

1 **Flavour enhancers** improve the taste of savoury food like crisps and meat pies. Monosodium glutamate (known as MSG) and salt are flavour enhancers.
2 **Food colours** do exactly what they say! Sweets often contain added colours.

Figure 1b.2 Wine gums contain food colours.

3 **Antioxidants** stop food reacting with oxygen in the air. Foods contain fats and oils (oils are fats that are liquid at room temperature). Fats and oils in food can react with oxygen. The products of such reactions make the food taste sour, or **rancid**. Adding an antioxidant stops this happening. Margarine and lard often have an antioxidant additive.
4 **Emulsifiers** help different ingredients in the food to mix together. The soya lecithin in the cereal bars is an emulsifier. Foods like tomato sauce, mayonnaise and salad cream contain water and oils. These do not naturally mix – they separate out and the oils float on the water. Adding an emulsifier will keep them mixed.

INGREDIENTS

INGREDIENTS: WATER, CHICKEN (2%), MODIFIED MAIZE STARCH, RAPESEED OIL, DOUBLE CREAM (FROM COWS MILK), SKIMMED MILK POWDER (FROM COWS MILK), POTATO STARCH, WHEAT FLOUR, SALT CHICKEN FAT, STABILISERS: DISODIUM DIPHOSPHATE, POLYPHOSPHATES, DISODIUM HYDROGEN ORTHOPHOSPHATE, ACACIA GUM; FLAVOURINGS, HYDROLYSED POTATO STARCH, YEAST, YEAST EXTRACT, GLUCOSE, PALM OIL, LACTIC ACID, SUGAR, ACIDITY REGULATOR: CALCIUM LACTATE, COLOUR: CAROTENES; LEEK POWER, FLAVOUR ENHANCERS: DISODIUM GUANYLATE, DISODIUM INOSINATE; GARLIC POWDER, CITRIC ACID, ONION POWDER, GARLIC EXTRACT, GLYCEROL, ANTIOXIDANTS: ALPHA TOCOPHEROL, ASCORBIC ACID, NATURAL TOCOPHEROLS.

PRODUCED IN THE UK USING UK CHICKEN

*CONTAINS MILK & WHEAT GLUTEN

INGREDIENTS: Milk chocolate (57%) (Sugar, Cocoa mass, Cocoa butter, Dried skimmed milk, Butterfat, Whey powder, Vegetable fat, Lactose, Emulsifier (Lecithin), Flavouring), Sugar, Wheat flour, Modified starch, Colours (E171, E104, E124, E110, E122, E133, E120), Glazing agents (Carnauba wax, Beeswax), Flavouring.

Figure 1b.3 Food ingredients for soup and Smarties.

1 Look at the food labels in the photographs in Figure 1b.3. For each food:

   a which ingredient does it contain most of?

   b which of the four types of additive does it contain?

   c what are the names of the additives in the food?

Hydrophobic and hydrophilic

Oil and water don't mix. The oil will float on the water, making the food look unattractive and also giving it an unappealing texture. The molecules of an emulsifier make the oil and water mix in a very clever way. One end of each emulsifier molecule is 'fat loving'. The correct word for this is **hydrophobic**; hydrophobic literally means 'water hating'. A substance that is hydrophobic doesn't mix with water but is good at mixing with fats. The hydrophobic end of each emulsifier molecule mixes in with the fats and oils in the food.

The other end of each emulsifier molecule is 'water loving'; the correct word for this is **hydrophilic**. A substance that is hydrophilic mixes with water but doesn't mix with fats. The hydrophilic end of each emulsifier molecule mixes in with the water in the food. In this way, the oil and water are made to mix because the emulsifier molecules join them together.

How emulsifiers work

The hydrophobic end of each emulsifier molecule is attracted to oil and fat molecules. This end of each emulsifier molecule forms **bonds** to oil and fat molecules. The hydrophilic end of each emulsifier molecule is attracted to water molecules. This end of each emulsifier molecule forms **bonds** to water molecules. The emulsifier molecule stops the oil molecules and water molecules from separating – it keeps them mixed by bonding to them.

Figure 1b.4 Are you an 'emulsifier'?

SAQ

2 Look at the cartoon strip in Figure 1b.4.

  a Why are Oliver and Walter able to go into town together even though they don't get on?

  b Compare the cartoon strip with the effect of adding an emulsifier to a mixture of oil and water. Make it clear which role each person plays.

Problems with additives

Most additives do a good job. If you live in Southampton, you would not enjoy a genuine Dundee cake made in Dundee if the butter used to make it had gone **rancid** and sour before the cake reached you. A little bit of added antioxidant stops this happening.

Sometimes, however, additives are a problem to some people. Tartrazine is a bright yellow–orange colour. It used to be used as a colouring in sweets and orange squash. Unfortunately, it sometimes caused itchy skin and hyperactivity, particularly in young children, so tartrazine is no longer used.

Salt is a flavour enhancer. You also need a little bit of salt in your diet in order to be healthy. However, too much salt can cause high blood pressure and possibly heart problems later in life.

Additives that no one wanted

In 2003, several cases came to light of food contaminated with Sudan Red 1 and related red dyes. Sudan Red 1 is a dye which is illegal in foods because it may cause health problems. It can legally be used to colour such things as petrol and shoe polish (yummy). It seems that the dye had been used to colour batches of chilli powder, which was brought into the UK as a food ingredient. The chilli powder found its way into many spicy ready meals and was also used to make Worcester sauce, which was then used to make many other foods. The foods that were contaminated with the dye were then sold all over the world. Food standards agencies around the globe reacted with impressive speed.

The South African Government Department of Health (SAGDofH) believed that 'Sudan Red 1 contains chemicals that could increase the risk of cancer if consumed over a long period or in large quantities'. SAGDofH also said that 'the withdrawal of certain foods should be seen only as a precautionary measure as the amounts used were extremely small'. SAGDofH instructed both the provincial health offices and the port health authorities to take samples from all red chilli and related seasoning products in order to ensure that there was no Sudan Red present.

The main worry in Australia and New Zealand was the possible presence of the dye in Worcester-sauce-flavoured crisps. These crisps are sold in Australia and New Zealand in specialist shops catering for homesick Brits. Food Standards Australia New Zealand (FSANZ) said, 'There is questionable evidence that Sudan red dyes may be associated with cancer formation in laboratory animals, but there is no evidence that they can cause harm in humans, particular at the low levels found in these foods'. FSANZ reported that it was unlikely that any contaminated food had found its way to Australia, but this important national agency continued to monitor the situation.

In the UK, over 150 products, ranging from flavoured crisps to curry powders, were recalled from the supermarket shelves and destroyed between June 2003 and June 2005. In addition, food standards laboratories have developed a screening test that gives a definite answer on whether or not a food contains Sudan Red 1 or its relatives Sudan Reds 2, 3 or 4. This will enable the UK to respond rapidly and with more certainty in the case of a similar scare in future.

Figure 1b.5 These jelly sweets contain β-carotene. This is a natural orange colouring that has replaced tartrazine.

3 Look at the photograph of jelly sweets in Figure 1b.5.

   a Why has the manufacturer decided to add colouring?

   b Why did the manufacturer use β-carotene not tartrazine?

Intelligent packaging

When you shop at a supermarket, you probably put the food in a carrier bag to take it home. The carrier bag makes the food easier to carry. Other packaging can do more than this. Such packaging is called **active** or **intelligent packaging**. Intelligent packaging can help to preserve the food, or it can make the food look more appealing, or it can heat or cool it for you. Three examples of intelligent packaging are shown in Figures 1b.6 to 1b.8.

Figure 1b.6 Hot coffee. This can holds the coffee inside it, but it also heats the coffee up. A special compartment inside the can contains chemicals which react together and produce this heat. Chemical reactions like this one that release heat are called exothermic reactions.

Figure 1b.7 Dried soup. Bacteria and fungi love soup! How can we keep soup in our cupboards at home for months without it going bad or mouldy? One answer is to take all of the water out of the soup by a process called **dehydration**. Dehydrated (or dried) soup will not have bacteria or moulds growing in it because these micro-organisms need water in order to grow. Removing the water makes it difficult for the bacteria or moulds to grow. The dried soup is then put into packets that stop water getting back into the food.

Figure 1b.8 Breathable film. Meat slowly turns brown unless oxygen can get to it. Brown meat looks unappetising and people are unlikely to buy it. If we leave meat open to the air it will stay pink but, if micro-organisms land on it, the meat can go bad. We need to let the oxygen get to the meat but keep out the micro-organisms. This is done by using packaging film with tiny holes in. The oxygen molecules are so tiny they can fit through these holes so the meat stays pink. Bacteria and moulds have cells that are too big to fit through the holes, so the meat doesn't go bad. The packaging film is called **breathable film**.

SAQ

4 Look at the cartoons in Figure 1b.9. Three of the people have a problem with some food or drink. Explain how each problem could have been avoided.

a

b

c

Figure 1b.9 Food problems.

Summary

You should be able to:

◆ state that everything in food is made of chemicals

◆ state the four types of food additive

◆ state the name of a food likely to have an antioxidant added to it and the name of a food likely to have an emulsifier added to it

◆ explain why antioxidants and emulsifiers are added to food

H ◆ explain how an emulsifier does its job

◆ give examples of active packaging

◆ explain why removal of water helps to preserve food

Questions

1 Explain each of the following statements.

   a 'All butter shortbread' contains an antioxidant additive.

   b Parents are advised to avoid giving their children ice-poles that contain tartrazine.

   c A man with a heart complaint is advised to cut down on the number of take-aways he is eating.

2 A manufacturer of mayonnaise uses vegetable oil and vinegar as its two main ingredients. It also used to add soya lecithin during the mixing process. In order to save money, the manufacturer decides to leave out the soya lecithin. Unfortunately, as the jars stand on the supermarket shelves, a layer starts to form on the top of the mayonnaise and people stop buying it. You are called in as a consultant to solve this problem. Compile a short report to the manufacturer. Your report should include:

   a what the layer is

   b why the layer separates out

   c the reason why soya lecithin used to be added

   d as full an explanation as possible of how the soya lecithin did its job

   e your recommendations to the manufacturer.

3 Give three examples of active or intelligent packaging. Explain how the packaging is good for the consumer or makes the food more likely to sell.

4 The list below describes three foods. The manufacturers might choose to blend one or more types of additive with the food in order to solve the problems described. You must decide which type or types of additive could be used and explain your choice. You should also discuss how necessary you consider the use of the additives to be.

   a A packet of dried low-fat chicken soup looks good when made up but doesn't have a strong enough chicken flavour.

   b A new line of fruit pastilles has five different fruit flavours but every pastille looks the same pale yellow.

   c A Mediterranean salad dressing uses olive oil and balsamic vinegar as its base. It looks good, tastes good and is in excellent condition when it leaves the factory but, unfortunately, the oil separates out into a top layer after a week's storage. The shelf life is also very short because the oil sometimes goes rancid.

Figure 1c.1 Problems with perfumes.

Perfumes

Many of us wear perfume now. You may not be the sort to dab little spots of very expensive 'parfum français' on your wrists or your neck, but you probably use a body spray, or you use perfumed soap, shampoo or deodorant. When these perfumes are breathed in through a person's nose, they have an effect on sense cells in the nose.

Like everything else that has a particular use, a perfume must have certain **properties**. A substance's 'properties' are all the ways in which it behaves. These have to be right for its use. If a substance is going to be used as a perfume then it must smell nice, but it is important that it has the properties described in Table 1c.1, too.

SAQ

1 Look at the three people in the cartoons in Figure 1c.1. What has gone wrong for each person? In each case, say which of the five desirable properties is missing from their perfume or spray.

Perfumes evaporate

Perfume is applied to the skin as a liquid. The perfume molecules are held to each other by weak attractive forces. The molecules are close to each other but they can move about. The perfume will be warmed up by the wearer's body, and the molecules will move more quickly as this happens.

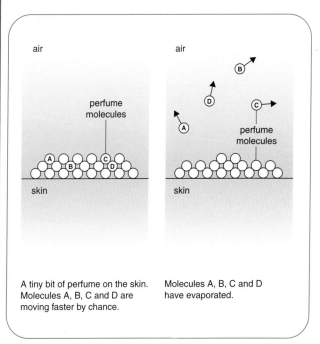

A tiny bit of perfume on the skin. Molecules A, B, C and D are moving faster by chance.

Molecules A, B, C and D have evaporated.

Property	Reason
Insoluble	It must not dissolve in water, otherwise it will wash off easily.
Non-toxic	It must not be **toxic** – this means it isn't poisonous.
Non-irritant	It must not be an **irritant**. Smelling nice is not so good if you come out in red blotches.
Unreactive	It must not react with water. A perfume that reacted with the water in your sweat might change into a substance that had no smell, or even into one that smelt nasty.
Volatile	It must evaporate easily. Otherwise, people passing nearby wouldn't know how lovely you smelt, which would be a waste of money.

Table 1c.1 Properties of perfume.

Figure 1c.2 The faster-moving molecules can evaporate.

H

A perfume molecule that happens by chance to be moving fast enough can break free of the weak attractive forces that hold it to its neighbours. It then moves away into the air as an independent molecule of gas. Lots of perfume molecules will be doing this at any one time. We say the molecules are **evaporating**. This is a **physical change**.

Perfumes need to be able to evaporate easily so that other people can smell them. A substance that evaporates easily is called a **volatile** substance. Perfumes must be made of substances with weak attractive forces between their molecules, so that they can evaporate easily.

SAQ

2   Why do substances with strong forces between the molecules make poor perfumes?

Natural perfumes

Many perfumes on the market today are **synthetic**. Another word for synthetic is **artificial**. These words both mean they are man-made substances. However, long before we were able to make perfumes in laboratories and factories, people were wearing perfumes. These perfumes came from natural sources. You will not be surprised to learn that rose water is perfumed with the scent from rose petals. The perfumes known collectively as musk are produced by animals such as the musk deer and a small wild cat called a civet.

Figure 1c.3 Civet cat.

A popular use of natural perfumes is to perfume a room by using an oil burner. A few drops of a natural perfumed oil like lavender oil or tea tree oil are put into a small puddle of water in the top of the oil burner, then a lighted candle

is put in the space underneath the water and oil. The heat of the flame helps give the oil molecules more energy so that they can evaporate.

Figure 1c.4 Using an oil burner.

Making esters

Some artificial perfumes can be made easily in a science lab. Chemicals called **esters** are easy to make. Esters have a strong smell, so some of them are used to make perfumes. You can make an ester in a school laboratory. The reactants used are dangerous so great care must be taken. An ester can be made by reacting ethanoic acid with an alcohol called ethanol. When an acid reacts with an alcohol, the products are an ester and water.

First you mix $5\,cm^3$ of ethanoic acid with $5\,cm^3$ of ethanol in a boiling tube, then you add three drops of concentrated sulphuric acid. You must now put the boiling tube in a warm water bath.

Half an hour later, some ester should have formed. Take the boiling tube out of the water bath and pour its contents into a beaker containing $50\,cm^3$ of cold sodium carbonate solution. The ester that formed will float on the sodium carbonate solution as an oily top layer. Some water was made in the reaction, too, but you won't notice this. You will notice immediately how smelly the ester is, not an obvious perfume perhaps, but other esters do smell much nicer.

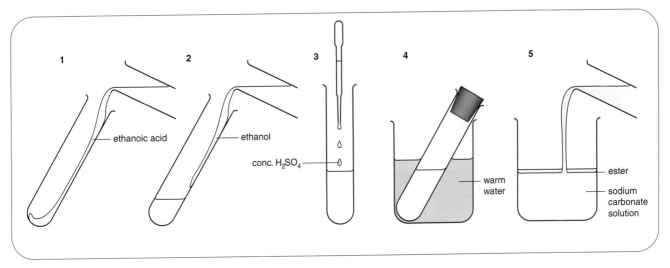

Figure 1c.5 Making an ester.

SAQ

3   In this reaction the name of the ester produced is 'ethyl ethanoate'.

   a   Write a word equation for this reaction.

   b   The concentrated sulphuric acid is acting as a *catalyst*. Find out what this word means.

   c   Why do you think the mixture was left in a warm place for half an hour?

   d   If you have done this practical yourself, write down all the safety precautions you had to take.

Nail varnish

Figure 1c.6 The lengths we go to…

When the nail varnish in Figure 1c.6 was first painted onto their nails, the person had to wait until it dried before doing anything with their hand. This is because the coloured nail varnish in the bottle is mixed in with a liquid called a **solvent**. After the nail varnish was painted onto each nail, the solvent had to evaporate; we say that it dried.

The nail varnish is said to be **dissolved** in the solvent. The coloured nail varnish (the substance that dissolves) is called the **solute**. The mixture that we get when a solute dissolves in a solvent is called a **solution**. Remember that, on page 156, you learnt that oil and water don't mix; if you shake them together, they separate. This is because oil and water don't form a solution. A solution *doesn't* separate if you leave it to stand.

SAQ

4   You put some instant coffee powder into a mug, add some boiling water and then stir it. Which word in **bold** from the paragraphs above describes:

   a   the instant coffee powder?

   b   the boiling water?

   c   what happened to the instant coffee powder when you stirred your mug of coffee?

   d   the mug of black coffee you ended up with?

Solubility

The solvent in nail varnish is not water. Water can act as a solvent: instant coffee dissolves in water and so does salt. However, nail varnish colours do not dissolve in water; if they did, the colours would wash off next time the hand was washed. Because nail varnish colours don't dissolve in water we say they are **insoluble in water**. Because nail varnish colours do dissolve in the solvent in the bottle, we say they are **soluble in that solvent**.

Figure 1c.7 Using nail varnish remover.

To remove nail varnish colours, a solvent similar to the one in the nail varnish bottle has to be used. This solvent is called nail varnish remover. The esters that are used to make perfumes can also be used as solvents.

Household glue often consists of a tiny bit of sticky stuff (called the **adhesive**) dissolved in a lot of solvent. When you use it to glue something, you have to 'wait until it dries'. This means you have to wait until the solvent evaporates, leaving the adhesive behind to stick the items together.

SAQ

5   Look at the data in Table 1c.2.

Solvent	Biro ink	Nail varnish	Wax	Salt
propanone	3	25	13	1
water	0	0	0	36
ethanol	40	1	2	20

Table 1c.2 Number of grams of various substances that will dissolve in $50\,cm^3$ of three solvents.

   a   In which solvent is wax most soluble?

   b   In which solvent is nail varnish insoluble?

   c   Which solvent would you use try to clean biro ink off a shirt?

   d   Does wax dissolve in water? How does the data tell you this?

   e   Which of the three solvents would be best as nail varnish remover?

H Intermolecular forces

Molecules of the same type attract one another in a pure substance. Molecules of the same type and molecules of different types attract one another in a mixture. All molecules attract one another. The forces of attraction between neighbouring molecules are called **intermolecular forces**.

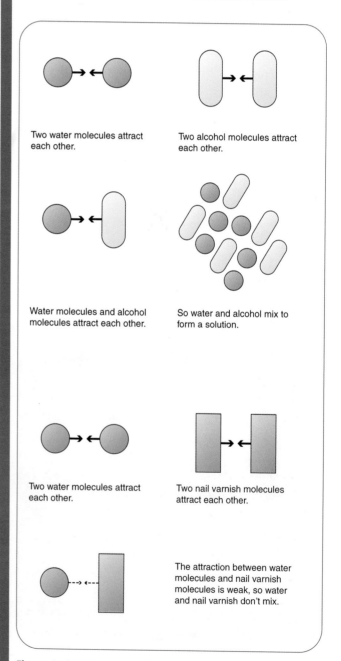

Figure 1c.8 Intermolecular forces.

H Intermolecular forces are not all equally strong. If you try to mix two pure substances, they will only mix if the intermolecular forces between the two different types of molecule are similar to the intermolecular forces in the two pure substances. If the intermolecular forces between the two different types of molecule are weaker than the intermolecular forces in the pure substances, the substances will not mix.

Nail varnish colours do not dissolve in water. The water molecules attract each other more strongly than the water molecules attract the nail varnish molecules. The nail varnish molecules attract each other more strongly than the water molecules attract the nail varnish molecules.

SAQ

6 Use these ideas to explain:

   a why alcohol dissolves in water

   b why oil does not dissolve in water

   c why nail varnish colours dissolve in nail varnish remover

   d why an ester does not dissolve in water.

For or against animal testing?

In 2004, in a small town in Hampshire, a group of people staged a piece of street theatre. A courtroom scene was presented, featuring a judge in a very elaborate wig, a sorry-looking defendant and a jury consisting entirely of people dressed as white mice. This street theatre wasn't just a piece of frivolous entertainment; the actors involved were bringing the public's attention to something they believed was deadly serious.

Modern law insists that nowadays, before cosmetics can be used and sold, they have to be tested to ensure they are safe. These tests have to be thorough and done over a long period of time. We must be sure the substance is not acting as a very slow poison.

How can new substances be tested? In the past they were always tested on large numbers of laboratory animals such as rats and rabbits. The substances were injected into the animals to see if they were poisonous. The substances were dripped onto the animals' eyes or skin to see if they were irritants. The advantage to this approach was obvious – humans didn't suffer.

Animal testing is now much less popular. New products may instead be tested on human volunteers or on human cells growing in an incubator. There are two main reasons for using less animal testing. Firstly, this reduces suffering caused to animals. Secondly, if a product is tested

Figure 1c.9 Many manufacturers now use their refusal to test on animals as a positive selling point.

and found to be safe on rats and rabbits, this does not always mean it is safe to be used by humans.

The opposition to animal testing has resulted in groups known as 'animal rights activists'. The street theatre in Hampshire was performed by such activists. The defendant was meant to represent the owner of a nearby laboratory. The laboratory they were campaigning against is alleged to test botox on white mice. Botox is designed to stop the user's face from wrinkling as they get older. The white mice had a chance to get their own back in the street theatre – they made up the jury. You can guess what the verdict was ...

Summary

You should be able to:

◆ name two perfumes obtained from natural sources

◆ explain the difference between *natural* and *synthetic* substances

◆ state that esters are used to make perfumes and act as solvents

◆ describe how an ester can be made in a laboratory

◆ list the properties of a good perfume

Ⓗ ◆ explain why some liquids are volatile

◆ use the terms *solution*, *solute*, *solvent*, *soluble* and *insoluble* accurately in the context of nail varnish and nail varnish remover

Ⓗ ◆ explain solubility using the concept of attractive forces between the molecules involved

◆ explain why cosmetics need to be tested and discuss the reasons for and against testing on animals

Questions

1 A manufacturer of chemicals and cosmetics called Global Synthetics (GS) develops a synthetic ester to use as a perfume.

   a What does *synthetic* mean?

   b What types of substance would be used to make the ester?

   c GS will need to make sure the ester has certain properties. What properties must it have?

   d One use of esters is for perfumes. Give another use.

2 GS tests the ester to make sure it has the right properties.

   a One of the directors of the company says the ester should not be tested on animals such as rabbits. What might be the reasons for this decision?

   b Another director disagrees with this. What might be their reasons for wanting to use animal testing?

   c GS decide not to test the ester on rabbits. They are also developing a new drug to treat diabetes in children. They have decided to test this drug on rabbits. Give at least one reason for this decision.

Ⓗ 3 GS also develop new solvents. Their latest solvent (called Trike®) is excellent for removing graffiti because 'permanent marker' ink that won't dissolve in water dissolves quickly in Trike. Use ideas about intermolecular forces to explain:

   a why permanent marker ink can't dissolve in water

   b why permanent marker ink dissolves in Trike.

Fossil fuels

The human race has been lucky. 300 million years ago, life teemed in the seas and on the swampy land. Many of the trees that died in the swamps were fossilised and eventually formed coal and gas. At the same time, the remains of some of the creatures that died in the seas were also fossilised, forming oil and more gas. Coal, oil and gas are **fossil fuels**, which are fantastically useful energy sources that have been powering our machines ever since the industrial revolution.

Fossil fuels have one major disadvantage. They are formed by very slow processes. We will soon use up the world's supplies, but meanwhile new supplies are not being formed fast enough to replace them. This means fossil fuels are **non-renewable** and that they are a **finite** resource. We are using them up very quickly; in particular, we are likely to run out of oil during your lifetime.

Figure 1d.1 This platform was used to drill down to a pocket of oil under the North Sea.

Crude oil

We find oil trapped deep in the Earth's crust. Next we drill down to it. The oil may then come to the surface because of its own pressure or we may have to pump it out. The oil that is obtained is not a pure substance but is a mixture of many different **hydrocarbon** compounds. A 'hydrocarbon' compound is a compound of hydrogen and carbon only. Because the oil is a mixture, it is called **crude oil**.

Oil spills

Much of the world's crude oil is found under the sea or is transported around the world in huge ocean-going supertankers. When there are accidents or spillages, the sea gets polluted. A large patch of oil called an **oil slick** may float on the sea or get washed up on beaches. Poisonous compounds in the oil then kill birds, fish and other marine life. The birds' feathers stick together so they can no longer fly. Oily beaches are dirty and unattractive.

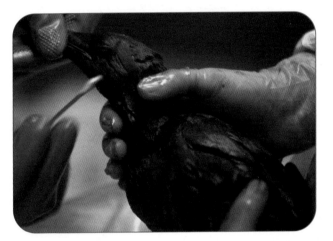

Figure 1d.2 Relief workers did their best to save the life of this guillemot. However, it could still have died if it had swallowed too much oil.

SAQ

1   **a** Name three fossil fuels.

b How did the fossil fuels in the Earth's crust originate?

c Fossil fuels are 'non-renewable'. Explain what this term means.

d Describe one environmental problem caused by our use of fossil fuel.

Oil politics

There are political problems associated with the extraction and exploitation of crude oil. Environmental groups have campaigned for greater controls to be put on the oil industry to make spills less likely, and to stop empty supertankers using seawater to wash out their tanks as they go home to fill up again.

H

In some of the world's nations, crude oil is their main mineral wealth. Pressure groups in these countries have wanted their governments to slow down the extraction of the oil in order to safeguard their nation's future. The oil-buying nations do not want them to do this.

Purifying crude oil

The crude oil in the Earth's crust is not useful itself because it is a mixture of very many different hydrocarbon compounds. If we purify this mixture, the crude oil becomes *very* useful. This is why it is so valuable. The crude oil is purified by **fractional distillation**.

The crude oil is shown being heated at the bottom left of Figure 1d.3. The useful products, called **fractions**, are shown coming out on the right of the fractionating tower. The fractionating tower is hottest at the bottom and coolest at the top. The hydrocarbons in the crude oil separate according to their boiling points. The hydrocarbons with lower boiling points leave nearer the top of the tower, where the temperature is lower. These

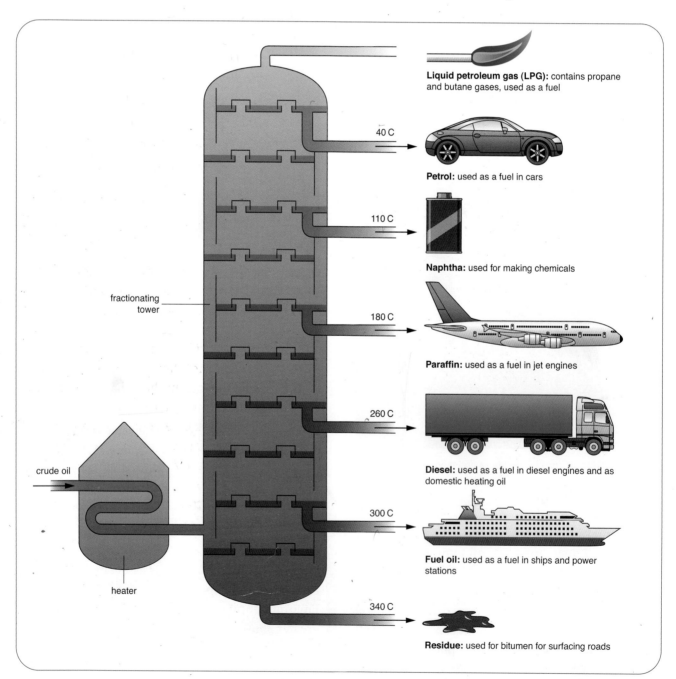

Liquid petroleum gas (LPG): contains propane and butane gases, used as a fuel

40 C

Petrol: used as a fuel in cars

110 C

Naphtha: used for making chemicals

180 C

Paraffin: used as a fuel in jet engines

260 C

Diesel: used as a fuel in diesel engines and as domestic heating oil

300 C

Fuel oil: used as a fuel in ships and power stations

340 C

Residue: used for bitumen for surfacing roads

fractionating tower

crude oil

heater

Figure 1d.3 Fractional distillation of crude oil.

hydrocarbons are made of smaller molecules. Their vapours condense at lower temperatures.

The hydrocarbons with higher boiling points leave nearer the bottom of the tower, where the temperature is higher. These hydrocarbons are made of larger molecules. Their vapours condense at higher temperatures. Each fraction consists of a mixture of hydrocarbons with similar boiling points. Although each fraction is still a mixture, it is now very useful and can be used for the purposes shown in Figure 1d.3.

SAQ

2 a How does a fractionating tower separate crude oil into fractions?

   b Is each fraction a pure substance?

   c Give the names of three of the fractions obtained from crude oil and give a use for each one.

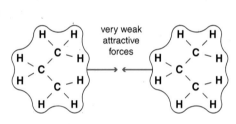

Small hydrocarbon molecules attract each other very weakly.

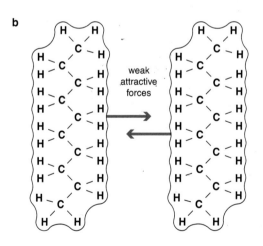

Large hydrocarbon molecules attract each other more strongly than small hydrocarbon molecules.

Figure 1d.4 a Small hydrocarbon molecules attract each other very weakly. **b** Larger hydrocarbon molecules attract each other more strongly than small hydrocarbon molecules.

Boiling and intermolecular forces

The atoms within each hydrocarbon molecule are held together strongly by strong **covalent bonds**. Hydrocarbon molecules are attracted to their neighbours by weak **intermolecular forces**. These intermolecular forces are broken when the substance boils. Larger hydrocarbon molecules have stronger intermolecular forces than smaller hydrocarbon molecules, so the hydrocarbons with larger molecules have higher boiling points than the hydrocarbons with smaller molecules.

SAQ

3 Copy the diagram of small hydrocarbon molecules in Figure 1d.4. Put the following labels on your copy:

   ◆ this is a strong covalent bond

   ◆ this is a weak intermolecular force

   ◆ this is broken when the hydrocarbon boils

   ◆ this is not broken when the hydrocarbon boils.

Cracking

Some of the fractions in crude oil are more useful than others. The petrol fraction is particularly useful. The oil refinery can sell the more useful fractions for a higher price. The oil refinery can make more profit if it can convert the less valuable fractions into the more valuable ones.

The hydrocarbons from the fuel oil fraction can be converted to petrol by a process called **cracking**. Cracking can be done in a school lab, using a kerosene-like substance called liquid paraffin in place of the fuel oil. Figure 1d.5 shows how this can be done.

The cracking reaction involves two types of hydrocarbon molecules. **Alkanes** have all their atoms held together by single covalent bonds, represented in Figure 1d.6 by a dash (i.e. C–C and C–H). **Alkenes** have most of their atoms held together by single covalent bonds, but two of the carbon atoms in the molecule are held together by a double covalent bond, represented in Figure 1d.6 by a symbol similar to an equals sign (i.e. C=C).

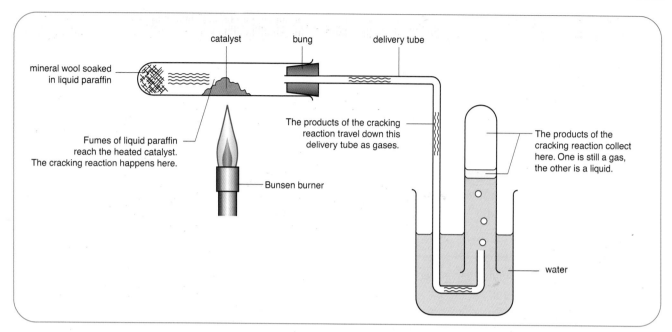

Figure 1d.5 Cracking liquid paraffin in the lab.

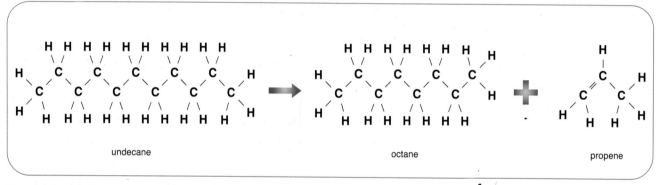

undecane   octane   propene

Figure 1d.6 Cracking.

Fuel oil and petrol are both mixtures of alkanes. The molecules of the alkanes in the fuel oil fraction are larger than the molecules of the alkanes in the petrol fraction. During cracking, the fuel oil is passed over a catalyst at a high temperature. The catalyst makes the molecules in the fuel oil break up into smaller alkane molecules and alkene molecules. The smaller alkane molecules are useful because they can be sold as petrol. The alkene molecules are useful because they can be used to make substances called polymers. Plastics such as polythene and polystyrene are polymers. In this way, the oil refinery can meet society's high demand for petrol while increasing its own profits.

SAQ

4 a Identify the reactants and the products in the cracking reaction shown in Figure 1d.6.

b Work out the molecular formula of each of the three substances in the reaction and state how many atoms there are in one molecule of each substance.

c Identify which of the three substances are alkanes and which one is an alkene.

d How did you decide which was the alkene?

e What will the propene that is produced be used for?

f What will the octane that is produced be used for?

g Why does cracking help an oil refinery make more money?

It could never happen again ...

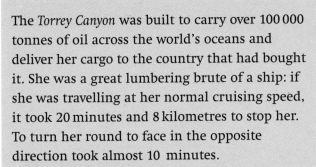

Figure 1d.7 Oil tanker disasters.

The *Torrey Canyon* was built to carry over 100 000 tonnes of oil across the world's oceans and deliver her cargo to the country that had bought it. She was a great lumbering brute of a ship: if she was travelling at her normal cruising speed, it took 20 minutes and 8 kilometres to stop her. To turn her round to face in the opposite direction took almost 10 minutes.

In March 1967, the *Torrey Canyon* was heading northwards through the gap between the Isles of Scilly and Lands End. This gap is 30 kilometres wide but the seas there are dangerous, with several reefs near the surface. And remember, the *Torrey Canyon* was not a nimble ship.

The *Torrey Canyon* hit the Seven Stones reef hard enough to rip open seven of her oil tanks. 30 000 tonnes of oil escaped immediately and drifted towards Cornwall. The world's first ever major oil tanker disaster had begun. The oil slick had a devastating effect. In the days after the tanker ran aground, thousands of sea birds were killed, marine life was poisoned and beautiful beaches became filthy and stinking.

The UK government tried desperately to minimise the impact on the environment and on the Cornish tourist industry. The RAF bombed the wreck to try to set light to the oil still inside her. Thousands of litres of detergent were sprayed onto the oil slick to try to disperse it. Unfortunately, the detergents caused environmental problems all of their own. It took years for the environment to return to normal. It was the first ever such disaster, but of course everyone was determined to learn from it and make it the last. However ...

In total, there have been over 30 major oil spills worldwide since the *Torrey Canyon* hit the Seven Stones reef and sank. We can tighten regulations and make improvements to our ships but the sea will always be a dangerous place. It seems that as long as we use crude oil on this scale we will need to use supertankers, and the environment will pay the price.

Summary

You should be able to:

◆ name three fossil fuels

◆ explain and use the term *non-renewable resource*

◆ describe crude oil as a mixture of hydrocarbons and describe how it is separated into useful products by fractional distillation

◆ explain how fractional distillation works

◆ name the fractions produced by fractional distillation of crude oil

H ◆ explain how intermolecular forces affect the boiling points of hydrocarbons

◆ describe the environmental impact of oil spills

H ◆ discuss the political problems associated with the exploitation of crude oil

◆ describe how molecular size affects the economic value of hydrocarbons

◆ label the laboratory apparatus used for cracking and explain how cracking can increase an oil refinery's profits

Questions

1 Three fractions obtained by heating crude oil are petrol, diesel and bitumen.

a Which of the three fractions has the lowest boiling point and which has the highest?

b Which of the three fractions has the largest molecules and which has the smallest? Explain your answer.

c Which of the three fractions comes out where the tower is coolest and which comes out where the tower is hottest?

d Which of the three fractions comes out of the tower highest and which comes out the lowest?

e Use your answers to parts **a–d** to make a general rule linking boiling point, molecular size, the temperature of the tower where the fraction emerges, and the part of the tower where the fraction emerges.

2 The formula of decane is $C_{10}H_{22}$. Cracking decane gives two products, octane C_8H_{18} and ethene C_2H_4.

a What conditions will be used for the cracking reaction?

b Suggest a use for each of the products of cracking decane.

continued on next page

Questions - *continued*

c Why might an oil refinery increase its profits by cracking decane?

d The laboratory apparatus in Figure 1d.5 could be used to crack decane like this. Copy Figure 1d.5 and add the labels 'decane goes here at the start', 'octane collects here' and 'ethene collects here' in the right places.

H 3 A poor nation has two main sources of foreign income, tourism and its oil wealth. Tourists come to enjoy the beautiful tropical beaches and to see the huge colonies of sea birds nesting on the cliffs of an island just off the coast. The level of tourism has now reached full capacity. At the current rate of extraction, the oil supplies will last 80 years. In order to improve the country's future, the government wants to invest heavily in education. This will be funded by increasing sales of oil fivefold. Recently, increased tanker traffic has resulted in two minor oil spills in the ocean near the country.

What are the advantages and disadvantages of increasing the sales of oil fivefold? Your answer should consider:

- the income the country will get from the oil, now and in the future

- the income the country will get from tourism, now and in the future

- the effect of investing in education, now and in the future

- the environmental effects of increased oil exports

- how other nations would react to their increased oil exports

- how 'green' pressure groups would react to their increased oil exports.

What advice would you give to the government?

Polymers

In our modern society, we use an enormous amount of the materials called **polymers**. The materials usually known as plastics are all polymers, and you will learn more about these materials in Item C1f *Designer Polymers*. Polymers consist of very large molecules called **polymer molecules**. These polymer molecules are in the shape of long thin chains. They are made by joining together many small molecules. The substance used to make a polymer therefore consists of small molecules. It is called a **monomer** and the small molecules are called **monomer molecules**.

Figure 1e.1 Polymers can be used to make children's toys.

Monomers are nearly all compounds of hydrogen and carbon only; such compounds are known as **hydrocarbons**. Hydrocarbons are obtained by the fractional distillation of crude oil. There are many different hydrocarbon compounds. In order to make them easier to remember and understand, they are grouped into families. Two important families of hydrocarbons are called the **alkanes** and the **alkenes**. The alkenes are the family of compounds used to make the monomers that polymers are made from.

SAQ

1 Explain the meaning of the term 'hydrocarbon'.

Alkanes

The alkanes are all compounds of hydrogen and carbon only, so they are hydrocarbons. Table 1e.1 contains some information about four members of this family: methane, ethane, propane and butane.

- The hydrogen atoms and carbon atoms shown in the molecular formula are joined together to form small groups called molecules.
- The molecular formula of each substance tells you how many carbon atoms and how many hydrogen atoms there are in each molecule of the substance. The molecular formula of methane, for example, is CH_4. This means every methane molecule consists of one carbon atom and four hydrogen atoms.

Name	methane	ethane	propane	butane
Molecular formula	CH_4	C_2H_6	C_3H_8	C_4H_{10}
Displayed formula				

Table 1e.1 Alkanes.

Figure 1e.2 Model of ethane.

- Every molecule made up like this is a methane molecule. Every molecule not made up like this is not a methane molecule.
- The displayed formula of methane shows the five atoms it is made of, shown as the letters C, H, H, H and H, joined together by bonds. The bonds are shown as the straight lines that join the atoms.
- The atoms in the substances around us are held together by bonds of several types. The type of bond that holds the atoms together in methane molecules is called the **single covalent bond**.

2 a What is the molecular formula of ethane?

   b What does this mean?

   c Look at the displayed formula of ethane.

     i How many atoms are there in one molecule of ethane?

     ii How many covalent bonds does each carbon atom make?

     iii How many covalent bonds does each hydrogen atom make?

   d Look at the displayed formula of propane.

     i How many atoms are there in one molecule of propane?

     ii How many covalent bonds does each carbon atom make?

     iii How many covalent bonds does each hydrogen atom make?

   e The molecular formula of pentane is C_5H_{12}. Draw the displayed formula of pentane.

Alkenes

The alkenes are all compounds of hydrogen and carbon only, so they are also hydrocarbons. Some information about three members of this family, ethene, propene and butene, is found in Table 1e.2.

Each displayed formula shows that each alkene molecule has one feature not seen in the displayed formula of any of the alkanes. This feature is shown in the displayed formula as C=C. This is called a **double covalent bond** and means there are two covalent bonds holding these two carbon atoms together.

Name	ethene	propene	butene
Molecular formula	C_2H_4	C_3H_6	C_4H_8
Displayed formula			

Table 1e.2 Alkenes.

Figure 1e.3 Model of ethene.

You should now have noticed two 'family differences' between alkanes and alkenes. Firstly, their names are different: alkanes have names that end –ANE, while alkenes have names that end –ENE. Secondly, alkanes consist of molecules held together by single covalent bonds only. Alkenes consist of molecules with at least one double covalent bond within each molecule.

SAQ

3 Figure 1e.4 shows 12 substances A to L. Every one is either an alkane or an alkene.

  a Which of these are alkanes?

  b Which of these are alkenes?

  c Which of these are hydrocarbons?

  d Which of these have a double covalent bond in each molecule?

  e Which of these have at least one single covalent bond in each molecule?

  f Which of these have only single covalent bonds in each molecule?

  g Which three letters represent descriptions of the same substance?

  h Which two letters represent descriptions of ethene?

  i Name the two different types of formula in **h**.

Single and double bonds

The molecular formula of ethane is C_2H_6. The molecular formula of ethene is C_2H_4. Ethane contains the maximum number of hydrogen atoms in each molecule for a hydrocarbon molecule with two carbon atoms. Because of this ethane is said to be a **saturated** hydrocarbon. All of the members of the alkane family are saturated

A	B	C	D
methane	H H \C = C/ H / \ H	butene	pentene
E	F	G	H
C_3H_6	H H \C/ / \ H H	CH_4	(structure: two-carbon branched structure with H's)
I	J	K	L
(structure: three-carbon chain with C=C double bond and H's)	hexane	C_2H_4	C_3H_8

Figure 1e.4 Substances A to L.

H hydrocarbons. Saturated hydrocarbons have molecules in which the atoms are joined together by single covalent bonds only.

Ethene contains less than the maximum number of hydrogen atoms in each molecule for a hydrocarbon molecule with two carbon atoms. Because of this, ethene is said to be an **unsaturated** hydrocarbon. All of the members of the alkene family are unsaturated hydrocarbons. Unsaturated hydrocarbons have molecules with at least one double covalent bond joining two of the carbon atoms together.

SAQ

4 Look at the substances A to L in Figure 1e.4.

   a Which of these substances are saturated?

   b Which of these substances are unsaturated?

   c What is meant by the term 'saturated hydrocarbon'?

   d What is meant by the term 'unsaturated hydrocarbon'?

Covalent bonds

Atoms consist of even smaller particles. One of the types of even smaller particle found within atoms is called the electron. To make a single covalent bond, the two bonded atoms share two electrons between them. A single covalent bond is

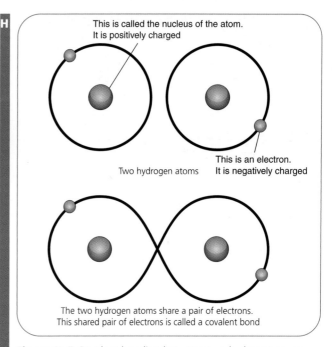

This is called the nucleus of the atom. It is positively charged

Two hydrogen atoms

This is an electron. It is negatively charged

The two hydrogen atoms share a pair of electrons. This shared pair of electrons is called a covalent bond

Figure 1e.5 Covalent bonding between two hydrogen atoms.

therefore known as a **shared pair** of electrons. To make a double covalent bond, the two bonded atoms share four electrons between them.

Using bromine water

It is possible to find out whether an unidentified hydrocarbon is unsaturated or not by shaking it with bromine water. (Bromine water is a dilute solution of the element bromine in water.) Bromine water is an orange colour. If you shake bromine water with a saturated hydrocarbon, no change is seen.

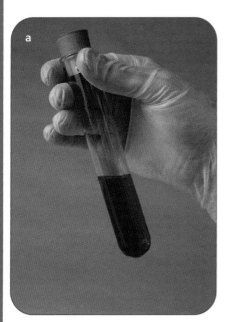
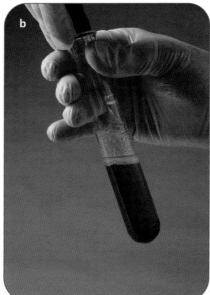

Figure 1e.6 a Some bromine water is put in a test tube. **b** Hydrocarbon is added. It floats on the top of the bromine water. **c** After the tube is stopped and shaken, the bromine water is decolourised.

H If you shake bromine water with an unsaturated hydrocarbon, the colour of the bromine water disappears. It changes from orange to colourless. We say it is **decolourised**. Only unsaturated hydrocarbons decolourise bromine water.

SAQ

5 Look at Figure 1e.6 and then answer questions **a** to **d**.

    a Was the hydrocarbon added in the figure an alkane or an alkene?

    b Was the hydrocarbon added in the figure saturated or unsaturated?

    c Explain why you decided on these answers.

Polymerisation

Alkenes are very useful substances because they are the raw materials used to make polymers. Alkene molecules are small but they can join together by a chemical reaction to make the long chain molecules that polymers are made of. The alkene molecules are called monomer molecules in this process.

The process of reacting the monomer molecules together to make the polymer molecules is called **polymerisation**. To make this reaction happen, the reaction vessel is pressurised and a **catalyst** is added. The catalyst speeds up the polymerisation reaction.

Look at Figures 1e.7 and 1e.8 showing polymerisation. You should be aware that these diagrams only show a small part of the process. During polymerisation, hundreds of monomer molecules link together to form each polymer molecule. Notice the names of the polymers. The monomer *ethene* produces the polymer *poly(ethene)*. The monomer *propene* produces the polymer *poly(propene)*. These polymers are commonly known as polythene and polypropene.

These two polymers are made by the adding together of monomer molecules without any other product forming. Because of this, these polymers are known as **addition polymers**. They have been made by a process called **addition polymerisation**.

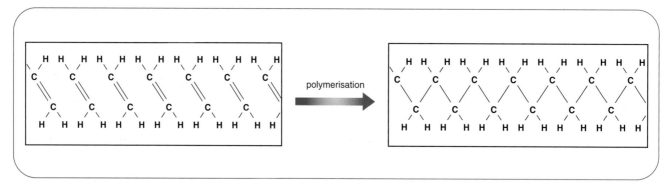

Figure 1e.7 Polymerisation of ethene.

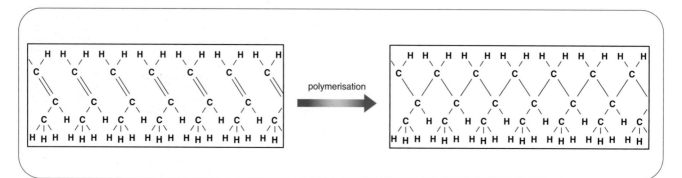

Figure 1e.8 Polymerisation of propene.

SAQ

6 a Copy and complete the two sentences that follow.

   i In Figure 1e.7, the monomer .... is shown making the polymer ....

   ii In Figure 1e.8, the monomer .... is shown making the polymer ....

b Complete the sentence using the three words in the brackets that follow.

Poly(styrene) is a .... made from a .... called styrene. It is made by .... polymerisation.

(monomer, addition, polymer)

c Complete the sentence using the five words in the brackets that follow.

The polymer .... is made from the monomer .... This process is called addition .... High .... and a .... are necessary.

(pressure, butene, polymerisation, poly(butene), catalyst)

Polymers are saturated

Look at Figures 1e.7 and 1e.8. You should notice that the monomers used are unsaturated while the polymers produced are saturated. This is always the case with addition polymerisation. A saturated polymer is made from an unsaturated monomer.

SAQ

7 a Using Figures 1e.7 and 1e.8 as examples to help you, draw a diagram of a short length of polymer molecule that would form from the four chloroethene molecules shown in Figure 1e.9.

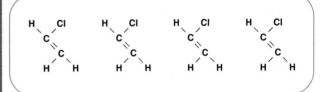

Figure 1e.9 Chloroethene molecules.

b Using Figures 1e.7 and 1e.8 as examples, draw the displayed formula of one of the monomer molecules that would be used to make the polymer molecule PTFE shown in Figure 1e.10.

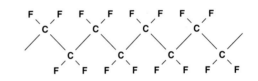

Figure 1e.10 PTFE.

8 What would happen if a polymer was shaken with bromine water? Explain your answer.

Engineering polymers

One of the greatest challenges facing human civilisation is global warming. The Earth's weather is changing and there is ever-increasing agreement among scientists that a major cause is the increase in CO_2 level in the atmosphere caused by burning fossil fuels. Travelling 100 km in a small car will require about 5 litres of petrol, which will release over 12 kg of CO_2 into the environment. However, reducing the mass of the car by 100 kg could reduce this figure by 10%. This is where polymers come in.

The properties of car parts made of polymers must meet three decisive criteria. They must be as strong as their metal equivalents but with lower mass. They must be cheap to make and to build into the car, thus helping to cut costs. Finally, they must make it possible for the manufacturers to incorporate new design ideas that will result in better cars.

Modern polymers fulfil all three of these criteria. These polymers are high-performance materials with extremely advantageous properties. They are known as 'engineering polymers'. Because engineering polymers deliver the all-important reduction in mass, they will help to reduce each car's CO_2 emissions. For example, polyphenylenesulphide (or PPS for short) can be used to make the air pipes in turbocharged engines; the pipes become 40% lighter than pipes made of the lightest metal alternative, which is aluminium.

Another engineering polymer is polyoxymethylene, known as POM. POM is described as having physical properties similar to cold-rolled steel, but it only has one-eighth of the density of steel. It is suitable for making gears, piston rings and the gaskets in fuel systems.

We already make up to 15% of a car from polymers, and this figure could rise to 20% over the next few years. A plastic car may not sound quite like what you want, but it will protect the environment and be superior to the steel one it has replaced.

Summary

You should be able to:

◆ use the following terms accurately: *hydrocarbon, polymer, monomer, polymerisation, covalent bond*

◆ describe the conditions necessary for polymerisation to take place

◆ describe the difference between an alkane and an alkene

◆ recognise alkanes and alkenes from their names

◆ represent an alkane molecule and an alkene molecule using displayed formulae

H ◆ use the terms *saturated* and *unsaturated* accurately to describe alkanes, alkenes, monomers and polymers

◆ describe a test to find out whether or not a hydrocarbon is unsaturated

◆ describe a covalent bond in terms of shared electrons

Questions

1   Use Table 1e.2 on page 175 to answer this question.

a   What is the molecular formula of ethene?

b   What does this mean?

c   Look at the displayed formula of ethene.

i   How many atoms are there in one molecule of ethene?

ii   How many covalent bonds does each carbon atom make?

iii   How many covalent bonds does each hydrogen atom make?

d   Look at the displayed formula of propene.

i   How many atoms are there in one molecule of propene?

ii   How many covalent bonds does each carbon atom make?

iii   How many covalent bonds does each hydrogen atom make?

2   a   Using suitable diagrams from this chapter explain the meaning of the following terms: *polymer*; *monomer*; *polymerisation*.

b   Alkenes are particularly useful hydrocarbons. Scientists are very interested in the properties of new alkenes that they are able to synthesise. Explain as fully as you can why these statements are true.

3   a   What do the terms *saturated hydrocarbon* and *unsaturated hydrocarbon* mean?

b   Is hexane a saturated hydrocarbon or an unsaturated hydrocarbon?

c   Is hexene a saturated hydrocarbon or an unsaturated hydrocarbon?

d   If you were given an unlabelled bottle of hexane and an unlabelled bottle of hexene, how would you find out which was which? Describe the test you would do and the observation you would make when you tested the hexane and the observation you would make when you tested the hexene.

Designer polymers

More about polymers

Some substances consist of large molecules made of very many small molecules linked in a chain. These substances are called polymers. You have learnt about them in Item C1e *Making polymers*. Their 'large molecules' can consist of thousands, even millions, of atoms. Some polymers are natural and some are **synthetic**. Natural polymers are important materials in all living things. Carbohydrates, proteins and DNA are all natural polymers. Synthetic polymers are also important materials. They include the artificial fibres and the many types of different substance known as plastics.

Synthetic polymers have been manufactured since the early part of the 20th century. One of the best known synthetic polymers, polythene, was first made in a laboratory in 1933. Polythene is used to make carrier bags and squeezy bottles. You will be familiar with the names of many other synthetic polymers, such as polystyrene, PVC, nylon and polyester.

Figure 1f.1 Plastics have many uses.

The right material

When a polymer is used for a particular purpose, it is important for it to have the right properties. Polymers have many useful properties.

● Most synthetic polymers are easy to melt and mould, easy to colour, and do not have sharp edges, so they are used to make many children's toys.

● Polythene is easily melted and made into thin sheets, so it is used to make carrier bags.

● When polystyrene is melted, it will form a stable solid foam if a gas like carbon dioxide is blown into it. This foam is called expanded polystyrene. Expanded polystyrene is used as packaging and also as heat insulation, for example in take-away soup cups.

● Nylon and polyester will form a thin fibre if the polymer is melted and forced through a narrow hole called a spinneret. The nylon and polyester fibres can then be spun into threads which are woven to make cloth and then used in clothing.

We use a particular material for a particular job according to the properties of the material. Our industries decide which materials to use to make each product. Their decisions are based on the properties and costs of the materials.

SAQ

1 **a** What is meant by the *properties* of a substance?

   b Which properties of polymers make them suitable for making children's toys?

   c State one property of nylon and polyester which makes them suitable for making artificial fibres.

Intermolecular forces

The molecules in a polymer consist of atoms joined to each other by **covalent bonds**. Covalent bonds are strong. The molecules can be joined to other molecules in two different ways.

If the molecules are joined to their neighbouring molecules by weak intermolecular forces then the polymer melts and softens easily

H when heated. The polymer will also stretch easily if a force is applied to it. Polymers like this are called **thermoplastics** and examples include polythene and polystyrene.

If the molecules are held to their neighbouring molecules by strong bonds then the polymer will have a high melting point and be rigid. The strong bonds can be covalent bonds that were formed when the polymer was first made. Polymers like this are called **thermosets** and examples include the formaldehyde resins used to make plugs, sockets and switches for domestic electricity. The covalent bonds that hold the molecules together are known as **cross-linking bridges**.

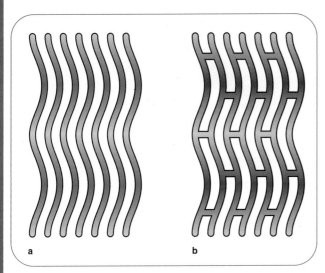

a   b

Figure 1f.2 The polymer molecules are shown as long curved lines. **a** These polymer molecules are held together by weak intermolecular forces. **b** These polymer molecules are held together by strong covalent bonds.

SAQ

2 Look at the two diagrams in Figure 1f.2.

   a Which one shows the molecules in a thermoplastic and which one show the molecules in a thermoset? Give reasons for your choice.

   b Which type of polymer would you use to make a plastic bag? Give reasons for your choice.

   c Why are electric fitments made of thermosets not thermoplastics?

   d Which type of polymer would you use to make kitchen work surfaces? Give reasons for your choice.

Waterproof clothing

Waterproof outer coats are a good example of how the desirable properties of polymers have improved the materials we use. In the 1970s, nylon began to be used to make coats called cagoules. They had many advantages over coats made from traditional materials. They were lighter, packed into a small space, were hard wearing, and could easily be made brightly coloured – very important for walkers and cyclists on busy roads. When the weather was bright and sunny, they provided good protection from the Sun's ultraviolet rays.

All of these advantages are consequences of the properties of nylon. Cagoules had one disadvantage though: water vapour could not get out through the nylon cloth as it could through traditional materials. Nylon is impervious. When the wearer sweated, the moisture was trapped inside the cagoule and made the wearer feel clammy.

Figure 1f.3 a Nylon cagoules. **b** The inside of a Gore-Tex® coat. The black membrane makes Gore-Tex breathable.

Many coats are now manufactured from a fabric called Gore-Tex®. Gore-Tex coats are waterproof, so rain can't get in. However, Gore-Tex is a **breathable** fabric. This means that when the wearer sweats the moisture can move through the fabric and away into the air. People engaged in active outdoor activities such as hill walking and cycling can sweat a lot, so breathable fabrics are very desirable. Gore-Tex has all the desirable properties of nylon and is breathable too – it is a better material.

SAQ

3 a Explain why nylon is a good material for making waterproof clothing.

   b Explain why Gore-Tex is better than nylon for this purpose.

How breathable fabrics work

Gore-Tex coats are made from a nylon fabric woven in such a way that it is not waterproof. A thin layer (called a membrane) made from polyurethane or PTFE is then coated onto the nylon.

   The membrane is waterproof, so liquid water cannot go through it, but it has tiny pores that molecules of water vapour can go through. The Gore-Tex stops rain getting in but allows sweat to evaporate out. The membrane must be used in combination with nylon because on its own it is too fragile for making coats. A material like Gore-Tex made by combining thin layers of other materials is called a **laminate**.

SAQ

4 a Why is Gore-Tex waterproof?

   b How does sweat escape from inside a Gore-Tex coat?

   c Why is it important that the nylon component of Gore-Tex is not waterproof?

   d What is a laminate?

   e What would be the problem with a coat made only from a PTFE membrane?

The problem with polymers

Polymers are excellent materials because they have many useful properties. Unfortunately, they have one bad property. They cannot rot naturally. If you throw away a newspaper or a bread roll, it will eventually rot away owing to the action of bacteria and fungi. We say that these substances are **biodegradable**.

   Most of the polymers used over the past 50 years are not biodegradable. As a society, we have a big problem dealing with our plastic bags, old toys and worn-out nylon clothes because the polymers they are made of are non-biodegradable.

Figure 1f.4 This rubbish will never rot away. It will stay until we clean it up.

The UK produces many thousands of tonnes of waste polymers every year. These all have to be dealt with. We have several answers to this problem.

1 We bury the waste. This is called **land fill**. Unfortunately, land fill sites waste valuable land.

2 We burn the waste in incinerators. This is not as straightforward as it sounds, because burning polymers can release **toxic** fumes.

3 We **recycle** the waste by melting it and remoulding it into new products. The different polymers in the waste have to be sorted out before recycling can be done. This sorting is often difficult to do, but even so the amount of polymer being recycled in the UK is increasing.

   The **raw materials** used to make polymers are **non-renewable**. This means they are running out and not being replaced. Disposing of waste polymers by burying or burning them is wasting a valuable resource. Recycling is a much better solution to the problem if it can be done efficiently.

The biomechanical mother and daughter

In the summer of 2004, at the age of 86, Mrs Thompson was able to stop hobbling and start walking normally. She had first found movement painful about 10 years earlier, particularly when getting up out of a chair or climbing stairs, and it had got steadily worse. The problem was her left hip joint. The joint had simply worn out and when she tried to move it there was a lot of friction, and this is what had caused the pain. Moving into a bungalow had helped solve the staircase problem, but finally she found any walking at all was very painful.

Her doctor said a hip replacement was the answer. Without this operation, it was likely that she would soon be completely housebound, or in a wheelchair. After careful consideration, she agreed to have the operation, and she went into Stoke Mandeville hospital in June 2004.

In a hip replacement operation, the surgeon uses a polymer called polytetrafluorethylene – PTFE for short. PTFE is used to make Gore-Tex because of its breathability. It is used when doing hip replacement surgery because it causes less friction than any other material.

After doing necessary repairs on Mrs Thompson's thigh bone and pelvic bones, the surgeon fitted the two bones with new surfaces made of PTFE. No friction means no pain! Mrs Thompson has made a full recovery after the operation. She is able to walk around the village where she lives and can even manage stairs, although she admits she doesn't look forward to them!

Mrs Thompson's daughter Catherine has also had a hip replacement operation. She had hers in 2003, just before her 50th birthday. Her doctor was reluctant to recommend the operation but decided in the end that it was the best choice. The problem is that the new joint only lasts about 20 years, so it is very likely that Catherine will need a second operation when she is older. Modern surgery certainly can work miracles, but it is interesting to think that nature can still make longer lasting hip joints than we can.

Figure 1f.5 a You can make a full recovery after a hip-joint replacement, like Mrs Thompson did. **b** This is an artificial hip joint. The black metal shaft ends in a silver metal ball embedded in a plastic socket that is coated with PTFE to reduce friction. During a hip-joint replacement operation, the shaft is cemented into a hole cut in the thigh bone, and the socket is cemented into the pelvis.

Figure 1f.6 Disposing of waste. A polymer incinerator (top) and a polymer recycling bin (bottom).

Scientists are now developing biodegradable polymers. You may have been given a carrier bag made of biodegradable polymer at a supermarket. In the long run, however, recycling waste polymers is a better solution to the problem.

SAQ

5 What are the problems with the methods of disposing of polymer waste shown in the photographs in Figure 1f.6?

Summary

You should be able to:

♦ state the names and uses of at least four polymers

♦ explain how the properties of a polymer make it suitable for a particular use

H ♦ explain how the intermolecular forces between polymer molecules affect the properties of the polymer

♦ compare the properties of nylon and Gore-Tex

H ♦ explain how Gore-Tex is made to be both waterproof and strong

♦ discuss the problems involved in dealing with waste polymers

Questions

1 a Give the names of three polymers.

   b Give a use for each of the polymers you have named.

   c Explain how at least one of the properties of the polymer makes it suitable for this use.

2 a What is meant by a *breathable* fabric?

   b What are the advantages of a waterproof coat made from nylon compared with a coat made from traditional fabrics?

   c What is the disadvantage of a coat made from nylon?

   d How does Gore-Tex solve this problem?

3 Thermoplastics and thermosets are two different types of polymer.

   a In what ways are the properties of thermoplastics and thermosets different?

   b Which type of polymer would you use to make a saucepan handle? Explain your choice.

   c Which type of polymer would you use to make the strings of a tennis racket? Explain your choice.

   d Which type of polymer has weak forces between its molecules?

   e Which type of polymer has molecules that are held to their neighbouring molecules by covalent bonds?

4 What problems are there when we have to dispose of an old nylon shirt compared with disposing of an old cotton shirt? Describe three ways the nylon shirt could be dealt with, stating one problem with each method and explaining which method of disposal should be used in your opinion.

5 Describe how the fabric called Gore-Tex is made and explain how it is able to be

   a hard-wearing

   b waterproof

   c breathable.

Better fuels

The hydrocarbons in **crude oil** have found many uses in our industrial society. Many of them are used as **fuels**. All industrialised countries rely to some extent on hydrocarbon fuels obtained from crude oil. As the developed world gets wealthier and people own more cars, enjoy more air travel and buy more manufactured goods, our use of these fuels is increasing.

Different crude oil fractions are used as fuels in different ways and in different types of engine, according to the particular properties of each fraction. When deciding which is the best fuel to use in a particular situation, several questions have to be asked.

1 How much energy is released when one gram of fuel is burnt? This is called the **energy value** of the fuel.
2 Is the fuel readily available?
3 Is the fuel easy to store?
4 How expensive is the fuel?
5 Is the fuel **toxic**? Gaseous fuels may be toxic if breathed in, liquid fuels may give off similarly toxic vapours or contain toxins that can be absorbed through the skin.
6 How much do the fuel's combustion products pollute the environment?
7 How easy is it to use the fuel?

SAQ

1 Look at Figures 1g.1, 1g.2 and 1g.3.

   a List the advantages and disadvantages of using fuel oil to power a car. Do you expect to see cars powered by fuel oil on our roads?

   b Your aunt and uncle are considering buying a methane gas fire for their sitting room. What advice would you give them on how to use the fire safely?

   c Would you like to own an LPG-powered car in future? Give your reasons.

Complete combustion

When a hydrocarbon fuel burns it is **reacting** with oxygen in the air. If the burning fuel has an adequate supply of oxygen the combustion products are carbon dioxide and water. This is known as **complete combustion**.

The carbon atoms in each molecule of the hydrocarbon combine with the oxygen from the

Figure 1g.1 This car uses LPG instead of petrol. The LPG is cheaper to buy than petrol but needs a special fuel tank in the car to store it. LPG cannot be bought at every service station. Most cars in the UK are petrol powered.

Figure 1g.2 This ship runs on fuel oil. It is cheap, readily available, easy to store, and has a high energy value. Fuel oil can release polluting combustion products but this might be seen as less of a problem out at sea, away from people. Fuel oil causes the ship to have a long start-up time but, because the crew know when they are going to set sail this is not a problem to them.

Figure 1g.3 This gas cooker is burning methane. Methane is not poisonous and is extremely easy to use in domestic heating systems and cookers. If the gas leaks into a room, the unburnt methane can cause explosions. Methane is difficult to store but, because it is delivered to houses in pipes, this is not a problem to the homeowner. Methane's combustion products are safe, as long as it is used in a well-ventilated room.

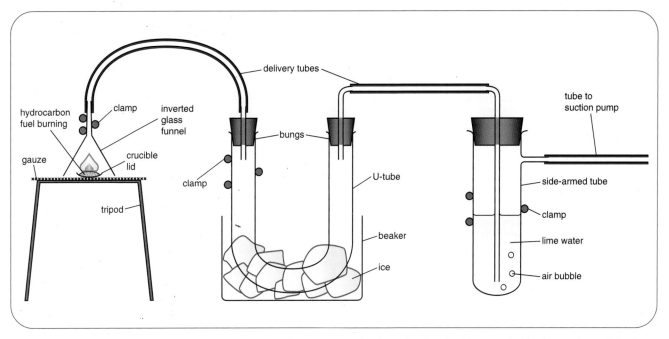

Figure 1g.4 After the hydrocarbon fuel has been burning for 2 minutes, droplets of a clear liquid appear inside the U-tube and the lime water turns cloudy.

air to produce carbon dioxide. The hydrogen atoms in each molecule of the hydrocarbon combine with the oxygen from the air to produce water. You can see evidence of the presence of carbon dioxide and water in the fumes from a burning hydrocarbon using the apparatus in Figure 1g.4.

SAQ

2 a Which product forms the droplets in the U-tube?

b Why is the U-tube being kept cold?

c Which product turns the lime water cloudy?

Incomplete combustion

If the burning fuel has an inadequate supply of oxygen, the combustion products may be carbon monoxide and water, or sometimes even carbon and water. This is known as **incomplete combustion**.

Incomplete combustion can also take place in the engine of a car, because car engines often don't take in enough air for complete combustion to take place. In this situation the burning fuel is still releasing useful heat energy, although less than it does during complete combustion, but the products produced cause environmental problems.

Carbon monoxide is poisonous, so starting a car in a confined space is dangerous. If carbon is produced, it is known as soot when it comes out of the exhaust. Soot is dirty and can cause breathing problems. Incomplete combustion can be a problem with domestic appliances that burn methane. The room where the methane is burning may not be adequately ventilated. If the

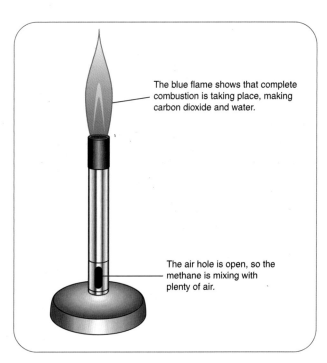

The blue flame shows that complete combustion is taking place, making carbon dioxide and water.

The air hole is open, so the methane is mixing with plenty of air.

Figure 1g.5 The blue flame transfers more chemical energy to heat energy per second than the yellow flame.

The yellow flame shows that incomplete combustion is taking place, making carbon monoxide, carbon (soot), and water.

The air hole is closed, so the methane has an inadequate supply of air.

Figure 1g.6 The yellow flame transfers less chemical energy to heat energy per second than the blue flame.

appliance is not regularly serviced, its air intakes can become blocked. In both these cases, poisonous carbon monoxide can be produced.

Incomplete combustion and complete combustion of methane can be observed using a Bunsen burner.

SAQ

3 **a** What colour is a Bunsen flame when complete combustion is taking place?

b What colour is a Bunsen flame when incomplete combustion is taking place?

c Give *three* advantages of using a Bunsen on a blue flame rather than a yellow flame.

H Complete and incomplete combustion

Balanced symbol equations

When methane undergoes complete combustion, it combines with oxygen to produce carbon dioxide and water. Representing these substances with their formulae, we would write this:

$$CH_4 + O_2 \rightarrow CO_2 + H_2O$$

H Notice that the formula of oxygen is O_2, which means that the oxygen atoms in the air bond together in pairs.

The equation above is not balanced. This means that it is not fully correct. It does not tell us exactly how many molecules of each substance are involved in each tiny 'reaction event'. To balance the equation, you must work systematically.

Worked example

Step 1: count the atoms on each side of the arrow.

On the left of the arrow, there are one carbon atom, four hydrogen atoms and two oxygen atoms. On the right of the arrow, there are one carbon atom, two hydrogen atoms and three oxygen atoms. The number of oxygen atoms is not the same on either side of the arrow. The number of hydrogen atoms is not the same on either side of the arrow. This means that the equation is not balanced.

Step 2: balance the equation for either oxygen or hydrogen.

It is easier to balance it for hydrogen, so we start there. The equation is balanced for hydrogen if we say that two water molecules are formed. The equation now looks like this:

$$CH_4 + O_2 \rightarrow CO_2 + 2H_2O$$

Step 3: balance the equation for oxygen.

The number of oxygen atoms is not the same on either side of the arrow. There are four oxygen atoms in the products so four oxygen atoms – two O_2 molecules – must react with each methane molecule.

$$CH_4 + 2O_2 \rightarrow CO_2 + 2H_2O$$

continued on next page

Worked example – *continued*

This is now a fully correct, balanced symbol equation for the complete combustion of methane. It is a correct statement of how many molecules of each substance are involved in each tiny reaction event. In each reaction event, one methane molecule meets two oxygen molecules and they become one carbon dioxide molecule and two water molecules.

A balanced symbol equation for the incomplete combustion of methane to give carbon is

$$CH_4 + O_2 \rightarrow C + 2H_2O$$

You should notice how the equation shows that incomplete combustion requires less oxygen – each methane molecule only reacts with one oxygen molecule not two. In each reaction event, one methane molecule meets one oxygen molecule and they become one carbon atom and two water molecules.

A balanced symbol equation for the incomplete combustion of methane to give carbon monoxide is

$$CH_4 + 1\tfrac{1}{2}O_2 \rightarrow CO + 2H_2O$$

This explains why incomplete combustion sometimes produces carbon and sometimes produces carbon monoxide. Carbon monoxide forms when the oxygen is in short supply. Carbon forms when the oxygen is in *very* short supply.

Although the equation

$$CH_4 + 1\tfrac{1}{2}O_2 \rightarrow CO + 2H_2O$$

is correct, it suggests that one methane molecule reacts with one and a half oxygen molecules. 'Half oxygen molecules' are extremely rare in the air.

It is better to double the whole equation to give

$$2CH_4 + 3O_2 \rightarrow 2CO + 4H_2O$$

In each reaction event, two methane molecules meet three oxygen molecules and they change into two carbon monoxide molecules and four water molecules.

Future fuels

Fossil fuels are non-renewable. At present, the vast majority of our road transport uses petrol or diesel obtained by the fractional distillation of crude oil. Our reserves of crude oil are dwindling fast. We estimate that petrol and diesel will not be on sale for use by the general public in 2040.

Ford, BMW, Toyota and the other businesses that manufacture our cars are not stupid. They know they must look to the future or go bust. Every one of them has been making preparations. This means that they have been researching, designing and building prototypes for the vehicles that will replace our petrol-powered cars.

Every car manufacturer has been thinking along similar lines. The new generation of cars will use hydrogen as a fuel. Some of these cars will have engines similar to our present-day cars, i.e. the hydrogen will be burnt in an internal combustion engine. Cars with this type of engine will be big, powerful and expensive.

It seems likely that most of the cars of the future will use their hydrogen fuel to generate electricity in fuel cells. These fuel cells will allow the hydrogen to react with the oxygen in the air but, rather than releasing heat, they will generate electricity, which will power the car via an electric motor.

There are many advantages to owning a car powered by a fuel cell. The electric motors will be quiet. They will not pollute the environment. When hydrogen combines with oxygen, the only product is water. The hydrogen fuel will be obtained from water, so we need not worry about it running out. The electric motors and fuel cells will be able to be fitted to small cars. The family runabout does have a future!

H **SAQ**

4 Write a balanced symbol equation for the complete combustion of ethene (C_2H_4) to give carbon dioxide and water.

Summary

You should be able to:

* assess the suitability of a fuel for a particular purpose

H * explain why the amount of fossil fuels being burnt is increasing

* describe the processes involved in the combustion of a fuel

* describe complete combustion and incomplete combustion of a hydrocarbon fuel, naming the products

* explain why the products of incomplete combustion can be dangerous

* describe the advantages of complete combustion over incomplete combustion

H * construct balanced symbol equations for complete combustion and incomplete combustion of a hydrocarbon fuel

Questions

1 Two students each boil $100\,cm^3$ of water in a beaker using a Bunsen burner, tripod and gauze. One student finds that the water boils in 3 minutes and the apparatus stays clean. The other student finds that the water boils in 5 minutes and the beaker and gauze get very dirty because of soot collecting on them.

a Give two ways in which the two students' experiments would have looked different as they started to heat the water.

b Explain as fully as you can why they got different results.

2 a Write a word equation for the incomplete combustion of methane, giving carbon and water as the two products.

b Write a word equation for the incomplete combustion of methane, giving carbon monoxide as one of the products.

c Write a word equation for the complete combustion of methane.

H **3** Write a balanced symbol equation for each of the following reactions:

a the incomplete combustion of propane (C_3H_8) to give carbon and water

b the incomplete combustion of ethane (C_2H_6) to give carbon monoxide and water

c the complete combustion of propane (C_3H_8) to give carbon dioxide and water

d the complete combustion of butane (C_4H_{10}).

Exothermic reactions

When fuels burn, energy transfers take place. Chemical energy stored in the fuel is transferred to the surroundings as heat and light. The reactants are giving out energy to the surroundings. We say they are releasing energy when this happens. A reaction that releases energy is called an **exothermic reaction**. The photographs on this page are examples of three different exothermic reactions.

Figure 1h.1 The wax and the oxygen in the air react, and heat and light energy are given out.

Figure 1h.2 This explosion is giving out heat, light and sound energy.

Figure 1h.3 The chemical energy in the batteries will be transferred to the CD player as electrical energy.

You are familiar with many exothermic reactions. Every burning reaction is exothermic and releases energy to the surroundings. The reactions taking place inside your body's cells between molecules of the food you have eaten and the oxygen you have breathed in are called respiration reactions. Respiration is an exothermic reaction, so you are releasing energy to your surroundings. The room where you are sitting is getting warmer because you are in it – its temperature is going up.

Endothermic reactions

Some reactions take in energy from their surroundings. We say that they are absorbing energy when this happens. A reaction that absorbs energy is called an **endothermic reaction**. If an endothermic reaction takes heat energy from the surroundings then the surroundings get colder – the temperature drops.

Endothermic chemical reactions are not as familiar to you from daily life as exothermic reactions are. The chemical reactions inside a rechargeable battery while it is being recharged are endothermic. The battery is absorbing energy from the electricity and storing it as chemical energy as it is recharged. This is an example of an endothermic reaction.

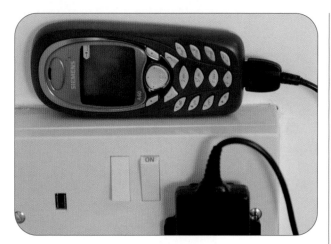

Figure 1h.4 The battery in this phone is being recharged. The chemical reactions taking place in the battery are endothermic.

Other examples are less familiar but can be observed in a school laboratory. If a spoonful of sodium hydrogencarbonate powder is added to a beaker of citric acid solution, the temperature of the beaker of solution will drop.

SAQ

1 a Explain what is meant by an exothermic reaction.

b Give an example of an exothermic reaction that releases heat energy to the surroundings.

c Give an example of an exothermic reaction that releases sound energy to the surroundings.

d Give an example of an exothermic reaction that releases light energy to the surroundings.

e Explain what is meant by an endothermic reaction.

f Give an example of an endothermic reaction.

H Bond breaking and bond making

All chemical reactions can be thought of as involving two stages. The first stage involves breaking the bonds in the reactants. This stage is endothermic. The second stage involves making the new bonds in the products. This stage is exothermic. The energy change of the reaction as a whole depends on which stage involves more energy. If the bond breaking stage involves more energy, the reaction is endothermic. If the bond making stage involves more energy, the reaction is exothermic.

H Worked example

When 2 g of hydrogen react with 16 g of oxygen, 18 g of water vapour are formed.

Breaking the bonds in the 2 g of hydrogen absorbs 436 kJ.

Breaking the bonds in the 16 g of oxygen absorbs 248 kJ.

The total energy absorbed is 436 kJ + 248 kJ = 684 kJ.

Making the bonds in the 18 g of water vapour releases 926 kJ.

More energy is released by the bond making than is absorbed by the bond breaking. (926 is bigger than 684.) Therefore the reaction is exothermic.

The amount of energy released when 18 g of water vapour are made from 2 g of hydrogen and 16 g of oxygen is 926 kJ − 684 kJ = 242 kJ.

SAQ

2 71 g of chlorine will react with 2 g of hydrogen. 73 g of hydrogen chloride are formed. The reaction between one chlorine molecule and one hydrogen molecule is shown diagrammatically in Figure 1h.5.

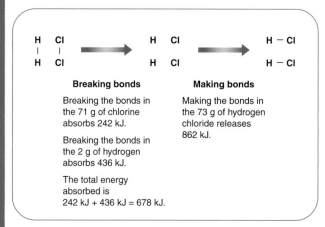

Figure 1h.5 Bond breaking and bond making.

a Is the reaction exothermic or endothermic?

b How many kJ of energy are absorbed or released when 73 g of hydrogen chloride are formed?

Measuring heat energy

It is possible to investigate the amount of energy released when a fuel burns, using the apparatus in Figure 1h.6.

As the fuel burns, heat is released. This heat will go into the water and its temperature goes up. This apparatus can be used to compare two different fuels. The experiment can be done with the first fuel then repeated with the second fuel. The results of the two experiments can then be compared.

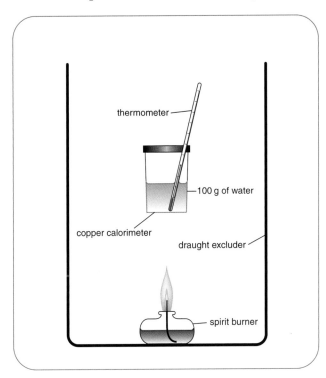

Figure 1h.6 Measuring heat energy.

For the comparison to be valid it is important that the test is **fair**. To ensure **fair testing**, everything should be the same in the two experiments except for the fuel used.
1 The same mass of water must be used.
2 The water must start at the same temperature.
3 The same mass of fuel must be burnt.
4 The height of the flame must be the same.
5 The calorimeter must be clamped at the same height.

Temperature is measured in °C, which is short for '**degrees Celsius**'. Energy is measured in **J**, which is pronounced '**joules**', or sometimes in **kJ**, which is pronounced '**kilojoules**'. One thousand joules equal one kilojoule. The fuel that makes the temperature of the water rise by the most degrees

Celsius has transferred the most joules of energy to the water.

SAQ

3 Draw apparatus for measuring the energy released by burning a tablet of solid fuel.

4 The apparatus in Figure 1h.6 was used to compare three fuels, ethanol, hexane and propanone. The results are as follows.

	Ethanol	Hexane	Propanone
Mass of fuel burnt	1 g	1 g	3 g
Mass of water in calorimeter	200 g	200 g	100 g
Starting temperature of water	21 °C	21 °C	19 °C
Final temperature of water	39 °C	50 °C	100 °C

Table 1h.1

   a The test to compare ethanol and hexane was fair. Which fuel released more joules of energy per gram when it burned?

   b Unfortunately, the test done with propanone was not a fair comparison with the other two fuels. Give three ways in which the test was not fair.

Calculating heat energy

The number of joules of heat energy transferred to the water from the burning fuel can be calculated using the equation

energy = mass of water in grams × 4.2
            × temperature change in °C

If this equation is used for the data for burning ethanol in Table 1h.1, the sum works like this:

energy = 200 × 4.2 × 18 J

energy = 15 120 J or 15.12 kJ

This is the number of joules transferred to the water per gram of fuel burnt.

SAQ

5 Use the data in Table 1h.1 to calculate the amount of energy transferred to the water from the burning hexane. Give your answer in joules and in kilojoules.

Energy value of a fuel

This method can be used to give a fair comparison even if the amount of fuel burnt is not exactly 1 gram. It is possible to calculate how many joules of heat would have been transferred to the water if 1 gram of fuel had burnt. This is done by calculating the number of joules of heat transferred to the water and dividing it by the number of grams of fuel that were burnt. The answer is known as the **energy value** of the fuel:

$$\text{energy value} = \frac{\text{joules of heat transferred}}{\text{grams of fuel burnt}}$$

Look at the diagrams in Figure 1h.7, which show the same experiment done with ethanoic acid (pure vinegar) in the spirit burner.

The energy transferred to the water is calculated in the same way:

energy = mass of water in grams × 4.2
× temperature change in °C

energy = 200 × 4.2 × 14 J

energy = 11 760 J or 11.76 kJ

Look at the mass of the burner plus fuel before and after the experiment. It has gone down from 213.10 g to 211.50 g. It has gone down by 1.60 g. The mass of fuel that burnt was 1.60 g.

Because 1.60 g of fuel transferred 11 760 J of heat to the water, 1 g of fuel would have transferred (11 760 ÷ 1.60) joules of heat to the water. The answer is 7350 J per gram of fuel.

SAQ

6 In an experiment, 0.75 g of fuel were burnt and the energy transferred to the water was 9600 J. How many joules of energy would have been transferred to the water if 1 gram of fuel had been burnt?

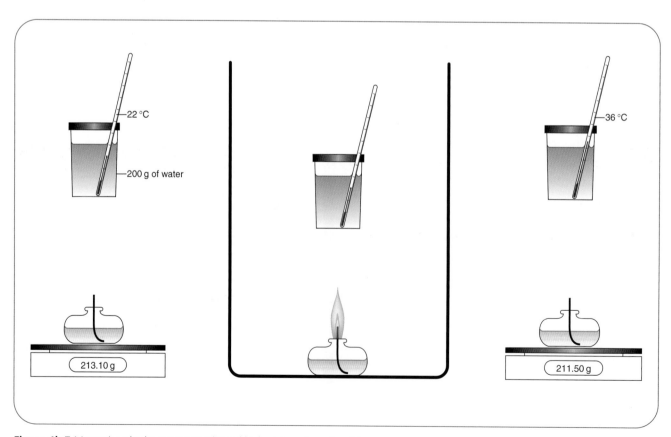

Figure 1h.7 Measuring the heat energy released by burning ethanoic acid.

Fuel + air + spark = BOOM!

Early in the morning of 11 December 2005, an explosion tore apart a hydrocarbon storage depot at Hemel Hempstead in Hertfordshire. Although this was the most recent acident of its kind in the UK, it wasn't the first.

One Sunday morning in July 1994, the oil refinery at Milford Haven in Wales experienced a severe thunderstorm. This set off an unfortunate chain of events and, 5 hours later, about 20 tonnes of flammable hydrocarbon liquid and vapour were released into the atmosphere from the fluidised catalytic cracking unit. There was an explosion as soon as the vapour reached a source of ignition just over 100 metres away. The plant was extensively damaged and the fires that followed could not be put out until over 48 hours later.

How much energy is stored in 'about 20 tonnes of flammable hydrocarbon liquid and vapour'? The answer is approximately a million million joules! If you find this hard to imagine then this is enough energy to:

- boil enough water to make a cup of tea for every person in Scotland
- blow 100 tonnes of solid rock 1000 kilometres into the air

- do almost 200 000 loads of washing in automatic washing machines
- light every light bulb in every house in the UK for 20 seconds (that may not be long but it's a lot of light bulbs)
- accelerate a fully laden Ford Mondeo containing the Watkins family, all their luggage and their dachshund Hermann up to a final speed of over 100 000 km per hour. At this speed, it would take them 24 minutes to circle the world.

Figure 1h.8 The Watkins family.

Summary

You should be able to:

- use and understand the terms *exothermic* and *endothermic*
- **H** understand how bond breaking and bond making contribute to the energy change of a reaction
- label the apparatus and describe the method used to compare the energy output of different fuels
- **H** calculate the amount of energy transferred to a known mass of water
- calculate the amount of energy released from burning one gram of a fuel

Questions

1 You are investigating three chemical reactions. You find that two of them are *exothermic* and one is *endothermic*.

   a What do the *italic* words mean?

   b How could thermometer readings help you to identify which reactions are exothermic and which is endothermic?

2 Harry has two different candles, one made of white wax and one made of red wax. He wants to see if the two different waxes release different amounts of heat energy when they burn. Harry performs two experiments and records the following results.

	White wax	Red wax
Mass of wax that was burnt	1.00 g	1.00 g
Volume of water that was heated	100 cm^3	100 cm^3
Starting temperature of water	19 °C	19 °C
Final temperature of water	42 °C	48 °C

   a Draw and label the apparatus that Harry might have used to perform his experiments.

   b How did Harry make sure he performed a fair test?

   c What did Harry conclude about the two waxes?

H 3 32 g of oxygen will react with 28 g of nitrogen. 60 g of nitrogen oxide are formed. The reaction between one oxygen molecule and one nitrogen molecule is shown diagrammatically in Figure 1h.9.

Figure 1h.9

Breaking the bonds in the 32 g of oxygen absorbs 496 kJ

Breaking the bonds in the 28 g of nitrogen absorbs 436 kJ

Making the bonds in the 60 g of nitrogen oxide releases 751 kJ

   a Is the reaction exothermic or endothermic?

   b How many kJ of energy are absorbed or released when 60 g of nitrogen oxide are formed?

continued on next page

Questions - *continued*

H **4** Three students do separate experiments to find out the amount of energy transferred as heat when 1 gram of a fuel called butan-1-ol is burnt. They used the apparatus in Figure 1h.7, although two of the students chose to put a different mass of water in the calorimeter. The three students got these results:

	Maria's results	Danielle's results	Sokina's results
Starting mass of spirit burner plus fuel	182.45 g	216.89 g	202.39 g
Final mass of spirit burner plus fuel	181.52 g	215.36 g	201.28 g
Mass of water in calorimeter	200 g	100 g	300 g
Starting temperature of water	20 °C	20 °C	20 °C
Final temperature of water	40 °C	76 °C	36 °C

a Calculate the number of joules of energy transferred to the water in each experiment.

b Calculate the energy transferred to the water per gram of fuel burnt from each student's results.

c Suggest why Danielle's result is well below those of Maria and Sokina.

Paints and pigments

Components of paint

Paints add colour to our lives and help to prolong the life of the materials we use. Paints therefore have two jobs – they decorate and they protect. A typical paint is made by mixing three main components.

1 The **pigment** is either a coloured compound or a white compound. The pigment hides the original colour of the painted surface and gives it its new colour.

2 The **binder** is usually a resin or polymer. When the paint has dried, the binder holds the pigment particles together and sticks them to the surface. The binder may also leave a shiny gloss finish when the paint is dry.

3 The **solvent** is added to dissolve the binder, so that the paint becomes a liquid. Liquids are much easier to apply to surfaces.

Figure 2a.1 Paint colours and protects.

SAQ

1 **a** What are the three components of paint?

 b Why is each component necessary?

Paints are colloids

A **colloid** is a mixture that will not settle out. This means that the two components of the mixture will not separate from each other. In a colloid, tiny particles of solid or liquid are **dispersed** in another liquid. The dispersed particles are not dissolved in this liquid. Paints are colloids.

Two types of mixture are used to make paints.

1 In **water paints**, the pigment is dispersed in water as tiny solid particles. The binder is also dissolved in the water. Water is a good solvent because it is non-toxic and non-flammable, but it does make the paints slower to dry than other solvents. Water paints are commonly called emulsion paints.

2 **Oil paints** use oil as the solvent or they use a solvent that dissolves oil. The pigment consists of tiny solid particles that are dispersed in the oil. The binder can be dissolved in the oil or the oil itself can act as binder. Linseed oil used to be used for oil paints but nowadays we use oils obtained from crude oil.

In both water paints and oil paints, the pigment is present as solid particles that do not separate out by settling. The solid particles are said to be **scattered**, or dispersed, in the liquid. They do not settle out because they are so small. More water paints are used than oil paints at present.

SAQ

2 **a** What do we mean by a *colloid*?

 b Explain why the pigment particles in paint do not settle out.

 c What is the difference between a water paint and an oil paint?

 d What is an emulsion paint?

How paints dry

Paints are applied as a liquid, so the paint then has to dry in order to form a hard protective finish. When a water paint dries, the water evaporates, leaving the binder and pigment behind on the surface.

Figure 2a.2 Another friendship ruined…

Oil paints dry differently. When an oil paint dries, the oil reacts with oxygen in the air to form the hard, glossy finish. This reaction is called **oxidation**. It involves several processes which you have already met in this book.

The oils used consist of molecules that include a hydrocarbon part. This hydrocarbon part has two or more carbon-to-carbon double bonds in it. It is unsaturated (see Item C1e *Making Polymers*). Without the double bonds, the hard, glossy finish will not form properly.

When the paint dries the oil is **oxidised**. This process involves the formation of cross-linking bridges (see Item C1f *Designer Polymers*). The oil molecules are polymerising (see Item C1e *Making Polymers*). The shiny protective film that forms consists of the polymerised oil molecules.

SAQ

3 Explain the relevance of the following terms to the making of plastics and to the drying of oil paints:

   a unsaturated

   b cross–linking bridges

   c polymerisation.

Dyes

Pigments have other uses besides paints. Pigments can be used to make glass and china colourful. Other substances are used to colour clothing. These substances are known as **dyes**. For

Figure 2a.3a Medieval peasants would have dressed like this.

Figure 2a.3b 21st century people.

centuries, people had to use natural dyes. A yellowy-brown dye can be made cheaply from onion skins, and stinging nettles can be used to make a green dye. In the Middle Ages, these were the colours the peasants wore. Only the very rich could afford to buy brightly coloured natural dyes, which is why red and purple are associated with kings and emperors.

Nowadays we can make many strongly coloured **synthetic** colourings and the world is a much brighter place for it. As well as ordinary dyes, we have learnt how to synthesise some with very unusual properties.

Thermochromic paints

Thermochromic pigments change colour when they get hot. These pigments can warn you when a mug contains a very hot drink, or when a kettle has just boiled, by changing colour. The

Figure 2a.4 Boiling water is being poured into this mug. One of the pigments used to colour the mug is thermochromic. The heat from the boiling water is actually making a single colour disappear. Where is this colour?

thermochromic pigment is usually coloured when cold and goes white above a certain temperature.

For example, a commonly used thermochromic blue changes to white at 27 °C. By mixing it with an acrylic paint, more colours can be obtained. The thermochromic blue mixed with yellow acrylic paint is green when cold but, above 27 °C, the blue disappears and the colour changes to yellow.

Phosphorescent paint

Phosphorescent pigments glow in the dark! These pigments can absorb and store energy when light shines on them. When the light is switched off, they release the energy slowly in the form of light, so they glow in the dark. These pigments have found many uses, including emergency exit signs, wrist watches with glowing numbers and hands, and glowing instrument panels for cars and aircraft.

Radioactive paints also glow in the dark. Most World War II aircraft had glow-in-the-dark instruments that were radioactive. Nowadays, phosphorescent pigments give us a very

Figure 2a.5 These instruments glow in the dark because of phosphorescent pigments.

effective and much safer alternative to using radioactive paint.

SAQ

4 a Explain what is meant by *thermochromic* and *phosphorescent* pigments.

b How have the discovery of thermochromic and phosphorescent pigments made life safer?

The Radium Girls

In the 1920s, a factory on the east coast of England used paint containing a radioactive metal called radium to paint the numbers on the dials of the clocks and watches they were making. They put dabs of the paint on the hands too. The radioactivity in the paint made the numbers and hands glow so that the wearer could still tell the time when it was dark. The workers who painted the dials were mostly young girls who had just left school. The factory paid well, so the work was popular. The painting was fiddly but, if the girls occasionally ran the bristles between their lips so that the brush kept a nice fine point, this made the work easier.

Ten years later, the doctors and dentists in the area began to notice more patients coming to them with cancer, especially cancers of the mouth and jaw. Many of the sufferers had worked at the dial company.

A similar episode took place at the Radium Dial factory in Ottawa, a small town in Illinois in the USA. Some of the girls were surprised that, when they sneezed, their hankies glowed in the dark, but none of them believed the paint was harming them. It is reported that some even painted their own teeth so as to surprise their boyfriends when they went to the cinema. I bet it worked, too!

Because Radium Dial paid over three times what the girls could earn elsewhere, there were few complaints. But soon a pattern of disease emerged that was the same as was seen in England. The radium paint was identified as the cause and, in 1926, some US compensation boards began making payments to the affected girls.

Summary

You should be able to:

◆ explain why we use paints

◆ name the components of paint and describe their functions

◆ describe paints as colloids

H ◆ explain the meaning of the term *colloid*

◆ describe oil paints and water paints

◆ explain how some paints dry by evaporation

H ◆ explain the role of oxygen in the drying of some paints

◆ describe how dyes are used and the difference made by using synthetic dyes

◆ know what thermochromic pigments are and describe some uses of them

H ◆ describe and explain how adding acrylic paints can give thermochromic pigments more colour possibilities

◆ know what phosphorescent pigments are and explain how they work

H ◆ explain why phosphorescent pigments are safer than radioactive alternatives

Questions

1 a What would be the disadvantage of using a paint that contained no binder?

   b Varnishes are made by mixing a binder with a solvent without adding pigment. Explain why this mixture is ideal for a varnish.

   c Paints are *colloids*. What does this word mean?

2 a When a brush is used to apply emulsion paint, it has to be cleaned in water. Explain why water has to be used as a cleaner.

   b When a brush is used to apply gloss paint, it has to be cleaned in white spirit. Explain why white spirit has to be used as a cleaner.

3 In the 1980s, manufacturers of clothing began making T-shirts that changed colour when worn. They soon stopped making them – they were unpopular because the shirt didn't all change colour at once. This made them look dirty.

   a Why did the shirts change colour?

   b Why didn't the whole shirt change colour at once?

4 Read the section on 'The Radium Girls' and answer the following questions.

   a Why were the watch faces and dials painted with paint that contained radium?

   b Why did the dial painters consume small amounts of radium regularly?

   c Why were the early signs of illness noticed by dentists as well as by doctors?

   d Why was the work popular?

   e What evidence is there that the plight of the Radium Girls was taken seriously in some US states?

   f Why is radium paint no longer necessary for painting watch and clock dials?

Materials from the Earth

We make the buildings where we live, work and study out of the materials we obtain from the Earth's crust. Many historical buildings and some more expensive modern buildings are made of blocks of **stone**. Most of the houses in the UK are built out of **bricks** and have roofs covered in **tiles**. Bricks and tiles are both made of baked clay. The bricks are joined to walls using mortar. Mortar is made by mixing water, sand and **cement**. Cement is made by heating together limestone and clay.

Many houses have doors and window frames made of **aluminium** which has been obtained from aluminium ore. The **glass** in the windows has been made from sand and other raw materials. Tall skyscrapers have a steel framework. The steel is an alloy containing over 98% **iron** which has been obtained from iron ore.

Figure 2b.1 A house.

1 Look at the picture in Figure 2b.1. List five construction materials that have been used to make the house. For each material, name one raw material from the Earth's crust that has been used to make it.

Concrete

One construction material is used more than any other. If we ranked all construction materials in order of how many tonnes are used per day, top of the list would be **concrete**. Concrete is used to make such different things as tower blocks, motorway bridges and the foundations of houses. It is made by mixing together water, cement and gravel. This makes a liquid mixture which sets into a hard rock-like substance.

Concrete is a hard material but it is not very strong. It can crack easily if subjected to bending forces. If liquid concrete is poured over a framework of reinforcing steel, the steel strengthens the concrete. When the concrete sets, we have a new material called **reinforced concrete**. This material is called a composite. A **composite** consists of two materials combined together. A composite material has improved properties compared with the materials that make it up. Reinforced concrete is an excellent construction material. Steel is a strong, flexible material that makes the reinforced concrete strong. Concrete makes it hard.

Figure 2b.2 Using reinforced concrete.

2 **a** How do the steel girders improve the properties of the concrete?

b Why is reinforced concrete an example of a composite material?

Scraping the sky

Figure 2b.3 Sears Tower.

Figure 2b.4 Petronas Twin Towers.

Until 1998, the tallest building on Earth was Chicago's Sears Tower, which measures a colossal 445 m. If there was a continuous staircase from the ground floor to the top floor, it would take you over 15 minutes to run up it! Then, in 1998, the record for tallest building was taken by the Petronas Twin Towers in Kuala Lumpur, Malaysia. These two stunning towers are 452 m high and are connected 170 m up in the air by the 58 m long 'sky bridge', a beautiful arc in the sky. The Petronas Towers are 7 m higher than the Sears Tower – enough for two more floors.

For the first ever high-rise building made of reinforced concrete, we have to go back to 1903. This was when the Ingalls Building was erected in Cincinatti, USA.

The Ingalls Building is 64 m high. Surprisingly, new buildings did not get significantly higher than this for over 50 years. It was then that ways of making much stronger reinforced concrete were developed and the race for the sky began. This is a common theme in the history of science and technology – a new material with better properties causes a leap forwards.

The builders of the Petronas Towers used composite structural systems. The main vertical components were cores, columns and sheer walls of exceptionally strong reinforced concrete. The main horizontal components were made of the highest quality structural steel. These have to resist the huge lateral and vertical forces of the building's own mass, and also the forces of the strongest possible wind.

Although the Petronas Towers is the tallest building in the world as this book is being written, it won't be for very long. Currently under construction, the Shanghai World Financial Centre will be 8 m taller at 460 m.

Making cement

One of the most important components of concrete is cement, which is made by heating limestone and clay together. Cement is also an important part of the mortar used by bricklayers. When water is added, the changes that take place in the cement cause the concrete or mortar to set hard. When cement is made, the limestone that is heated with the clay undergoes a chemical change. Limestone is an impure form of calcium carbonate. When it is heated, it decomposes to form calcium oxide and carbon dioxide.

The word equation for this reaction is

$$\text{calcium carbonate} \rightarrow \text{calcium oxide} + \text{carbon dioxide}$$

Figure 2b.5 Limestone and gypsum (a naturally occurring clay) are being heated together in this rotary kiln to make cement.

H The symbol equation for this reaction is

$$CaCO_3 \rightarrow CaO + CO_2$$

SAQ

3 a What is the chemical formula of calcium oxide?

b What is the chemical formula of carbon dioxide?

c What is the chemical formula of calcium carbonate?

d How many atoms are there in one formula unit of calcium carbonate?

e Is the symbol equation above balanced? Explain your answer.

Thermal decomposition

The reaction that happens when calcium carbonate is heated is called **thermal decomposition**. It is *thermal* decomposition because heat is needed to make it happen. It is thermal *decomposition* because the calcium carbonate breaks down into two simpler substances. *Decomposition* always means that a substance breaks down into at least two simpler substances.

The thermal decomposition of calcium carbonate is an **endothermic reaction**, because heat energy has to be put in when it happens. Cement contains calcium oxide. When water is added to cement, the calcium oxide reacts with the water and heat energy is given out in this reaction. The reaction that occurs when cement sets hard is therefore an **exothermic reaction**.

SAQ

4 Explain the following statements as fully as you can.

a The reaction that occurs when calcium carbonate is heated is an *endothermic* reaction.

b The reaction that occurs when water is added to calcium oxide is an *exothermic* reaction.

The limestone that is used to make cement is a mineral. We have no shortage of limestone in the UK. There is plenty of it and it is dug out of the ground all over the country. Where the limestone is near to the surface, it is easy to dig out by quarrying. Quarrying is much cheaper than deep mining.

Figure 2b.6 Quarrying for limestone in Derbyshire, UK.

Quarrying and the environment

Unfortunately, limestone quarrying has environmental consequences. The quarries occupy land that could otherwise be used for housing or farming. The rock in quarries is often loosened by blasting with dynamite and then carried away in trucks. Quarrying is therefore noisy and dusty, and creates additional traffic. When the rock in the quarry has been removed, an ugly pit is left behind. This can be changed into an environment for wildlife and human recreation by flooding. The quarrying has changed the environment, but it can still be beautiful afterwards.

Figure 2b.7 Thrapston gravel pits.

SAQ

5 Figure 2b.7 shows a flooded gravel pit near Thrapston, Northants. The flooded pit is now an important environment for herons and many other water birds.

  a Describe the sequence of events that has taken place at Thrapston:

    ● start when the gravel deposit was first discovered

    ● finish at the present day.

  b How does the photograph show Thrapston gravel pits to be an environment for wildlife and recreation?

  c What might the gravel dug out of the quarry at Thrapston have been used for?

Limestone, granite and marble

Limestone is also an important building stone. The Houses of Parliament and the tower of Big Ben are built of limestone. Granite and marble are two more rocks much used as building stone. These different rocks have different properties. Marble is harder than limestone and granite is harder still.

Figure 2b.8a The Houses of Parliament are made of limestone.

Figure 2b.8b Florence Cathedral is made of marble.

Figure 2b.8c Mareschal College in Aberdeen is made of granite.

Sedimentary, igneous and metamorphic rocks

Limestone, marble and granite have different hardnesses because the three rocks are formed in very different ways. Limestone is a **sedimentary rock**. It formed many millions of years ago from the shells of dead sea creatures. These shells formed deep layers at the bottom of prehistoric oceans which were then covered by other sediments so that the shells were compressed and stuck together to form limestone.

Marble is a **metamorphic rock**. It is formed when great heat and pressure act on limestone deep in the Earth. In Item C2c *Does the Earth move?*, you will find out that huge sections of the Earth's crust are moving very slowly. These movements can cause limestone to be found deep in the Earth.

Granite is an **igneous rock**, formed from a substance called **magma**. Magma is molten rock from below the Earth's crust. When magma rises into the crust through cracks in the crust, it can cool and set to form granite.

SAQ

6 a Describe the different ways by which granite, limestone and marble are formed.

b Describe the difference in hardness between granite, limestone and marble.

Summary

You should be able to:

- name some construction materials manufactured from resources in the Earth's crust

- state that limestone, marble and granite are used to make buildings

- rank these three rocks in order of hardness

- explain the hardness of these three rocks with reference to their origins

- discuss the effects of quarrying on the environment

- describe thermal decomposition and write a word equation for the thermal decomposition of calcium carbonate

- write a balanced symbol equation for the thermal decomposition of calcium carbonate

- describe how cement and concrete are made

- describe how concrete is reinforced and understand the term *composite material*

- explain why reinforced concrete has superior properties as a building material

Questions

1 a What is meant by a *composite* material?

   b Explain why reinforced concrete is classified as a composite material.

   c Explain why Gore-Tex® is classified as a composite material.

2 The raw materials listed below are all obtained from the Earth's crust. Explain why each one is important in the construction industry:

   a iron ore

   b aluminium ore

   c clay

   d sand

   e limestone

   f sand and gravel.

3 When limestone is heated, a chemical reaction occurs that gives a solid and a gas as the two products.

   a What name is given to this type of reaction?

   b Write a word equation or balanced symbol equation for this reaction.

   c How could you prove the identity of the gas that is produced?

   d Why is this reaction important to the construction industry?

4 Explain what is meant by a sedimentary rock, a metamorphic rock and an igneous rock, giving one example of each.

5 Imagine the following situation.

   A high-quality gravel deposit has been found in an area of outstanding natural beauty. This area is also near to an urban area with high unemployment.

   a Discuss the arguments for and against quarrying in the area.

   b Make your own recommendation and explain your decision.

Does the Earth move?

The layered Earth

The Earth is not the same all the way through. It consists of three layers, called the **crust**, the **mantle** and the **core**.

The crust

The layer we live on is called the **crust**. The crust is a thin layer compared with the other two layers. It is made of rock but it is not one continuous solid piece of rock. It is broken up into pieces of various sizes called **tectonic plates**. Imagine dropping a hard boiled egg so that the shell broke but the pieces stayed in their places. This is what the Earth's crust is like.

The tectonic plates are not all of equal thickness. The plates that make up the Earth's land masses are thicker; they are called continental plates. The plates that make up the Earth's ocean floors are thinner and denser; they

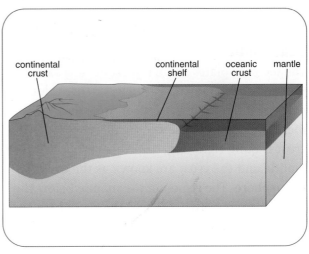

Figure 2c.1 A small section of the Earth's crust with the mantle underneath.

are called oceanic plates. The tectonic plates that the crust is made of move very slowly, at a speed of a few centimetres per year. This movement may seem too slow to matter but it has some very

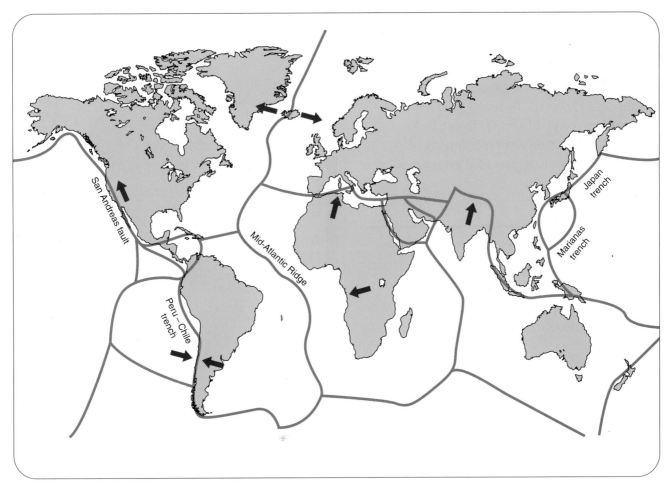

Figure 2c.2 World plate boundaries. The arrows show the direction of movement of some of the plates.

dramatic effects. The movement causes volcanoes, earthquakes and tsunamis, and these effects are mainly seen where the plates meet each other.

The mantle

The tectonic plates sit on top of the next layer (the mantle) because they are less dense than the mantle. The mantle consists of extremely hot rock. The region of the mantle directly under the crust is quite rigid. Together, this part of the mantle and the crust are known as the Earth's **lithosphere**. The region of the mantle below the lithosphere is hotter still, and the rocks here are sufficiently molten to allow them to flow.

The core

The centre of the Earth is called the core. The Earth's core is made mostly of iron and is extremely hot and dense. The core causes the Earth to have a magnetic field, and this makes a compass needle point northwards.

Finding out about the layered structure of the Earth has not been easy. It can't be done by digging down and taking a look, because the lower part of the crust is far too hot for humans to survive. It is too hot for measuring instruments to function. However, the waves produced by earthquakes travel differently through the different layers. Careful measurements of earthquake waves have provided evidence for the existence of the layers.

SAQ

1 Whereabouts in the Earth are the mantle, crust and core? Make one statement about each part of the Earth.

2 When geologists study the Earth's tectonic plates, the most interesting places to look at are the edges of the plates. Why can this be a dangerous thing to do?

3 What is the Earth's lithosphere?

⒣ Evidence for plate tectonics

Geologists are now agreed that the Earth's crust is broken into tectonic plates which move slowly relative to each other. The evidence for this is now overwhelming.

- The coastlines of South America and Africa fit together like two slightly damaged jigsaw pieces. This is because they were once a single land mass that has broken apart.
- The layers of rock on the east coast of South America and on the west coast of Africa are very similar.
- The same fossils are found in these two areas of the world.
- **Plate tectonics** explains the existence of great rift valleys, such as the East African rift valley. The continent is slowly tearing apart as two plates move in opposite directions.
- Satellites can observe and measure the plate movement. London and New York get a few centimetres further apart every year. Los Angeles and Tokyo get a few centimetres closer together.

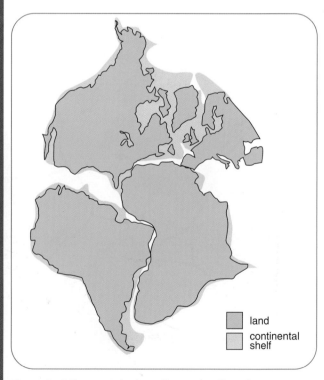

land
continental shelf

Figure 2c.3 The tectonic plates fit together like a jigsaw.

Convection currents in the mantle

The theory of **plate tectonics** was first suggested in 1912 by a German scientist called Alfred Wegener. The theory was not accepted at first, mainly because no-one could explain *why* the plates moved.

We now believe that the Earth's tectonic plates move because the deeper parts of the mantle are liquid and can flow. The temperature in this part of the mantle is uneven and this causes very slow moving currents (called convection currents). These convection currents transfer heat energy from the hotter lower mantle to the cooler upper mantle. The convection currents cause the tectonic plates in the Earth's crust to move, slowly.

Types of plate movement

In some places, the plates move away from each other. This sort of movement is causing the East African rift valley to form. It is also happening down the middle of the Atlantic ocean: as the plates move apart, molten lava is rising up through the crack. There is a line of underwater volcanoes down the centre of the Atlantic, called the Mid-Atlantic Ridge. Thousands of years from now, these volcanoes will form a new chain of islands.

Figure 2c.4 When enough lava has spewed out, an underwater volcano reaches the surface and a new island is seen. This island near Iceland, called Surtsey, appeared in 1963.

In other places, the plates move towards each other and collide. This can cause the Earth's crust to crumple, forming new mountain ranges. The Alps and the Himalayas formed in this way.

When an oceanic plate collides with a continental plate, the thinner, denser oceanic plate is usually pushed underneath the continental plate into the mantle. This is called **subduction**. The oceanic plate then undergoes partial remelting in the mantle to form molten rock called magma.

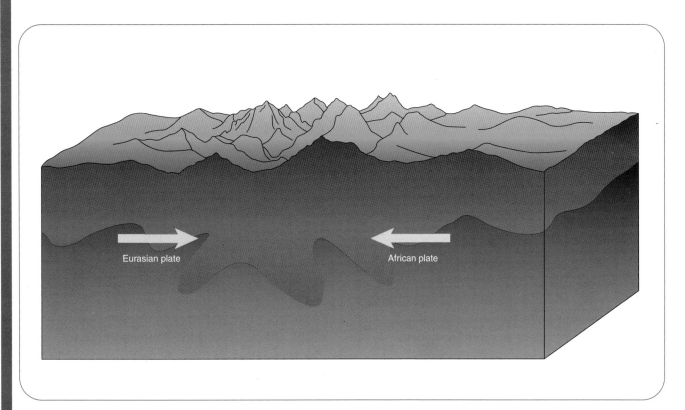

Eurasian plate

African plate

Figure 2c.5 This is how the Alps formed.

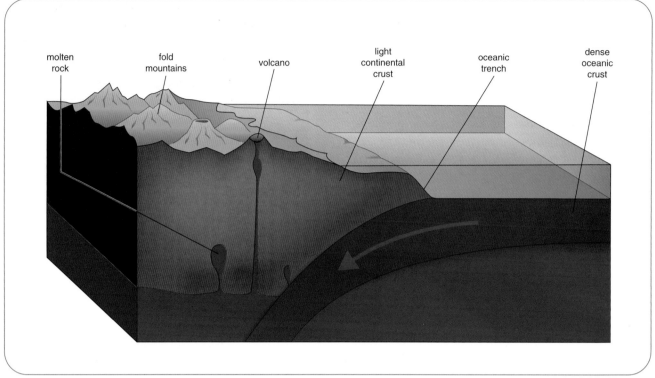

Figure 2c.6 Subduction at colliding plates.

SAQ

4 a How old is the theory of plate tectonics?

 b Why was this theory not accepted at first?

 c Give four bits of evidence for plate tectonics.

Rising magma

Molten rock in the mantle is called **magma**. The magma is under great pressure. If the pressure on the magma is released, for example by the cracking of the crust above it, the density of the magma drops. The cracks in the Earth's crust, for example at places where two plates meet, causes the magma to rise to the Earth's surface. The magma comes out onto the Earth's surface when a volcano erupts. When the molten rock comes out of the volcano, its name changes to **lava**.

Not all lava is the same. Some volcanic lava is thin and runny, and this sort of lava flows easily out of the volcano. Some volcanic lava is much thicker and is more likely to result in sudden, violent volcanic eruptions. Geologists study volcanoes to give them a better understanding of the structure of the Earth's crust. It also helps them

Figure 2c.7 A typical volcano.

to predict when and where a volcano is likely to erupt, although not with complete certainty. The lava moves up through weaknesses in the Earth's crust. It then cools and solidifies into new rock. The type of rock formed when lava cools and solidifies is called **igneous** rock. Areas of igneous rock are popular with farmers, because the soil that forms there is usually very fertile. Most igneous rocks are crystalline. The crystal size depends on how fast the lava cooled. If the lava cools and solidifies underground, the cooling is slow, which gives large crystals. If the lava cools and solidifies above ground, the cooling is fast, which gives small crystals.

The silica content of lava is important in determining the type of igneous rock that will form. If lava that is poor in silica but rich in iron cools slowly, it forms a coarse rock containing large crystals; this rock is called **gabbro**. If the same lava cools quickly, it forms a finer grained

rock containing small crystals; this rock is called **basalt**. Similarly, if lava rich in silica cools slowly, it forms a coarse rock containing large crystals; this rock is called **granite**. If the same lava cools quickly, it forms a finer grained rock containing small crystals; this rock is called **rhyolite**.

SAQ

5 **a** What is meant by an igneous rock?

  b Why do some people choose to live in areas where the underlying rock is igneous?

  c Explain with examples why some igneous rocks have large crystals and some have small crystals.

Climate change can be natural

Volcanoes have been causing climate change on Earth for billions of years. The climate change worrying scientists at the moment is global warming – the world's average temperatures are getting higher because of the increasing amount of carbon dioxide in the atmosphere. The amount of carbon dioxide is increasing because of our reliance on burning fossil fuels.

Despite pouring out huge amounts of red-hot molten lava, volcanoes don't make the world warmer – instead, they cool it down. The huge amounts of ash and dust spewed into the air by an erupting volcano can partly shade the land

from the Sun's light and heat. Shading the Earth from the Sun has a *much* bigger effect than the hot lava, causing lower temperatures.

When Mount Tambora in Indonesia erupted, in 1815, this effect was so strong that the following year was known in Europe and North America as 'the year without a summer'. In Europe and North America, it is estimated that over 90 000 people died of starvation because their crops wouldn't grow in the dimmer light and colder temperatures.

Much more recently, in 1991, Mount Pinatubo in the Philippines erupted. It released enough dust into the air to cause a temperature reduction of up to 0.5 °C. In addition to this, a huge cloud of sulfur dioxide gas was released. Amazingly, this cloud of gas had spread right round the world in only three weeks.

Sulfur dioxide in the atmosphere has an effect on the colours of the sky. One consequence of the Pinatubo eruption was that many parts of the world had unusually beautiful sunsets during this period. The great British artist J. M. W. Turner lived from 1775 to 1851. Turner loved painting sunsets. Was he inspired by beautiful sunsets caused by the Tambora eruption in 1815?

Figure 2c.8 *The Fighting Temeraire* by J. M. W. Turner.

Thick and thin lava

The composition of the lava has an effect on how violent a volcanic eruption is. Lava which is poor in silica but rich in iron is runny and so flows out of the volcano comparatively smoothly and freely. When it solidifies it produces basalt. Silica-rich lava is much thicker. When the silica-rich lava

Figure 2c.9 The explosion of Mount St Helens – a very violent volcanic eruption.

solidifies, it produces rhyolite. Sometimes, it solidifies inside the volcano, blocking the gap through which the lava is flowing. The pressure of rising lava inside the volcano exerts huge forces on the blockage.

When the blockage (or some other part of the volcanic mountain) suddenly gives way, it can cause an extremely violent eruption. This sort of explosive eruption can produce a rain of volcanic ash and can throw huge boulders into the air. These boulders are called volcanic bombs. A type of porous rock called **pumice** is also formed in this sort of violent eruption.

Graded bedding

The boulders and big rocks hit the ground first then the ash and dust settle on top. The large stones therefore form a rock layer beneath the layer containing the ash and dust. The formation of rock layers like this, with heavier particles below lighter particles, is called graded bedding. The presence of **graded bedding** and pumice in igneous rocks is evidence that the original volcanic eruption was violent.

Summary

You should be able to:

◆ describe the layered structure of the Earth

◆ describe the Earth's lithosphere, including a description of the theory of tectonic plates

◆ state two consequences of plate movements

◆ describe the mantle

◆ explain the theory of plate tectonics and its development

◆ describe magma and explain why magma can sometimes rise from the mantle to the surface

◆ describe the igneous rock that forms when lava cools down, referring to the crystal sizes present

◆ describe how the thickness of lava can affect the violence of a volcanic eruption

◆ describe the composition of a thick lava and a thin lava

◆ describe the consequences of a violent volcanic eruption

◆ give some reasons why geologists study volcanoes

Questions

1 Gabbro, basalt, granite and rhyolite are different types of igneous rock.

   a What type of lava produces each of these rock types?

   b How does the rate at which the lava cooled affect these rocks?

   c What causes these rocks to cool at different rates?

   d How would these rocks appear different from each other if you examined them closely?

H 2 **a** Explain what is meant by *subduction*.

   b Use the ideas of subduction and the formation of the Mid-Atlantic Ridge to explain why the Earth's crust stays the same size.

   c Explain why Mount Everest is gradually increasing in height.

3 Draw a cross-section of the Earth and put on the following labels:

this part contains iron	mantle
lithosphere	this part consists of rigid magma
crust	core
this part consists of non-rigid magma	this part consists of tectonic plates

4 **a** Why do some people choose to live near volcanoes?

   b Why do scientists choose to study volcanoes?

   c Why is it more dangerous to be near a volcano if the lava that comes from it is thick?

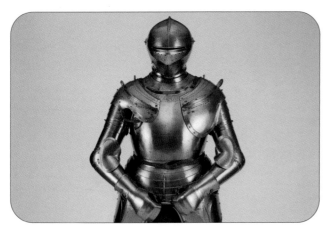

Figure 2d.1a Steel has been used for centuries for its strength.

Figure 2d.1b Aluminium's low density made it first choice for the structure of the Spitfire.

Figure 2d.1c Lead is flexible and doesn't rust.

Figure 2d.1d Silver is one of the most beautiful of metals.

Metals in the service of humans

The earliest humans lived in the Stone Age. Next came the Bronze Age, then the Iron Age. The early history of the human race is divided into three periods which are defined according to the materials people used. Materials are of crucial importance. When scientists discover how to synthesise a new material, a new technology that uses it follows soon after. Many of the most important materials used by humans are metals, including mixtures of metals called **alloys**. Different metals have different properties, so we use them for different jobs.

Metal ores

Tiny specks of pure gold can occasionally be found in the earth, but most metals are found in the ground as **ores**. Pure metals like gold are **elements**. An element is made of one type of atom only. Ores contain metal **compounds**, usually mixed with other unwanted materials such as sand. A compound is made of at least two different types of atom joined together by chemical bonds. The ore is called a mixture because its components, the metal compound and the sand, are not joined together by chemical bonds.

To get the pure metal from the ore is not always easy. Sometimes, this is because very high temperatures are needed. Sometimes, it is because the ore has to be melted and then has to have electricity passed through it. Copper is a comparatively easy metal to get from its ore. This is why it was one of the very first metals obtained from its ore by early humans.

Copper

Figure 2d.2a This 2p coin is made of steel but has a thin outer coating of copper to improve its appearance.

Figure 2d.2b These electrical wires are made of pure copper.

Figure 2d.2c Copper pipes are used for domestic central heating.

One of the best-known copper ores is called malachite. Malachite is an impure form of a compound called copper carbonate.

Figure 2d.3 The green malachite (top) is an ore containing copper. The black cassiterite (bottom) is an ore of tin.

The formula of copper carbonate is $CuCO_3$. If copper carbonate is heated, it decomposes to form copper oxide (CuO) and carbon dioxide (CO_2). If the copper oxide is heated with carbon, they react together to form copper (Cu) and carbon dioxide (CO_2). This process is called **reduction**. When humans first did this, charcoal was used as the source of carbon. Now we use coke, which is obtained from coal.

SAQ

1 **a** Copper is an element. Copper carbonate is a compound. Explain the difference between an element and a compound.

   b The formula of copper carbonate is $CuCO_3$. How many different elements are present in one formula unit of copper carbonate?

   c Name the elements present in copper carbonate.

   d How many atoms are there in one formula unit of copper carbonate?

2 **a** What is the name for the type of chemical reaction that happens when copper carbonate is heated?

   b Write a word equation for this reaction.

   c Write a balanced symbol equation for this reaction.

   d Identify the reactant and the products in this reaction.

3 Copper oxide reacts with carbon if they are heated together.

   a Write a word equation for this reaction.

   b Write a balanced symbol equation for this reaction.

   c Identify the reactants and the products in this reaction.

Purifying copper

Copper is the 'conductivity metal'. All of our houses use copper wires to conduct electricity from room to room. The copper produced by smelting is impure, and is known as boulder copper or blister copper. To get the best electrical conductivity, pure copper is needed. The purification of copper involves electricity too. The process is called **electrolysis**.

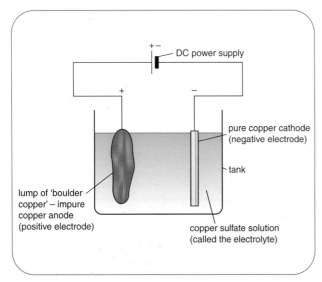

Figure 2d.4 Copper purification by electrolysis.

When the electricity is switched on, the copper from the impure anode dissolves into the copper sulfate solution. Copper will not do this unless an electric current is used. The copper from the solution turns back to solid copper on the cathode. Only pure copper builds up on the cathode. The impurities in the anode either stay dissolved in the solution or form a solid sludge at the bottom of the tank.

You might expect something with such an unattractive name as 'solid sludge' to be thrown away. However the impurities in the boulder copper include gold and silver. When the copper is purified these two metals end up in the sludge. Even sludge can be valuable!

SAQ

4 Copy the diagram in Figure 2d.4 but don't include the labels. Look at the diagram hard, learning the labels. Close this book and add the labels to your diagram.

H 5 Copper is purified by electrolysis.

a Which substance is the anode made of?

b What does 'anode' mean?

c Which substance is the cathode made of?

d What does 'cathode' mean?

e Is the power source AC or DC?

f Which substance is used as the electrolyte?

g Find out what 'electrolyte' means.

Recycling copper

Our supplies of copper ore are non-renewable. This means they are not being re-formed in the Earth's crust. As we extract copper ores by mining and quarrying, the remaining supply becomes less until, one day, it will run out. We believe this will happen sometime between 2250 and 2300 if we carry on extracting copper ore at the present rate. Copper ore reserves will last longer if we recycle used copper instead of making new copper. A typical car contains between 15 kg and 25 kg of copper. All of this can be recycled when the car is scrapped.

It is cheaper to recycle copper than to make new copper. Unfortunately, recycling copper scrap is not straightforward. Copper is used in many

Figure 2d.5 This copper mine is at Parys on Anglesey in Wales. Copper was discovered here in 1768 and it was one of the world's largest copper mines. Even this huge mine, with over 6 million tonnes of ore still remaining, will eventually run out.

different alloys. If we collected all the copper alloys from metal recycling projects and melted them together, the resulting mixture would be useless. The mixture wouldn't be the same as any of the alloys we started with.

We could sort out the scrap. If we only melt alloys of the same type together, this problem would be avoided. Unfortunately, sorting the scrap is expensive. Items made of pure copper, for example domestic central heating pipes, are a very useful source of recycled copper.

SAQ

6 a Why is recycling copper a good thing to do?

  b What problems are involved in recycling copper?

Alloys

Extracting ores from the Earth's crust enables us to make over 60 pure metals. These 60 pure metals include many you will have heard of, like copper, iron and aluminium. You are less likely to

Figure 2d.6 Solder is an alloy of tin and lead. It melts at quite a low temperature. It is used to join electric components because of this. The heat needed to melt the solder will not damage the electrical components.

Figure 2d.7 Brass is an alloy of copper and zinc. Almost all metals are 'sonorous' – this means they can be made to vibrate and produce musical notes. Brass is particularly good at this. Brass has an attractive gold-like colour and doesn't corrode easily. These properties have led to its widespread use in making musical instruments.

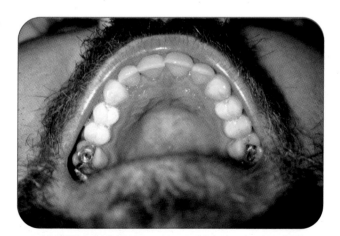

Figure 2d.8 Amalgam is an alloy of the liquid metal mercury and other metals such as tin, silver and copper. A short while after mixing the mercury and the other metals, the amalgam sets solid. It expands slightly as it solidifies. This makes it perfect for dental fillings. While it is setting, it can be packed tightly into the cavity in the tooth. When it expands, it grips the inside of the cavity firmly and so won't fall out. Or at least, it shouldn't fall out …

Figure 2d.9 Steel is the name for any iron-based alloy. The car body shown above is made of mild steel, which is over 99.75% iron, the remainder being carbon. Mild steel is strong, so it is used to reinforce concrete. Mild steel is ductile and **malleable**. Ductile metals can be drawn out to form wires. Malleable metals can be made to change shape easily. Many car body panels are made out of a flat sheet of mild steel by a single stamping process. **Stainless steel** is made by alloying iron with nickel and chromium. Stainless steel doesn't rust and is used to make cutlery.

Figure 2d.10 Bronze is an alloy of copper and tin. It has an attractive dark brown colour and is easily cast into shapes. 'Casting' means pouring the alloy as a liquid into a mould. About 5000 years ago, people used bronze to make the first metal weapons. Now we use bronze to make statues and medals.

have heard of some of the others, like rhenium and gadolinium. All of these metals show typical metal properties, such as conducting heat and electricity. In addition, each metal has its own distinctive properties. For example, they vary in hardness, strength and appearance.

We are not limited to using only pure metals. If we mix a metal with another substance (usually another metal), we produce an **alloy**. Every new alloy has its own distinctive properties. These properties are often different from the metals used to make the alloy (its **constituents**). Some alloys have properties that make them very useful.

Figure 2d.11 Nitinol is an alloy of **ni**ckel and **ti**tanium first made in 1962 by the US **n**aval **o**rdnance **l**aboratory (**ni** + **ti** + **n** + **o** + **l** = nitinol). If you bend a nitinol object, it always returns to its original shape, like the spectacle frames shown above. It is called a 'memory metal' and has also been used to make parts on the space shuttle and underwiring for bras.

SAQ

7 Copy the table below. In column one write the names of six *alloys*. In column two write what the alloy is made of – its *constituents*. In column three write one useful *property* of the alloy. In column four write one *use* of the alloy that depends on the property you have written.

Alloy	Constituents	Property	Use

Table 2d.1 Alloys.

With a little bit of luck …

Wanta was fed up. Their small tribal enclosure had been raided for the third time since the last full moon. The raiders had more strong adults than her tribe, and they always came armed with stone axes and pieces of bone. There was no real point fighting back. Her father had decided it was best to let them take a few goats and hope no-one in their tribe got hurt.

She was fed up too because her brother Zento wasn't helping. Again. Wanta and Zento were meant to be making fires for cooking food later that evening. She was having to do it all herself. Doubtless her brother was engaged in some of the idiocy he called 'testing'. Last week he wasted hours soaking the leather of his catapult in goat's blood to see if it became any more or less stretchy. All it did was smell.

Wanta built a circle of stones. The diameter of the circle was the length of her forearm and was three stones high. She filled the circle with small pieces of dry wood and then went to collect a piece of burning wood from the fire her father guarded. She dropped it into her pile of wood, which caught fire. She put in larger pieces of wood, making a fire that would burn all evening.

As she put the wood into the fire, her bracelet fell in too. Wanta was frightened and cursed her luck. The bracelet was made of a leather strip. Threaded on it were six polished green stones, each with a hole carefully bored in it using a sliver of bone. It was the mark of her membership of the tribe. To lose it was meant to bring terrible bad luck.

Next morning, she made sure Zento couldn't run off. She was standing by him when he woke, and she insisted he clean up from last night's cooking. Zento muttered to himself, but he cleaned the pottery bowls they had used for cooking and then dismantled the stone circle and cleared away the burnt wood. He found six little round beads in the ashes. They were a strange brown colour and, when he hit them with a stone, they didn't break, they just flattened a bit. He had never seen anything like them before. Wanta told him how she had lost her bracelet.

Over the next few weeks Zento learnt a lot about the little brown beads, constantly annoying his sister as he failed to help her with their duties around their home. He found out how to make lots more beads – all he needed was fire, burnt wood and broken bits of green stone. He found out how to use smooth stones to hammer the beads into little knives and arrow heads. He noticed that they stayed sharp when he used them.

Figure 2d.12 Building a fire.

Figure 2d.13 Binding on an arrow head.

continued on next page

With a little bit of luck ... - *continued*

Zento's father began to take notice. He worked with Zento. They tried using other stones instead of the green stones to see if they would make other beads. Nothing worked, until they took some shiny black stones and put them in the fire. In the ashes of the fire they found, not brown beads, but grey ones that shone strangely in the light.

One day, they learnt how to make an especially hot fire, by blowing air in at its base using hollow bones. They put a pottery bowl on the fire and put their beads in the bowl, where they melted and ran together, and shone with the colours of the evening sun.

They made more brown beads, and more grey beads, and they melted them and mixed them together. They made clay moulds in the shape of axe heads and in the shape of wooden spears. They poured their molten mixture into the moulds, and made axe heads and spear points.

They discarded their stone axe heads, and bound their new metal axe heads to the wooden handles. They tied their spear points to their

Figure 2d.14 Casting bronze.

wooden spears. Compared with stone, wood and bone, the new weapons were fearsome and powerful. The axes were heavy and would easily break the skull of an animal. The spears went straight through the thickest animal skin. Next time the raiders came, they would fight back. The Stone Age was over.

Summary

You should be able to:

- ◆ state how copper is extracted from its ore

- ◆ label the apparatus used to purify copper

- Ⓗ ◆ name the anode, cathode and electrolyte used when purifying copper

- ◆ explain why copper is recycled and describe the problems that are encountered

- ◆ explain the meaning of the term *alloy*

- ◆ give the names, constituents and uses of five alloys

- ◆ explain how the properties of an alloy can make it useful

- Ⓗ ◆ explain how the properties of nitinol have made it suited to particular uses

Questions

1 Table 2d.2 describes the composition, hardness, tensile strength and ductility of copper and various types of brass. The higher the hardness number, the more difficult it is to scratch the material. The higher the tensile strength, the more force is required to snap a piece of the material. The higher the ductility, the greater the length the metal can be drawn out to without cracking.

% copper	% zinc	Hardness	Tensile strength	Ductility
100	0	53	15	45
90	10	60	18	55
80	20	62	20	65
70	30	65	21	70
64	36	70	22	60
60	40	85	27	45
55	45	90	28	20

Table 2d.2

a Which two metals are mixed to make brass?

b What is the composition of the strongest alloy in the table?

c Plot a graph of ductility (vertical axis) against percentage of zinc (horizontal axis).

d What happens to the ductility of brass as the percentage of zinc increases from 10% to 45%?

e What percentage of zinc gives the *most* ductile alloy?

f Plot a graph of hardness (vertical axis) against percentage of zinc (horizontal axis). As the percentage of zinc increases, does the hardness increase in a regular fashion?

2 *Copper* can be produced from an ore called *malachite*. Malachite contains *copper carbonate*.

a Which of the three emphasised substances is an *element*? Explain what the term *element* means.

b Which of the three emphasised substances is a *compound*? Explain what the term *compound* means.

c Which of the three emphasised substances is a *mixture*? Explain what the term *mixture* means.

3 The following processes are all involved in making pure copper from malachite. Describe what is taking place in each process.

a Thermal decomposition. b Reduction. c Electrolysis.

4 When copper oxide is heated with carbon, it is possible for the products to be copper and carbon monoxide.

a Write a balanced symbol equation for the reduction of copper oxide by carbon to give copper and carbon monoxide.

b What safety issues are there when performing this reduction? *continued on next page*

Questions – *continued*

5 In each of parts **a** to **e** that follow, explain how the properties of the alloy make it suited to the stated use.

 a Steel is used to make car body panels.

 b Amalgam is used to make dental fillings.

 c Brass is used to make musical instruments.

 d Solder is used to join electrical components.

 e Nitinol is used to make spectacle frames.

6 Reread the story describing an episode in the life of a Stone Age tribal group. Identify and explain the *chemistry* of the episode described. For example, you should name the stones used to make Wanta's bracelet and identify the material in the brown beads Zento found next morning. You should try to explain the chemical reaction that produced the brown beads from the green stones. Symbol equations will add to the quality of your answer. Then go on to explain the rest of the story.

7 **a** Write a balanced symbol equation for the thermal decomposition of copper carbonate to copper oxide and carbon dioxide.

 b How would you show in a laboratory that carbon dioxide gas is made here? You should draw and label the apparatus you would use and say what changes you would observe.

Cars for scrap

Making a car

When you are 17, one of your first priorities may well be to pass your driving test and then get your first car. Many different materials are used to make a car. Some of the materials used are listed below.

Figure 2e.1 Many different materials are used to make a car.

SAQ

1 Look at Figure 2e.1. Name five materials that are visible in the photograph. For each material, give one of its properties that makes it suitable for the job.

Steel and aluminium

The body panels of the car are usually made from a strong metal such as steel or **aluminium**. This metal provides the main structural strength of the car.

Plastics

Plastics are cheap and are used for many of the articles inside the passenger compartment. So-called engineering plastics are now being developed for use as engine parts. These new plastics can withstand high temperatures and lower the mass of the car.

Glass

The **glass** used for windows is transparent and is also strong enough to contribute to the strength and stiffness of the car.

Copper

Copper is an excellent conductor of electricity and so it is used for the wiring in the car. It is also a good conductor of heat, so it is used to make the car's radiator, which has to transfer heat out of the engine quickly.

Fibres

The car seats are usually covered in textile **fibres**, either natural or synthetic. Fabric seats are cheap and easy to manufacture. Synthetic fibres have the disadvantage of making the driver and passengers feel sticky and sweaty on hot days. Leather seats don't have this problem. They are more expensive but are popular owing to their 'luxury' feel and smell.

Rusting

Unfortunately, cars have a relatively short lifespan, because many of the materials they are made of slowly deteriorate as they wear out or rust. Car manufacturers are trying to make cars last longer by using improved materials and by preventing rusting.

Figure 2e.2 From this … to this.

The main metal used to make most car bodies is steel. Steel is an alloy which consists mostly of iron with a small amount of carbon. Steel is used to make cars instead of iron for several reasons. Steel is harder and stronger than iron. Steel rusts more slowly than iron, but unfortunately it does still rust.

Corrosion

For iron or steel to rust, it must be in contact with both oxygen and water. Rust looks ugly but, more importantly, it is weak and crumbly. Rust flakes off the iron. This exposes more iron to the rusting process. We say the iron is **corroding**. This means it is being eaten away by a chemical reaction.

As a car body rusts, it loses its strength, which was the main reason for making it out of steel in the first place. Road and weather conditions can make the rusting faster. We put salt on our roads in winter to help to melt ice, but salty water makes iron rust more quickly. Rain that has become acidic because of air pollution can make iron rust faster, too.

Figure 2e.3 This lorry is spreading grit and salt on a road in January. Why is it doing this and what problem does it cause at the same time?

Oxidation and reduction

When copper is obtained from copper oxide, the oxygen is *removed* from the copper oxide. This is called a **reduction** reaction. When iron rusts, oxygen is *added* to the iron. Rusting is an **oxidation** reaction. Oxidation and reduction reactions are *opposites*. When iron rusts, it reacts with oxygen and water to form a compound

called hydrated iron (III) hydroxide. Hydrated iron (III) hydroxide is therefore the chemical name for rust. The word equation for this reaction is

iron + oxygen + water
$$\rightarrow \text{hydrated iron (III) hydroxide}$$

SAQ

2 a Give two reasons why we don't want our cars to rust.

b What does iron react with when it rusts?

c Give two examples of road and weather conditions that can make iron rust more quickly.

Properties of metal

How could we avoid the rusting away of cars?

Aluminium and iron are both metals, so they have certain properties in common.

1 They are both good conductors of electricity.
2 They are both malleable, which means they can both be pressed into new shapes without breaking.

Because aluminium and iron are *different* metals, they have some properties that are not the same.

1 Iron is magnetic but aluminium is not.
2 Iron is denser than aluminium. One cubic metre of iron has a mass of 7.83 tonnes. One cubic metre of aluminium has a mass of 2.70 tonnes.
3 When iron is in contact with water and oxygen, it rusts and corrodes away. Aluminium doesn't corrode when in contact with water and oxygen.

Protection by aluminium oxide

Aluminium does in fact react with the oxygen in the air. This forms a very thin layer of aluminium oxide, which is **impervious**. This means it won't let any more oxygen through to the aluminium underneath. The layer of aluminium oxide is **tenacious**. This means that it sticks onto the aluminium and doesn't flake off. The aluminium oxide layer **protects** the metal underneath it.

The rust layer that forms on iron is not impervious. Water and oxygen can get through it and the iron underneath rusts. Iron rust is not tenacious. It flakes off. Iron and steel objects will corrode away if they are in contact with water and oxygen.

SAQ

3 **a** If you have a steel drinks can and an aluminium drinks can, how could you tell them apart?

b How many times denser is iron than aluminium?

c Draw a circuit diagram to show how you could prove that iron and aluminium both conduct electricity.

4 Explain why aluminium doesn't corrode.

Aluminium has obvious advantages over steel as a material for car body panels. Aluminium doesn't corrode. Aluminium is less dense so the car will be lighter. Unfortunately, aluminium is much more expensive to manufacture from its ore than iron is. This is why most car bodies are made of steel.

Figure 2e.4 The body panels of this car are made of aluminium.

Aluminium cars may be cheaper to run

Aluminium cars are more expensive to buy than steel cars because aluminium car bodies are more expensive to manufacture. The consideration of cost should not stop there, however. Because the aluminium car won't corrode, it should last longer. It will be less likely to need expensive welding work done on it to strengthen it after a few years. Because the aluminium car is lighter, it will use less petrol – we say it has better **fuel economy**. The lighter car is also likely to have a higher top speed.

Aluminium – on the Moon and in your car

In the world of metals, aluminium is a newcomer. Its very existence was not suspected until 1807, and scientists did not succeed in making the first tiny pieces of aluminium until 1825. Scientists took a further 20 years to establish some of the properties of the metal. It was the very low density of aluminium that made people sit up and take notice.

People became very excited about the new 'wonder metal', but there was one huge snag. Although aluminium compounds were plentiful on Earth, getting the aluminium out of the compounds was very difficult. Aluminium was very hard to extract from its ore, so it was very expensive. In 1880, the price of aluminium was £220 000 per tonne.

The problem was solved in 1886. Two young scientists hit on the same solution at the same time. Neither knew of the other's existence. Both built their own makeshift laboratories. In America, Charles Hall worked in his father's woodshed. In France, Paul Héroult worked in the backroom of a local tannery.

Figure 2e.5a Charles Hall.

Figure 2e.5b Paul Héroult.

continued on next page

Aluminium - on the Moon and in your car - *continued*

Paul and Charles both discovered that, if aluminium ore was mixed with a mineral called cryolite, the mixture would melt at 850 °C. If electric current was passed through this molten mixture, aluminium was obtained. Because of the Hall–Héroult process (as it became known), by 1910 the price of aluminium had fallen to £85 per tonne.

Our society uses more aluminium every year. The biggest user of aluminium is the transport sector. Aluminium alloys were used to make the Saturn V rocket that helped land the first man on the Moon. The strength and low density of aluminium alloys means most passenger aircraft are made from them. The amount of aluminium used in making an average family car increased from 100 kg in 2000 to 150 kg in 2005. But this doesn't mean the cars got heavier. The aluminium replaced *over* 50 kg of other materials. The use of aluminium made the cars lighter.

Figure 2e.6 A Boeing passenger aircraft.

Recycling cars

Unless the owner spends a lot of time and money looking after a car, it will, in the end, become unfit to use. Perhaps the car 'fails its MOT' – this means it is judged to be unsafe for one reason or another. Perhaps it has rusted so badly that the body is no longer strong enough. Perhaps the moving parts in the engine have just worn out. It's time for a new car and the old one has to be scrapped.

Many of the resources used to make cars are not renewable. The ores used to make the metals and the crude oil used to make the plastics are non-renewable. The available supply of these natural resources can be prolonged by **recycling**. When a car is scrapped, its body panels can be melted down. The copper in the radiator and electrical wiring can be melted down.

The metals obtained can be used to make new cars. Recycling also prevents mountains of rusting cars building up in scrap yards outside our towns. Recycling is better than other means of disposing of them.

Recycling difficulties

There are problems involved in recycling old cars. Many small parts are made of different metals and other materials. These parts all need to be removed from the car, sorted and recycled separately. It is no good melting everything together into one horrible mixture. You should recall that this is a problem involved when recycling polymers too. Recycling must give us the *pure* materials we use to make new products. If it doesn't, it is of little use to us.

We need to overcome these problems in order to conserve resources and our environment. In various countries governments are now passing laws that insist that a certain minimum percentage of every scrapped car must be recycled. Car manufacturers are designing and making new cars in order to make this easier to do.

Figure 2e.7 All of these cars contain parts that are recyclable.

Summary

You should be able to:

◆ list the materials used to make cars, linking their properties to their uses

◆ state what causes iron to rust and what accelerates it

◆ state that aluminium doesn't corrode and explain why

[H] ◆ describe rusting as an oxidation reaction and write a word equation for the rusting of iron

◆ describe the similarities and differences between the properties of iron and aluminium

◆ describe two properties of steel which make it superior to pure iron

[H] ◆ explain the advantages and disadvantages of making a car body from aluminium or steel

◆ describe and explain the advantages and disadvantages of recycling the materials used to make cars

◆ understand how new laws will increase the level of recycling

Questions

1 Steel can be protected in various ways to help prevent rust.

   a Why does painting a steel car slow down its rate of rusting?

   b Why does oiling a chisel after you've used it stop the steel rusting?

   c Why is paint an unsuitable way to prevent a chisel rusting?

2 a Which is stronger, iron or steel?

   b Which rusts faster, iron or steel?

   c What does *corroding* mean?

   d Explain why iron corrodes but aluminium doesn't.

[H] e Write a word equation for the rusting of iron.

3 A friend wishes to buy an aluminium-bodied car but is put off by the high price. Write a short piece explaining to him or her:

   a why aluminium cars have a high initial cost

   b why there is a possibility of saving money over a period of time.

4 a Summarise the advantages and difficulties of recycling old cars after they cease to be useful.

   b Give *three* reasons why the recycling of materials from old cars should increase in the near future.

Atmosphere

The Earth has a layered structure. It has a core containing iron, a mantle composed of molten and near-molten magma, and a crust of rocky tectonic plates (see Item C2c *Does the Earth move?*). But there is also a fourth layer that is essential to life. Stretching upwards for over 700 km from the surface of the Earth is a complex mixture of gases – the **atmosphere**.

The Earth formed 4.5 billion years ago. It was so hot that it was molten. The outside cooled and the first crust formed. At this stage, there was much gas dissolved in the molten interior of the Earth. The Earth continued to cool, and some of the gas escaped to the surface of the Earth through volcanoes and cracks in the crust. The first atmosphere formed. It consisted mostly of carbon dioxide.

Figure 2f.1 The gases that belch out of volcanoes still contain large amounts of carbon dioxide.

The atmosphere has evolved

Three billion years ago there was life on Earth – cyanobacteria. Cyanobacteria were an early form of simple plant life. These early plants, like modern-day plants, were able to **photosynthesise**. They used photosynthesis to make sugars, taking in carbon dioxide and giving out oxygen. The level of carbon dioxide in the atmosphere fell and the level of oxygen rose. At the same time, the amount of nitrogen gas in the air increased.

SAQ

1 a How high is the Earth's atmosphere?

  b Has the Earth's atmosphere always been the same?

  c Name one gas present in the Earth's first atmosphere.

  d What has changed the composition of the atmosphere?

  e How has the composition of the atmosphere changed over time?

Ⓗ Nitrogen and the oceans

The release of dissolved gases from the molten Earth is known as **degassing**. Degassing released carbon dioxide into the early atmosphere. Water vapour and ammonia (NH_3) gases were released too. As the Earth cooled, the water vapour condensed to form liquid water, and the early lakes and seas formed. The first life evolved in these seas. After the appearance of the first plants, oxygen levels rose due to photosynthesis. Some of the oxygen reacted with the ammonia, producing nitrogen (N_2) and water. Because nitrogen is very unreactive, it does not get consumed in any large scale process, so nitrogen levels steadily increased.

SAQ

2 a Write a word equation for the reaction between ammonia and oxygen.

  b Write a balanced symbol equation for the reaction between ammonia and oxygen.

Respiration and combustion

The oxygen produced by photosynthesis in plants was used by plants and animals. The plants and animals needed the oxygen for the process called **respiration**. When living things respire, they take oxygen from the air and put carbon dioxide back into the air. Most recently, humans evolved on Earth and learnt how to burn fuels. This process called **combustion** also takes oxygen from the air and puts carbon dioxide back into the air.

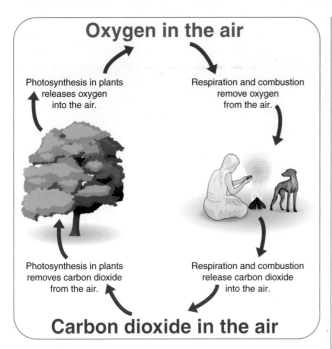

Figure 2f.2

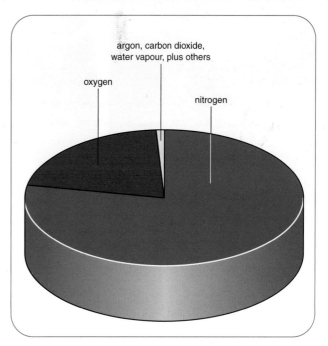

Figure 2f.3 The composition of the air today.

SAQ

3 a Name two processes that remove oxygen from the atmosphere and release carbon dioxide into it.

b Name a living organism that performs one of these processes.

c Name the process that removes carbon dioxide from the atmosphere and releases oxygen into it.

d Name a living organism that performs this process.

The atmosphere today

For a long time now, the atmosphere on Earth has consisted mostly of two gases. It is 78% nitrogen and 21% oxygen. Nitrogen and oxygen make up 99% of the atmosphere. The final 1% contains small amounts of many gases, including carbon dioxide, argon and water vapour. Some of these gases have been present in the atmosphere for a long time, but others are there as pollutants.

Compared with nitrogen and oxygen, the amount of carbon dioxide is tiny – only 0.035%. Fortunately, this is enough to enable plants to keep photosynthesising and to stay alive. The percentages of nitrogen, oxygen and carbon dioxide in the atmosphere today are fairly constant.

The effect of humans

Every compound has a precise composition that never varies. Water, for example, is always 11.1% hydrogen and 88.9% oxygen. The air is not a compound. Air is a mixture. This means that its composition is not exactly the same everywhere and at every time. The composition of the air can change. For example, the percentage of water vapour changes according to the weather. If a rain shower is followed straight away by a hot sunny period, the puddles evaporate and the percentage of water vapour in the air increases.

The living things on planet Earth have maintained the balance between the gases in the atmosphere for millions of years. Photosynthesis in plants has taken in carbon dioxide and given out oxygen. Respiration in animals and plants has taken in oxygen and given out carbon dioxide. In recent years, this balance has begun to be upset by humans. The composition of the air is being affected by humans in three main ways.

Combustion

Since the industrial revolution, we have burned more and more wood and fossil fuels every year in order to satisfy our need for energy.

H

Figure 2f.4 a Combustion. **b** Deforestation. **c** Respiration.

Deforestation

People need agricultural land to grow food on. Huge areas that were once covered in forest have been cleared to provide this land. This is called **deforestation**. Deforestation has been much in the news recently owing to the dwindling size of tropical rainforests. Sometimes the trees are cut down and sold as timber, but sometimes they are burnt.

Respiration and human activity

The human population is growing at an alarming rate. Every new human being respires in order to provide its body with energy. Every new human being increases the need to burn fuels to cook its food, and to provide it with goods.

SAQ

4 Look at the three photographs in Figure 2f.4. They illustrate three ways in which humans are changing the composition of the atmosphere. Consider the photographs one at a time.

a Which gas will be removed from the air and which gas added to the air by each activity? Answer as fully as you can.

b What will be the overall effect of these three activities on the composition of the atmosphere?

Pollution

The air we breathe contains small amounts of **carbon monoxide, oxides of nitrogen** and **sulphur dioxide**. These gases are naturally present in the air – for example, sulphur dioxide is released by volcanoes. However, natural processes only account for a really *tiny* percentage of each of these gases. The amounts of these gases in the air today are well above their natural levels. These gases are present at higher levels because of human activity. They are **pollutants**.

Carbon monoxide

Carbon monoxide is **poisonous**. It is released into the air from the exhaust pipes of cars, buses and trucks. It is produced by the incomplete combustion of petrol and diesel in the engines of these vehicles. People must be very careful not to start their cars in enclosed spaces, for example a garage with the doors closed. The build-up of carbon monoxide could be fatal.

Figure 2f.5 Exhaust fumes.

Sulfur dioxide

Sulfur dioxide causes **acid rain**. Rain naturally has a pH of around 5.5. Acid rain has a pH of around 4.0. Acid rain kills plants, kills the aquatic life in lakes, rivers, ponds and streams, and erodes stonework. Sulfur is naturally present in coal and oil. When we burn these fossil fuels, or fuels made from oil like petrol and diesel, the sulfur combines with oxygen and makes sulfur dioxide. Coal-burning power stations are responsible for much of the sulfur dioxide in the air. Petrol and diesel engines are also responsible for a significant amount.

The word equation for the formation of sulfur dioxide is:

sulfur + oxygen → sulfur dioxide

Figure 2f.6 This coal-fired power station is releasing sulfur dioxide into the air.

Oxides of nitrogen

Oxides of nitrogen also cause acid rain. In addition, oxides of nitrogen can take part in a complex series of chemical reactions to produce other pollutants. Some of these pollutants cause **photochemical smog**. This horrible yellow mixture of smoke and fog is caused by the action of light on the chemical pollutants. ('Photo' means light.) Photochemical smog and other pollutants involving oxides of nitrogen irritate the lungs of humans and other animals.

Oxides of nitrogen are released into the air from the exhaust pipes of petrol powered cars. In the internal combustion engine nitrogen and oxygen from the air react to produce oxides of nitrogen.

SAQ

5 a How do oxides of nitrogen get into the air?
  b What problems do they cause?
  c How does carbon monoxide get into the air?
  d What problem does it cause?
  e How does sulfur dioxide get into the air?
  f What problem does it cause?

Catalytic converters

The unpleasant nature of the problems caused by these three pollutants shows how important it is to control air pollution. If the problem with acid rain carries on getting worse, plant and water life will be very seriously affected. Farmers will find it even harder to earn their living and outdoor pleasures like rambling and fishing will be less enjoyable.

You should have noticed that car engines, known as internal combustion engines, can release all three pollutants. Some of the pollutants in the exhaust fumes can be removed by fitting a catalytic converter to the exhaust system under the car. The fumes go through the converter, which changes carbon monoxide to carbon dioxide. Carbon dioxide is not poisonous. The converter also removes oxides of nitrogen from the fumes.

Figure 2f.7 Catalytic converter.

One of the oxides of nitrogen present in car exhaust fumes is nitrogen monoxide (NO). It reacts with carbon monoxide on the surface of the catalyst. The products are carbon dioxide and nitrogen. A balanced symbol equation for this reaction is:

$$2CO + 2NO \rightarrow N_2 + 2CO_2$$

Flue gas desulfurisation

Power station emissions contain sulfur dioxide. This gas can be removed by **flue gas desulfurisation** – FGD for short. FGD involves making the sulfur dioxide react with either calcium carbonate or calcium hydroxide before it can leave the chimney. In both these reactions calcium sulfate is produced. Calcium sulfate is used to make plasterboard.

SAQ

6 How can we reduce the amounts of sulfur dioxide, carbon monoxide and oxides of nitrogen being released into the atmosphere?

Is carbon dioxide harmless?

Figure 2f.8 Rising global temperatures are making the West Antarctic ice shelf break up.

Figure 2f.9 The warmer winters in the west of the UK are due to the North Atlantic Drift, allowing plants like this to grow.

In 1859, a British scientist called John Tyndall studied how effective different gases were at trapping heat. He took careful measurements with many gases. He came to the conclusion that oxygen and nitrogen were very poor at trapping heat, but his results for carbon dioxide told a different story. Carbon dioxide was good at trapping heat.

John Tyndall's results showed that carbon dioxide was partly responsible for keeping the Earth's surface at its current temperature. The idea of the greenhouse effect was born. Without greenhouse gases like carbon dioxide, the average global temperature would be −18 °C. The presence of greenhouse gases means the actual value is +14 °C

The greenhouse effect is not a constant, because it depends on the composition of the atmosphere. Human activity is increasing the percentage of carbon dioxide in the Earth's atmosphere. This is causing the average temperature at the Earth's surface to rise. This is known as 'the enhanced greenhouse effect' or 'global warming'.

Weather scientists find it difficult to agree on the exact numbers, but it looks like a warming of over 0.5 °C has taken place since 1970. Half a degree doesn't sound much but, because a rise of 4 °C may well prove to be disastrous, it is a significant start.

'Global warming' might make the UK get colder! The UK climate depends partly on a warm water current that brings heat to its western shores. The current is called the North Atlantic Drift, commonly known as the Gulf stream. It depends on very cold water off the coast of Norway sinking downwards. This cold water then travels southwards through the deep ocean until it reaches the tropics. Meanwhile warm water travels northwards along the surface to replace it.

If global warming heats the seas near Norway, there will be no cold current to move south, so the hot current will stop flowing north and the UK will get colder.

Early in the 21st century, the leaders of many of the world's industrialised nations entered into negotiations designed to reduce the carbon dioxide levels in the atmosphere. These may be the most important decisions they will ever make.

Summary

You should be able to:

♦ describe air as a mixture and state the percentages of the main gases

♦ state that the percentages of the main gases are approximately constant

♦ describe how photosynthesis, respiration and combustion affect the percentages of oxygen and carbon dioxide in the air

♦ describe how the present-day atmosphere evolved

H ♦ evaluate the effects of human activity on the composition of the air

♦ state the common pollutants found in the air, their origins and the problems each pollutant causes

♦ state how a catalytic converter helps to reduce atmospheric pollution

H ♦ use a balanced symbol equation to describe how a catalytic converter lowers carbon monoxide emissions from cars

Questions

1 In photosynthesis, carbon dioxide and water are used by a green plant to make oxygen and glucose.

   a Write a word equation for photosynthesis.

H **b** The formula of glucose is $C_6H_{12}O_6$. Write a balanced symbol equation for photosynthesis.

   c Why does the plant need to photosynthesise? (Use the Index to find the answer in the biology section of this book.)

   d What effect does photosynthesis have on the composition of the Earth's atmosphere?

2 In respiration, oxygen and glucose are used by both plants and animals. Carbon dioxide and water are produced.

   a Write a word equation for respiration.

   b Write a balanced symbol equation for respiration.

H **c** Why do animals and plants need to respire? (Use the Index to find the answer in the biology section of this book.)

   d What effect does respiration have on the composition of the Earth's atmosphere?

continued on next page

Questions - *continued*

3 The burning of a fuel such as methane is called *combustion*. When oxygen and methane react together, carbon dioxide and water are produced.

H

a Write a word equation for the combustion of methane.

b Write a balanced symbol equation for the combustion of methane.

c Why do humans burn (combust) hydrocarbons like methane?

d What effect does the combustion of fuels like methane have on the composition of the Earth's atmosphere?

4 Copy the table below. In the first column, add the names of three air pollutants. In the second column, explain how each pollutant gets into the air. In the third column, describe the environmental problems caused by that pollutant.

Pollutant gas	Cause	Environmental impact

H

5 Describe the evolution of the Earth's atmosphere as a series of illustrations, like a comic strip. Frame one should be of a lifeless Earth with a volcano belching out carbon dioxide, ammonia and water vapour. Gases will need to be identified using labels. Your final frame should show the Earth and its atmosphere as it is today.

6 It is expected that supplies of petrol and diesel for road transport will run out sometime in the middle of this century. How will this affect the air in the centre of a big city like New York, Tokyo or London? Go into as much detail as you can.

H

7 Look at this balanced symbol equation:

$$2CO + 2NO \rightarrow N_2 + 2CO_2$$

a Rewrite the equation as a word equation.

b Close this book. Using your word equation, rewrite the balanced symbol equation in your exercise book. Do this by working it out from your word equation, not by memorising the balanced symbol equation above.

c Confirm the equation is balanced by counting the atoms present on each side of the arrow.

d In this reaction, carbon is oxidised and nitrogen is reduced. Use the balanced symbol equation or word equation to explain why this is so.

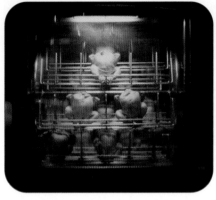

Figure 2g.1 How long does each change take?

Rate of a chemical reaction

The speed at which something happens is called its **rate**. Some changes have a very high rate – the explosion of a firework is over in a fraction of a second. Other changes take longer – for example cooking a joint of meat can take 2 hours. Some changes have a very low rate – it takes years for a piece of iron on a scrap heap to rust away completely.

SAQ

1 **a** Name the three chemical reactions shown in the pictures in Figure 2g.1.

   b Put the reactions in order of their rate, starting with the slowest.

Colliding particles

A chemical reaction between two or more substances happens when the reactant particles **collide** with each other. 'Particles' here means atoms and molecules. Look at the photograph of the rusting car in Figure 2g.1. Every little particle of rust on the car formed when some iron atoms were hit by some water molecules and some oxygen molecules. Substances will only react with each other when their particles collide with each other.

Moving particles

Every substance is made of particles (atoms and molecules) that are moving. Solids are made of closely packed particles that stay in the same place but vibrate from side to side. Liquids are made of closely packed particles that move slowly throughout the body of liquid. Gases are made of widely spaced particles that are free to move throughout the gas at high speed.

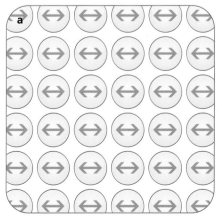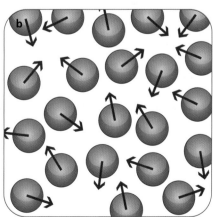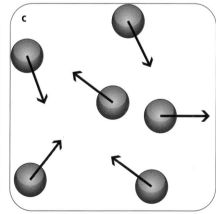

Figure 2g.2 The movement of particles in a solid (**a**), a liquid (**b**) and a gas (**c**).

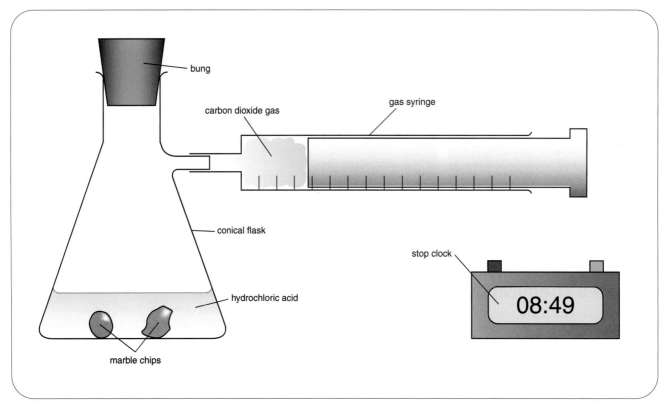

Figure 2g.3 Apparatus used to measure the rate of a reaction.

2 Summarise the difference in movement between the particles of a solid, a liquid and a gas.

Measuring the rate of a reaction

Calcium carbonate reacts with hydrochloric acid. Calcium carbonate is readily available in lumps called 'marble chips'. It is easy to measure the rate of this reaction, because one of its products is a gas. The word equation for this reaction is:

calcium carbonate + hydrochloric acid
→ calcium chloride + water + carbon dioxide

The gas produced is carbon dioxide. If you measure the amount of carbon dioxide given off, and measure the time it takes, you can tell the rate of the reaction. The apparatus you would use is shown in Figure 2g.3.

Reaction rates can vary

Chemical reactions happen at a variety of rates. Some tend to be fast, some tend to be slow. In addition, the rate of a particular reaction can be speeded up or slowed down. This involves particle collisions.

● Any change that makes the reactant particles collide **more frequently** will speed up a reaction.
● Any change that makes the reactant particles collide **harder** will speed up a reaction.

The explanations that follow refer to the rate of the reaction between hydrochloric acid and calcium carbonate.

More collisions

If you vary the **concentration** of the acid, the rate of the reaction will change. The acid being used here is a solution of hydrochloric acid in water. 'Concentration' here refers to the relative amounts of hydrochloric acid and water. If the acid contains less hydrochloric acid dissolved in a certain volume of water, it is said to be **less concentrated**. If the acid contains more hydrochloric acid dissolved in the same volume of water, it is said to be **more concentrated**.

Concentration is a measure of the number of particles of solute dissolved in a stated volume of solvent.

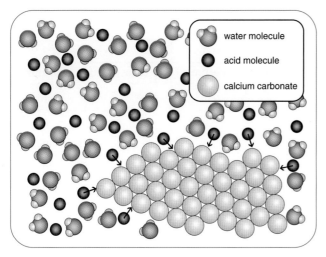

Figure 2g.4a Higher concentration of acid, more collisions.

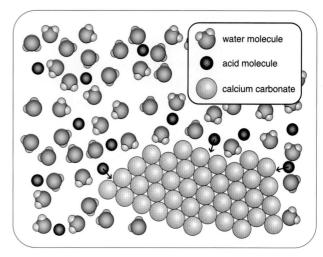

Figure 2g.4b Lower concentration of acid, fewer collisions.

In order to react, the acid particles must collide with the calcium carbonate particles. In more concentrated acid, the acid particles are closer together. This means that they collide with the calcium carbonate particles more often. This makes the reaction faster. More concentrated acid reacts faster with calcium carbonate than less concentrated acid does.

SAQ

3 a Explain what is meant by 'more-concentrated' acid.

  b Explain what is meant by 'less-concentrated' acid.

  c Explain why more-concentrated hydrochloric acid reacts more quickly with calcium carbonate.

  d Explain why less-concentrated hydrochloric acid reacts more slowly with calcium carbonate.

Gas pressure

The same effect can be seen if one (or both) of the reactants is a gas. To increase the concentration of a gas, you must increase its **pressure**. If gases are pressurised, the gas particles become closer together. This makes them collide with other reactants more frequently. This makes the reaction faster.

Harder collisions

If you vary the **temperature** of the acid, the rate of the reaction will change. In order to react, the acid particles must collide with the calcium carbonate particles. In hotter acid, the acid particles are moving more quickly. This makes them collide harder with the calcium carbonate particles. This makes the reaction faster. Hotter acid reacts faster than colder acid with calcium carbonate.

SAQ

4 Explain why hotter acid reacts faster than colder acid with calcium carbonate.

Successful and unsuccessful collisions

When a piece of calcium carbonate is immersed in hydrochloric acid, the acid particles are constantly colliding with the calcium carbonate particles. At lower temperatures, most of these collisions do not cause a reaction. In such collisions, the colliding particles don't have enough energy. The acid particles just bounce off the calcium carbonate particles without changing. These collisions are described as **unsuccessful collisions**.

At higher temperatures, more of the collisions cause a reaction. In such collisions, the colliding particles have enough energy so that, when the acid particles collide with the calcium carbonate particles, new products form. These collisions are described as **successful collisions**. Because more successful collisions happen at a higher temperature, the rate of the reaction is faster.

Using graphs

How fast is the reaction?

If you record the volume of gas collected at the end of each minute after starting the reaction, you are collecting **data** which tells you about the rate of the reaction. The apparatus in Figure 2g.3 was used to do two experiments.

In experiment 1, 50 cm^3 of acid was put in the conical flask and 1 g of calcium carbonate was added. The volume of gas in the syringe was written down after each minute.

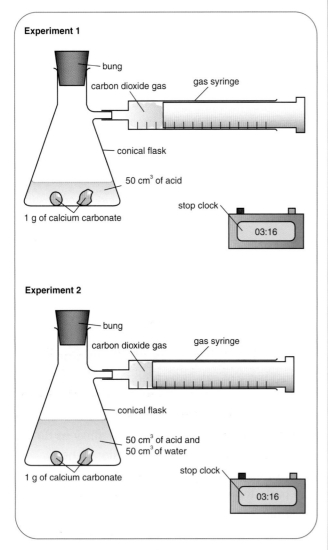

Figure 2g.5 Experiments 1 and 2.

In experiment 2, 50 cm^3 of the same acid was put in the conical flask and 50 cm^3 of water was added to it. Then 1 g of calcium carbonate was added. The gas volume in the syringe was recorded in the same way.

The results were recorded in two tables.

Experiment 1		Experiment 2	
Time (min)	Volume of gas (cm^3)	Time (min)	Volume of gas (cm^3)
0	0	0	0
1	20	1	10
2	32	2	18
3	38	3	25
4	41	4	30
5	42	5	34
6	42	6	37
7	42	7	39
8	42	8	40
9	42	9	41
10	42	10	42
11	42	11	42
12	42	12	42

Table 2g.1 Experiments 1 and 2.

Experiment 1 produced gas more quickly because the acid was more concentrated than the acid in experiment 2. Graphs are an excellent way to present this data.

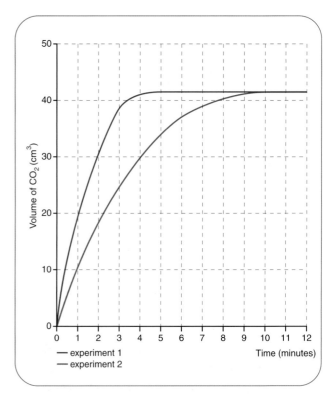

Figure 2g.6 Results from experiments 1 and 2.

These graphs show us in two ways that experiment 1 reacted faster.

1 The graph of experiment 1 goes up more steeply. The steeper the graph, the faster the reaction.

2 Experiment 1 finished first. When the graph runs parallel to the bottom axis, it means that no more gas is being collected. The reaction has finished. Experiment 1 finished after 5 minutes. Experiment 2 finished after 10 minutes.

The graphs also show that both reactions were at their quickest right at the beginning. Each line is steepest at the start; this means the reactions were quickest then.

Each reaction finished when one of the reactants was completely used up. In the situations we have described here, each reaction finished when all the acid was used up. The calcium carbonate was not all used up. The reactant that is not all used up is said to be in **excess**.

SAQ

5 a How much gas was collected after 2 minutes:

   i in experiment 1?

   ii in experiment 2?

  b Why did the chemicals in experiment 1 react faster than the chemicals in experiment 2?

  c Explain your answer to **b**. Use the idea of collisions between particles in your answer.

  d How do the graphs show that experiment 1 reacted faster than experiment 2?

  e Why did the two reactions stop?

Changing the temperature

The apparatus in Figure 2g.3 was used to do two more experiments.

In experiment 3, 60 cm^3 of acid was put in the conical flask. The acid was at 20 °C. 1 g of calcium carbonate was added. The volume of gas in the syringe was written down after each minute.

In experiment 4, 60 cm^3 of the same acid was put in the conical flask and then heated to 40 °C. Then 1 g of calcium carbonate was added. The gas in the syringe was recorded in the same way.

The results were recorded in two more tables.

Experiment 3		Experiment 4	
Time (min)	Volume of gas (cm^3)	Time (min)	Volume of gas (cm^3)
0	0	0	0
1	24	1	41
2	38	2	48
3	44	3	51
4	47	4	51
5	49	5	51
6	50	6	51
7	50	7	51
8	51	8	51
9	51	9	51
10	51	10	51
11	51	11	51
12	51	12	51

Table 2g.2 Experiments 3 and 4.

SAQ

6 a Use the data to plot graphs of the results of experiments 3 and 4.

  b What volume of gas was collected in each experiment after 1.5 minutes?

  c Why did the two reactions stop?

  d What happened to the rate of reaction in experiment 3 during the first 5 minutes?

7 a How do your graphs show that experiment 4 reacted faster than experiment 3?

  b Why did experiment 4 react faster than experiment 3?

  c Explain your answer to **b**. Use the idea of collisions between particles in your answer.

Experiments 3 and 4 both produced the same final volume of gas: 51 cm^3. This happened because both experiments involved exactly the same amounts of reactants. Changing the temperature changes the rate at which the products form, but the amounts of products that form depend only on the amounts of reactants used.

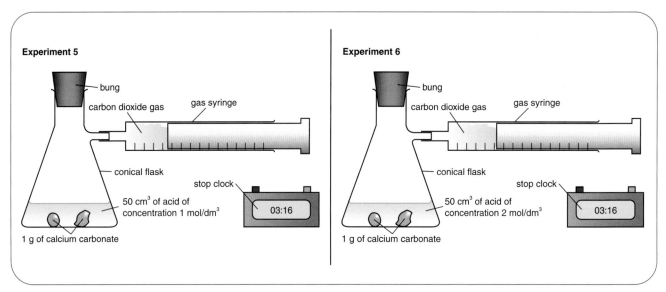

Figure 2g.7 Experiments 5 and 6.

Dresden

Figure 2g.8 At least 25 000 people died in the Allied firebombing raid on Dresden.

On the evening of 13 February 1945, British and American bombers attacked the German city of Dresden. The bombing may have caused more deaths than the atomic bombs caused in Hiroshima and Nagasaki added together. The raid on Dresden involved conventional high explosives, but it caused such massive damage and loss of life because it resulted in a firestorm.

The bombs that rained down on Dresden were firebombs. The fires that they started increased the temperature of the air so that it rose upwards in a massive convection current. More air moved in from the area around Dresden to replace the rising air, bringing with it fresh oxygen as it did so. The heat of Dresden's own fire provided the supplies of oxygen that it needed to keep burning. The flames grew hotter, the air currents got stronger and soon the whole city was nothing but a mass of flame.

The firestorm in Dresden happened because all of the ingredients were there for a very rapid chemical reaction. There was a large supply of reactants. The fuel for the fire came from the bombs and the material of the city itself, and fresh supplies of oxygen were constantly brought in by the wind. This meant the *concentration* of oxygen stayed high. There was a very high *temperature*. The burning reactions were exothermic, so the more the city burnt the more heat was released, and the temperature climbed higher and higher.

It is not certain how many people died in Dresden that night. The city's normal population of 600 000 had probably doubled in recent days owing to refugees fleeing the advancing Russian Army. The official estimate in 1945 was that well over 25 000 people lost their lives, but some authorities believe the true figure to be many times higher.

Amount of product

The apparatus in Figure 2g.3 was used to do two more experiments.

In experiment 5, $50\,cm^3$ of acid was put in the conical flask and $1\,g$ of calcium carbonate was added. The volume of gas in the syringe was written down after each minute.

In experiment 6, $50\,cm^3$ of acid that was twice as concentrated was put in the conical flask. Then $1\,g$ of calcium carbonate was added. The gas in the syringe was recorded in the same way. Each reaction finished when the acid had all reacted.

The units used to measure concentration are mol/dm^3. The sketch graph in Figure 2g.9 shows the amount of gas collected after each minute for experiments 5 and 6.

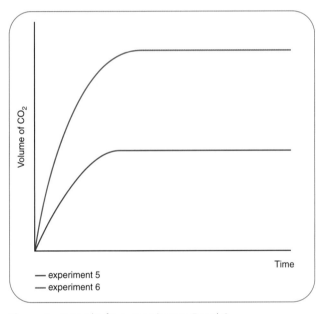

— experiment 5
— experiment 6

Figure 2g.8 Results from experiments 5 and 6.

You should notice two things.

1 Because experiment 6 used acid that was twice as concentrated, the rate of the reaction is faster. The line for experiment 6 is therefore steeper than the line for experiment 5.

2 Both experiments used $50\,cm^3$ of acid. Because experiment 6 used acid that was twice as concentrated, there is twice as much acid dissolved in the water. When the acid was used up, there was twice as much gas produced as there was in experiment 5.

Calculating reaction rates

A **rate of reaction** can be calculated by dividing the amount of product collected by the time taken.

Worked example

Experiment 2 produced $40\,cm^3$ of gas during the first eight minutes. Calculate the average rate of reaction over this period.

Rate of reaction
        = volume of gas collected ÷ time taken

Rate of reaction = $40\,cm^3$ ÷ 8 minutes

Answer: Rate of reaction = $5\,cm^3$ per minute

This method gives us the average rate of reaction over the specified time period. A rate of reaction can also be found at a particular time using a graph. To show you how this can be done, we are going to use the graph of the results of experiment 2 again.

Using a tangent

The rate of a reaction is given by the steepness of the graph. This is also called the **gradient**. The gradient of the graph in Figure 2g.10 at 3 minutes can be found by drawing on a line at this point called a **tangent**. The tangent should exactly match the steepness of the graph at this point. To find the gradient of the tangent, first use the tangent to draw a right-angled triangle (blue and green lines). Then divide the length of the green line ($36\,cm^3$) by the length of the blue line (6 minutes).

    $36\,cm^3$ ÷ 6 minutes = $6\,cm^3$ per minute

This is the rate of the reaction at 3 minutes. It has been found by working out the gradient of the graph at 3 minutes.

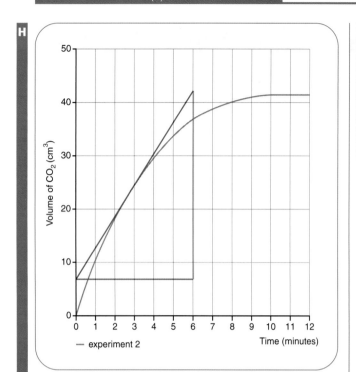

— experiment 2

Figure 2g.10 Using a tangent.

8 Use the gradient method to find the rate of reaction of experiment 2 at the following times:

a 2 minutes

b 5 minutes

c 11 minutes.

Summary

You should be able to:

◆ distinguish between a fast and a slow reaction

◆ explain that chemical reactions require collisions to occur between particles

◆ describe how increasing temperature or reactant concentration affects the rate of a reaction

◆ use collision theory to explain why increasing temperature or reactant concentration affects the rate of a reaction

◆ explain why a reaction stops

◆ label the apparatus that is used to investigate the rate of a reaction

◆ interpret data involving the rate of a reaction

◆ draw sketch graphs to show how temperature or reactant concentration affects the rate of a reaction and the amount of product formed

◆ calculate the rate of a reaction from the gradient of a graph

Questions

1 Four students investigated the rate of the reaction between hydrochloric acid and marble. They all used two similar marble chips but otherwise the experiments were all set up differently. They got the following results.

Sokina collected $24\,cm^3$ of carbon dioxide in 8 minutes.

Charlotte collected $15\,cm^3$ of carbon dioxide in 3 minutes.

Darryl collected $16\,cm^3$ of carbon dioxide in 4 minutes.

Richard collected $10\,cm^3$ of carbon dioxide in 5 minutes.

a Charlotte's experiment happened at the fastest rate. Whose was second fastest and whose was third fastest?

b Suggest *two* possible differences between the conditions used in Charlotte's experiment and in Richard's that made her experiment react faster than his.

2 Malachite reacts with sulfuric acid to make carbon dioxide gas. Leila investigated the rate of this reaction by adding an excess of malachite to a small amount of sulfuric acid. She measured how much gas was produced every minute. Her results are shown in Table 2g.3.

a Draw and label the apparatus Leila would have used to perform the experiment.

b Plot a graph of Leila's results.

c How long did it take the reaction to stop?

d Why did the reaction stop?

e When was the rate of the reaction highest?

f Did the reaction speed up or slow down as time went on?

g Why did it do this?

Time (minutes)	Volume of carbon dioxide (cm^3)
0	0
1	23
2	40
3	52
4	60
5	66
6	70
7	73
8	75
9	77
10	78
11	79
12	79
13	80
14	80
15	80

Table 2g.3 Leila's experiment.

continued on next page

Questions – *continued*

3 Draw sketch graphs to compare the gas given off in the following reactions. In each case the calcium carbonate is in excess. You must decide how quickly the gas is produced, i.e. how steeply the line goes up. You must decide how much gas is produced, i.e. at what height the line levels off.

   a 50 cm³ of acid at 'normal' concentration and 1 g calcium carbonate.

   b 50 cm³ of acid at 'twice normal' concentration and 1 g calcium carbonate.

   c 100 cm³ of acid at 'normal' concentration and 1 g calcium carbonate.

4 You will need a graph of the results from question 2 for this question.

   a Find the rate of the reaction at 4 minutes by drawing a tangent at this point.

   b Find the average rate of the reaction over the first 4 minutes.

   c Why is your answer to b greater than your answer to a?

5 Explain why the chemical reactions inside you (body temperature 37 °C) happen faster than the chemical reactions inside a newt (body temperature 10 °C). Use the idea of collisions between particles in your answer.

6 The fertiliser industry needs to make a lot of a strong-smelling gas called ammonia. The ammonia is made by reacting nitrogen gas and hydrogen gas. Unfortunately, this reaction is very slow.

   a Suggest two ways in which this reaction could be speeded up.

   b Explain why these changes would speed up this reaction. Use the idea of collisions between particles in your answer.

7 a Sketch the arrangement of the particles in an ice block, the water vapour you are breathing out and a drop of condensation on a window pane.

   b Comment on the movement of the particles in each situation.

8 Write a balanced symbol equation for the reaction between calcium carbonate and hydrochloric acid. Use the word equation on page 240 to help you. You will need to use the following formulae:

hydrochloric acid, HCl
calcium chloride, $CaCl_2$
calcium carbonate, $CaCO_3$.

Faster or slower (2)

Surface area

Calcium carbonate, in the form of marble chips, will react with hydrochloric acid. The rate of the reaction can be found by measuring the volume of carbon dioxide given off every minute. If the marble chips are crushed into small pieces, the rate of the reaction increases. If they are ground into powder, the rate of the reaction becomes even higher. Solid reactants will react faster if they are in smaller pieces.

When a large marble chip is broken into pieces, the inside surface of the chip is exposed. The chip now has a greater **surface area**. If the broken pieces are put into hydrochloric acid, there is now a greater surface area for the acid particles to collide with. This greater surface area means that the acid particles collide with the calcium carbonate particles more frequently. This increases the rate of the reaction.

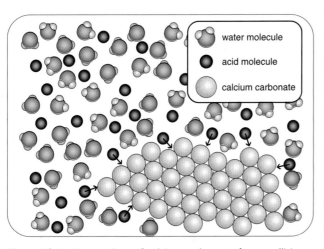

Figure 2h.1a Larger piece of calcium carbonate, fewer collisions.

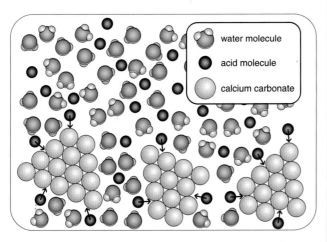

Figure 2h.1b Smaller pieces of calcium carbonate, more collisions.

SAQ

1 Explain why grinding a marble chip up into a powder makes it react faster with hydrochloric acid.

2 Experiments 1, 2 and 3 were performed by putting 1 g of marble into 50 cm^3 of hydrochloric acid.

Experiment 1 was done with a single marble chip.

Experiment 2 was done with the chip broken into small pieces.

Experiment 3 was done with the chip ground to a fine powder.

The volume of gas collected was recorded after each minute, as shown in Table 2h.1.

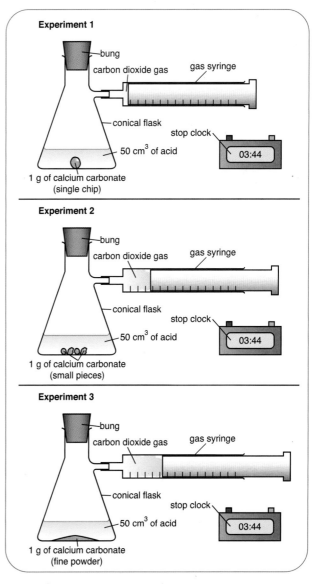

Figure 2h.2 Experiments 1, 2 and 3.

Experiment 1	
Time (min)	Volume of gas (cm^3)
0	0
1	2
2	4
3	6
4	8
5	10
6	11
7	13
8	15
9	17
10	18
11	20
12	21

Experiment 2	
Time (min)	Volume of gas (cm^3)
0	0
1	17
2	27
3	34
4	39
5	44
6	47
7	50
8	51
9	52
10	53
11	53
12	53

Experiment 3	
Time (min)	Volume of gas (cm^3)
0	0
1	47
2	51
3	53
4	53
5	53
6	53
7	53
8	53
9	53
10	53
11	53
12	53

Table 2h.1 Experiments 1, 2 and 3.

a Plot a graph of these results. The time axis of your graph should go as far as 20 minutes to allow you to answer part **d**.

b Which experiment had the fastest reaction and which had the slowest?

c Give two ways in which your graph could have helped you to answer part **b**.

d Look at the line you have plotted for experiment 1. Decide how this line is likely to have continued if readings had been taken for a further 8 minutes. Sketch the continuation of the line on your graph.

3 a Experiments 1, 2 and 3 were performed to investigate the effect of grinding the marble. What other factors should have been kept unchanged in order to make this a fair test?

b What could the experimenters have done to ensure their data was reliable?

4 a Experiment 1 was allowed to continue. What total volume of gas do you expect would have been collected 16 minutes after it started?

b How many minutes do you think it took experiment 1 to produce $30\,cm^3$ of gas?

c Find the gradient of each of your graphs from SAQ 2 at 2 minutes and hence state the rate of each reaction at this time.

Catalysts

A **catalyst** is a substance that increases the rate of a reaction and is unchanged itself at the end of the reaction.

Hydrogen peroxide solution decomposes. When hydrogen peroxide molecules collide with each other, they can react to produce water and oxygen. It is a useful way to make oxygen gas in a laboratory, but the hydrogen peroxide decomposes very slowly. Unless we can speed up the reaction, it is not very useful.

SAQ

5 a Write a word equation for the decomposition of hydrogen peroxide.

b The formula of hydrogen peroxide is H_2O_2. The formula of oxygen is O_2. Write a balanced symbol equation for the decomposition of hydrogen peroxide.

The rate of the reaction can be increased by using very concentrated hydrogen peroxide solution or by heating it. Unfortunately, neither of these changes speeds up the reaction enough to make it useful. If we add a catalyst, the reaction proceeds at a very good speed – useful amounts of oxygen can be collected in just a few minutes. A black powder called manganese dioxide will act as a catalyst and speed up the decomposition of hydrogen peroxide.

Figure 2h.3 The conical flask on the left contains hydrogen peroxide solution and manganese dioxide powder. The gas collecting on the right is oxygen.

SAQ

6 a Why does making hydrogen peroxide solution more concentrated or hotter make it decompose more quickly? Use the idea of collisions between particles in your answers.

  b What is a catalyst?

  c Name a substance that acts as a catalyst to increase the rate of decomposition of hydrogen peroxide.

Reusing a catalyst

When manganese dioxide is added to hydrogen peroxide solution, it acts as a catalyst and increases the rate of decomposition of the hydrogen peroxide. The reaction finishes when the hydrogen peroxide is used up. The catalyst is not used up. When the reaction in Figure 2h.3 has finished, the catalyst powder can be recovered from the flask on the left by filtering it out. If this catalyst powder is added to another flask of hydrogen peroxide solution, it will increase its rate of decomposition too.

Even a very small amount of catalyst is effective at speeding up the reaction of very large amounts of reactants. The same piece of catalyst can be used over and over again. This is why the catalyst in a car's catalytic converter can keep on working. This catalyst speeds up the reaction that converts carbon monoxide into carbon dioxide, but the catalyst itself is unchanged.

Catalyst specificity

Although a catalyst speeds up a particular reaction, it may not speed up any other reaction. If a catalyst speeds up only one reaction, it is said to be **specific** for that reaction. The biochemical reactions happening inside you are speeded up by catalysts. These catalysts are proteins called enzymes. Most enzymes only speed up one reaction. Such enzymes are said to be specific for the one reaction they speed up.

SAQ

7 a Why can a small amount of catalyst act on a large amount of reactants?

  b Which reaction is speeded up by the catalytic converter in a car?

  c How could you prove the catalyst in Figure 2h.3 is not being used up?

  d Most catalysts are specific. What does 'specific' mean here?

Figure 2h.4. This catalyst gauze speeds up one of the reactions used to make nitric acid. This circle of gauze contains platinum and is worth over a quarter of a million pounds. It's a good job that catalysts aren't used up.

Explosions

A very rapid reaction that produces gaseous products is called an **explosion**. 'Very fast' here means a reaction that is over in well under a second.

You are already familiar with one explosion in your school lab – the 'pop test' for hydrogen involves a very small explosion. The hydrogen mixes with oxygen from the air and this mixture explodes when you put a flame near it. The gaseous product is water vapour, which condenses to form water. The reaction only makes a little pop because you only use small amounts. In large amounts, a hydrogen–oxygen explosion can be very dangerous.

Figure 2h.5. The *Hindenburg* airship was filled with hydrogen gas. It caught fire on 6 May 1937. 36 people died in this disaster.

TNT is a well known 'high explosive'. The full chemical name for TNT is trinitrotoluene (TriNitroToluene). The explosion of TNT is a very rapid decomposition. TNT decomposes to make carbon, carbon monoxide, nitrogen and water vapour.

SAQ

8 **a** What do we mean by an 'explosion'?

   b Write a word equation for the decomposition of TNT.

   c Which gases form when TNT explodes?

Powder explosions

If a substance can react with oxygen, it can be dangerous in very fine powder form. The fine powder has a large surface area for oxygen molecules to collide with, so the reaction can be rapid enough to cause an explosion. Fine sulfur powder can explode in this manner. These explosions are called powder explosions or **dust explosions**.

Figure 2h.6 Large pieces of aluminium won't burn in a normal Bunsen flame. This is what happens when fine aluminium powder is sprinkled into the flame.

When sulfur reacts with oxygen the product is sulfur dioxide, which is a gas. Factories that handle fine sulfur powder have to do so very carefully.

You may not expect icing sugar, custard powder or flour to be dangerous, but they are. They are all compounds of carbon, hydrogen and oxygen. When they react with oxygen in the air, the products are gases, for example:

sugar + oxygen → carbon dioxide + water

Because this reaction is exothermic, it releases a lot of heat. The water that is produced is released as water vapour, which is a gas. Factories that make icing sugar, custard powder or flour have to be very careful. It is almost impossible for them to prevent the fine powders from floating in the air. To keep safe, they have to make sure there is nothing that could set off an explosion. Naked flames and sparks have to be avoided at all costs.

SAQ

9 Why can icing sugar cause an explosion but sugar lumps can't?

Boring old job?

Figure 2h.7 A grain storage facility.

A grain elevator is a storage facility for wheat or some other cereal crop. A moving conveyor belt is used to lift the grain up to the top of a huge silo. The silo is the vessel where the grain is stored. As long as you keep clear of the moving parts, it sounds like a very safe place to work. Boring even.

Was that what the workers thought at the grain elevator at Wichita in the USA when they went to work on the morning of 8 June 1998? Later that day, a grain dust explosion ripped through the facility, killing seven and injuring ten, three of them seriously. Almost a kilometre length of the elevator was damaged.

In fact, the workers at Wichita probably knew that grain dust explosions were a distinct possibility where they worked. There had been 43 in the USA during the previous 3 years, and the average was running at 13 per year. Any industrial facility working with wheat and cereal grains is going to produce a fine, flour-like dust simply through friction. It then only takes a spark, possibly when a metal link on a conveyor built rubs over another piece of metal, to set off the explosion.

It's not even new. In 1785 in Turin, a lamp in a bakery is reported to have set off a flour explosion. Never underestimate the effect of a large surface area on a combustible fuel.

Summary

You should be able to:

- describe a catalyst and state how a catalyst affects the rate of a reaction
- **H** recognise that a catalyst is specific to a particular reaction
- explain why powders react faster than lumps
- **H** explain how an increase in surface area affects the frequency of collisions
- describe an explosion and give examples of explosive reactions
- explain the dangers of combustible powders in industry
- interpret data involving the rate of a reaction
- draw sketch graphs to show how temperature or reactant concentration affects the rate of a reaction and the amount of product formed
- **H** calculate the rate of a reaction from the gradient of a graph

Questions

1 Explain the following observations:

a Malachite reacts much more quickly with sulfuric acid if large pieces of malachite are ground to a powder first.

b When sulfuric acid is being made sulfur dioxide has to react with oxygen. This reaction happens a lot faster if divanadium pentoxide is present. The divanadium pentoxide itself is unchanged at the end of the reaction.

c Food is much easier to digest if it is chewed up into small pieces.

d Yeast can make alcohol from sugars. The chemicals the yeast uses to do this do not get used up in the process.

e Lumps of iron don't burn but fine iron filings do.

continued on next page

Questions - *continued*

2   Richard investigated the rate of decomposition of hydrogen peroxide. He used the apparatus in Figure 2g.3 (page 240). He put 30 cm³ of hydrogen peroxide solution in the flask and began to take measurements. Nothing happened at first so he added some manganese dioxide and continued to take measurements. He drew a graph of his results which looked like this (Figure 2h.8).

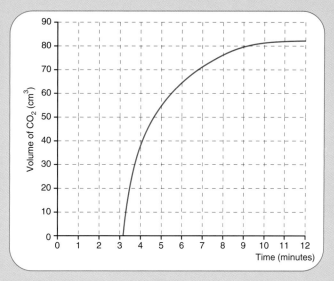

Figure 2h.8

a   What volume of gas was in the syringe at 7½ minutes?

b   When did Richard add the manganese dioxide?

c   What role does manganese dioxide play in this reaction?

d   When was the rate of the reaction fastest?

e   How long did the reaction take to finish?

f   Why did the reaction finish?

g   Which gas did Richard collect and measure in the gas syringe?

continued on next page

Questions - *continued*

3 Tamsin did two experiments involving calcium carbonate and hydrochloric acid. In both experiments, she used $500\,cm^3$ of the same hydrochloric acid, which was in excess. In one experiment, she used 2 g of large marble chips. In the other experiment, she used 1 g of very small marble pieces.

The graphs of her results looked like this (Figure 2h.9).

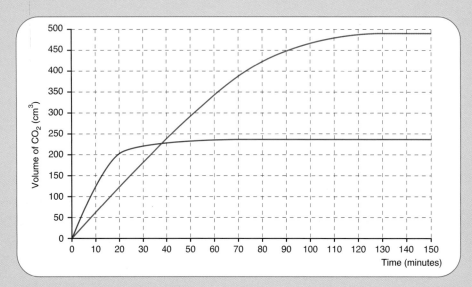

Figure 2h.9

a Which experiment started off more quickly?

b Which line shows the results of the experiment in which Tamsin used large marble chips?

c After what time had both experiments produced the same volume of gas?

d Why did both experiments eventually stop producing gas?

e Why did the experiment shown by the blue line eventually produce twice as much gas as the experiment shown by the red line?

f Copy Figure 2h.9 and calculate the rate of each reaction at time 20 minutes.

4 Why does increasing the surface area of a solid reactant speed up a reaction?

5 The chemical formula of TNT is $C_7H_5N_3O_6$. Write a balanced symbol equation for the explosion of TNT. (Hint: use your answer to SAQ 8b.)

Thermograms – picturing temperature

The photo shows a rather unusual view of the London Eye. The photograph is an example of a **thermogram**, taken with a thermal imaging camera. This camera detects the **temperature** of the different parts of the scene. The colours tell you which parts are hottest (white, yellow) and which are coldest (purple, black).

Passengers travel around the wheel in the pods attached to the rim. These pods are heated to a higher temperature than their surroundings, so they appear yellow in the thermogram. The main structure of the wheel appears orange and purple because it is colder.

The emergency services use thermal-imaging cameras when they are searching for someone at night. A person is usually warmer than their surroundings, so it is possible to spot them even on a dark night.

Figure 1a.1 Thermogram of the London Eye.

SAQ

1 You can see a streetlight in the centre of the thermogram of the London Eye. What can you say about its temperature?

Medical thermography

Thermograms have uses in medicine. A part of the body which is inflamed is likely to be warmer than normal, so it will show up. In the photo in Figure 1a.2, areas of the patient's shoulders where arthritis has set in appear warmer than the rest of the body.

It might seem that this is a great, modern way of using technology. In fact, the ancient Greeks had their own way of finding 'hotspots' in their sick patients 2400 years ago. Here is what they did.

The patient was stripped naked and then their whole body was painted with runny mud. As the mud dried out, the doctor looked to see if the mud covering any part of the body dried more quickly. If so, this was likely to be the seat of the infection and the doctor could then recommend suitable treatment. Perhaps this seems like a crude technique but, like any form of medical diagnosis, it took skill to apply it. The doctor needed to know which parts of the

body were more likely to dry off quickly in a healthy patient, so that they were looking for *differences* in the rate of drying, compared to a healthy person.

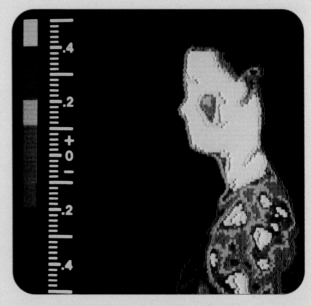

Figure 1a.2 Thermogram of a person suffering from arthritis.

Cooling down

The pods on the London Eye have to be continuously heated because **heat energy** escapes from them all the time. This is because the air around them is at a lower temperature, and energy flows from hotter places to colder places. It is the same with your home in winter. Energy is always escaping to the surroundings so, if you turned off the heating, the temperature would start to drop. Eventually your home would be as cold as the outside.

We human beings (and other mammals) are usually warmer than our surroundings and heat energy escapes from us because of the temperature difference. We lose energy faster in the winter because the temperature difference is greater. Our food is our energy supply; we generate heat energy through the process of respiration (reacting glucose with oxygen) and so we may need to eat more food in the winter than in the summer.

Heating up

If you want to heat something up, you have to supply it with energy. Figure 1a.3 shows a pan of water being heated on a stove. Energy must travel from the flame or hotplate through the metal base of the pan to the water. There is a temperature difference between the two sides of the metal and this is what makes the energy flow.

Now imagine putting a lump of ice in a warm drink. What happens? The ice warms up and melts. The drink cools down and becomes pleasantly cold. This happens because heat energy travels from the warm drink into the cold ice. Once the drink and the melted ice are at the same temperature, energy stops flowing.

Temperature and heat energy

It's easy to become confused about temperature and heat. Here's how you can think of them.

- The **temperature** of something tells you about its hotness (how hot it is). A thermometer measures the temperature of something by detecting its hotness. Temperature is measured in degrees Celsius ($°C$).

- If one object is at a higher temperature than another, **heat energy** will flow from the hotter to the cooler. Heat energy is the form of energy which flows when there is a temperature difference. Like any other form of energy, it is measured in joules (J).

Take care! There is no such thing as 'cold energy'. We can't talk about 'coldness' flowing from place to place. It's always heat that flows, from a hotter place to a colder place.

SAQ

2 Sketch a copy of Figure 1a.4. Add arrows to show how heat energy will flow.

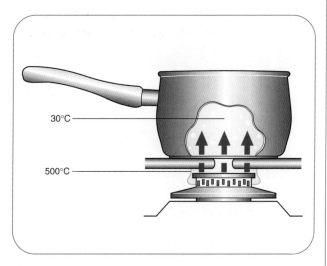

Figure 1a.3 Heating a pan on a stove.

Figure 1a.4

H Scales of measurement

0 °C is the 'zero' of the Celsius scale of temperature. It is the temperature of pure melting ice. But don't forget that there can be negative temperatures, too. If you take some ice from a domestic freezer, its temperature is likely to be about −20 °C.

This shows that you can still remove energy from something, even when its temperature is below 0 °C. In fact, you can go on removing energy until the temperature reaches −273 °C, the absolute zero of temperature.

SAQ

3 Test your understanding of these ideas.

a What would be the final temperature if you mixed two equal amounts of water, one at 20 °C and the other at 50 °C?

b What would be the final temperature if you mixed 2 kg of water at 20 °C with 1 kg at 50 °C?

Specific heat capacity

To make a hot drink, you need to bring some water to the boil. The cold water from the tap may be at a temperature of 20 °C; you have to provide enough energy to heat it to 100 °C. Its temperature must increase by 80 °C. The amount of energy you need to supply to boil the water depends on two things:

● the mass of the water

● the desired increase in temperature.

In order to calculate how much energy must be supplied to boil a certain mass of water, we need to know one other fact:

● it takes 4200 J to raise the temperature of 1 kg of water by 1 °C.

Another way to express this is to say:

● the **specific heat capacity (s.h.c.)** of water is 4200 J/kg °C (we say '4200 joules per kg per degree C').

The meaning of s.h.c.

Some materials are easier to heat up than others. We can compare different materials by considering a standard amount (1 kg), and a standard increase in temperature (1 °C). Different materials require different amounts of energy to raise the temperature of 1 kg by 1 °C. In other words, they have different specific heat capacities. Table 1a.1 shows the values of s.h.c. for a variety of materials.

From the table, you can see that there is quite a wide range of values. The s.h.c. of steel, for example, is one-tenth that of water. This means that, if you supplied equal amounts of energy to 1 kg of steel and 1 kg of water, the steel's temperature would rise ten times as much. A material with a high s.h.c. holds a lot of energy when it is at a high temperature.

The s.h.c. of water

Water is an unusual substance. As you can see from Table 1a.1, it has a high s.h.c. compared with

Type of material	Material	Specific heat capacity (J/kg °C)	Type of material	Material	Specific heat capacity (J/kg °C)
metals	steel	420	non-metals	glass	670
	aluminium	910		nylon	1700
	copper	385		polythene	2300
	gold	130		ice	2100
	lead	130			
liquids	water	4200	gases	air	1000
	seawater	3900		water vapour	2020 (at 100 °C)
	ethanol	2500		methane	2200
	olive oil	1970			

Table 1a.1 Specific heat capacities of a variety of materials.

most other materials. This has important consequences.

● It takes a lot of energy to heat water up.
● Hot water takes a long time to cool down.

The consequences of this can be seen in our climate. In the hot months of summer, the land warms up quickly (low s.h.c.) while the sea warms up only slowly. In the winter, the sea cools gradually while the land cools rapidly. People who live a long way from the sea (in the continental interior of North America or Eurasia, for example) experience freezing winters and roasting summers (see Figure 1a.5). People who live in coastal areas (such as western Europe) are protected from climatic extremes because the sea acts as a reservoir of heat in the winter, and stays relatively cool in the summer.

The high s.h.c. of water makes it useful in central heating systems. Water is heated in the boiler and then pumped around the radiators,

carrying heat energy with it. If we used a different liquid, it would need to be heated to a higher temperature or we would need to pump it around faster to carry the same amount of energy around the house.

SAQ

4 Look at Table 1a.1. Which of the materials listed has the greatest specific heat capacity? In general, do metals or non-metals have greater values of s.h.c.?

5 In an experiment, 5 kg blocks of two different metals are heated at the same rate. The temperature of metal A rises at twice the rate of that of metal B. Which metal has the greater specific heat capacity?

Calculations involving s.h.c.

Here is the formula we can use to work out the energy needed to raise the temperature of a mass of any material:

energy required = mass × specific heat capacity
× increase in temperature

The worked example shows how to use this formula in more detail.

Worked example

A domestic hot water tank contains 200 kg of water at 20 °C. How much energy must be supplied to heat this water to 70 °C? (Specific heat capacity of water = 4200 J/kg °C.)

Step 1: calculate the required increase in temperature.

Increase in temperature = 70 °C − 20 °C = 50 °C

Step 2: write down the other quantities needed to calculate the energy.

Mass of water = 200 kg

Specific heat capacity of water = 4200 J/kg °C

Figure 1a.5 In places far from the sea, summers are very hot and winters are very cold.

continued on next page

Worked example - *continued*

Step 3: write down the formula for energy required, substitute values and calculate the result.

Energy required = mass × specific heat capacity
$\qquad$ × increase in temperature

$$= 200\,kg \times 4200\,J/kg\,°C \times 50\,°C$$

$$= 42\,000\,000\,J$$

$$= 42\,MJ$$

So 42 MJ are required to heat the water to 70 °C.

SAQ

6 Do this in your head. The specific heat capacity of gold is 130 J/kg °C. How much energy is needed to raise the temperature of 2 kg of gold by 10 °C?

7 10.5 MJ of energy are supplied to 20 kg of water. By how much will its temperature rise? (Specific heat capacity of water = 4200 J/kg °C.)

8 When a dishwasher has finished washing, its contents are very hot. The china plates cool down very slowly, the metal cutlery more quickly. Use the idea of specific heat capacity to explain this. What other factors might affect the rate at which the different items cool down?

Latent heat

Energy and steel

Figure 1a.6 Handling liquid steel in a steelworks. A lot of energy must be supplied to the blast furnace to produce molten steel. Then the energy is released as the steel cools and solidifies. In an efficient modern works, some of this energy is recovered to heat the next batch of steel.

In a steelworks, people have to work with red-hot molten steel. The glowing colour of the steel in the photograph tells us that it is very hot, over 1400 °C. It takes energy to heat the steel up to such a high temperature and it takes more energy to melt it. Energy costs money, so the people who manage steelworks have to think carefully about how to avoid wasting energy.

The liquid steel is transported in metal flasks or beakers. Before it cools down and solidifies, it must be poured into moulds of the desired shape. Once the material reaches its freezing point, the temperature remains constant for a while. Energy is still being lost as the material solidifies. So, when molten steel is poured into a mould, it remains partly liquid until it has lost sufficient energy to become completely solid. (You may have noticed the same thing if you have ever filled an ice tray and put it in a freezer. The ice forms gradually, because it takes a while for energy to be removed from the water.)

The energy which comes out of a liquid such as steel when it freezes is the energy that was put into the solid material in the first place to make it liquid. In a well-designed steelworks, this energy is recycled to heat the next batch of steel. One way to do this is to blow air over the solidifying steel. The hot air which results can be used to heat up the walls of the blast furnace where more steel is being made. This is a process which cannot be 100% efficient – some energy escapes along the way – but saving a fraction of the energy in the molten steel can bring big savings to the steel manufacturer.

Boiling water

Imagine that you have put a pan of water on the stove to boil. Its temperature rises to 100 °C, at which point it is boiling. You forget the pan and it continues to boil – but its temperature does not rise. You are supplying energy to the water but this energy is needed simply to turn it from liquid to gas, without its temperature rising (see Figure 1a.7). Eventually, all of the water will have boiled away, taking the energy with it.

Similarly, if you melt ice, its temperature does not increase until all of the ice has melted.

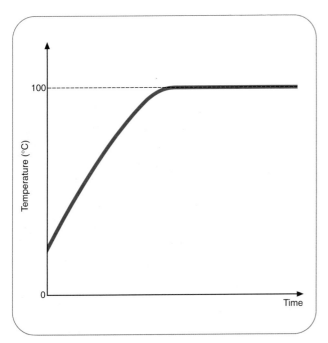

Figure 1a.7 When water is heated, its temperature rises until it reaches 100 °C. After that, its temperature remains constant until no water remains.

The energy which you put into water to turn it from hot water into steam, or to turn ice into water, is called **latent heat**. (The word 'latent' means 'hidden'; the heat energy is hidden because it has been put into the water or ice but it doesn't show up as an increase in temperature.) When steam condenses, or water freezes, that energy must come back out again.

Table 1a.2 shows that it takes a lot of energy to melt 1 kg of ice. It takes even more energy to boil 1 kg of water (that is, to turn all of it into steam). The amount of energy needed to melt or boil 1 kg of a substance is known as its **specific latent heat (s.l.h.)**.

The s.l.h. of ice is high compared with most other solids.

You may have used ice packs in a cool box to keep food and drinks cold. These packs contain a liquid with a high s.l.h. A lot of energy must be extracted from them when they are cooled in the freezer; it then takes a lot of energy to warm them up, so they (and the food) stay cool for longer. Athletes who have pulled a muscle are sometimes recommended to cool it using a pack of frozen peas. Because the peas are largely composed of water, they can absorb a lot of energy before they melt. They are convenient because the pack can be moulded around the person's body.

Energy required to melt 1 kg of ice	330 000 J
Energy required to boil 1 kg of water	2 260 000 J

Table 1a.2 Specific latent heat values for water.

SAQ

9 Which requires more energy, melting 1 kg of ice or boiling 1 kg of water?

Energy and particles

Energy is needed to raise the temperature of any material and to make it melt or boil. Where does this energy go?

When a material gets hotter, the energy supplied increases the **kinetic energy (KE)** of the particles of the material. In solids, they vibrate more. In gases, they move about faster. In liquids, it is a bit of both.

When a material melts or boils, the energy put in is needed to overcome the forces (break the bonds) between molecules. They don't move any faster (their KE doesn't increase) but they can move more freely.

It takes less energy to melt 1 kg of ice than to boil 1 kg of water (Table 1a.2). This is because, when ice melts, only one or two bonds are broken for each molecule of water; when water boils, all the remaining bonds are broken. (It is the weak **intermolecular bonds** *between* neighbouring water molecules which are broken, not the strong bonds that hold the individual molecules together.)

Latent heat calculations

If we know the s.l.h. of a substance, we can calculate the amount of energy needed to melt or boil any amount of that substance.

Energy required = mass × specific latent heat

Specific latent heat is measured in joules per kilogram (J/kg) or kilojoules per kilogram (kJ/kg).

Worked example

Figure 1a.8 Silver metal is often melted in order to shape it into jewellery.

A silversmith melts small quantities of silver. How much energy is needed to melt 10 g of silver at its melting point (960 °C)? The specific latent heat of silver is 100 kJ/kg.

Step 1: express the mass of silver in kg.

10 g = 0.01 kg

Step 2: calculate the energy required.

Energy required = mass × specific latent heat

= 0.01 kg × 100 kJ/kg

= 1 kJ

So 1 kJ of energy is required to melt the silver. (If we were given the specific heat capacity of silver, we could also work out the energy needed to heat it to its melting point.)

SAQ

10 Calculate the energy required to melt 100 kg of ice at 0 °C.

11 What mass of silver could be melted if 250 kJ of energy is supplied? (The specific latent heat of silver is 100 kJ/kg.)

12 Energy must be supplied (latent heat) to melt a metal such as gold. Use the idea of particles to explain where this energy goes. When liquid gold solidifies, what happens to the energy?

Summary

You should be able to:

◆ describe how heat energy flows from hotter places (higher temperature) to cooler places (lower temperature)

◆ use the terms *temperature* and *heat energy* correctly, and understand the difference between them

◆ state that the specific heat capacity of a substance is the amount of energy required to raise the temperature of 1 kg of the substance by 1 °C

◆ use the equation
energy required
= mass × specific heat capacity × increase in temperature

◆ state that the specific latent heat of a substance is the amount of energy required to melt or boil 1 kg of the substance

◆ use the equation
energy required
= mass × specific latent heat

Questions

1 What quantities are being described here?

   a A measure of the hotness of something.

   b The energy needed to melt or boil 1 kg of a substance.

   c The energy needed to raise the temperature of 1 kg of a substance by 1 °C.

2 A pan of water is boiling on the stove. Which of the following statements is/are true? For those which are untrue, provide a correct version.

   a The temperature of the water rises as it boils.

   b Energy must be supplied to keep the water boiling.

   c The water boils when it reaches the same temperature as the flame which is heating it.

3 Look at Table 1a.1 on page 259.

   a Which has the greater specific heat capacity, ice or water?

   b If you were to dissolve sodium chloride (common salt) in water, would this increase or decrease the specific heat capacity?

H 4 The specific heat capacity of steel in 420 J/kg °C. How much energy is needed to raise the temperature of a 20 kg steel bar from 20 °C to 500 °C?

5 A 100 g block of ice at 0 °C is added to 100 g of boiling water at 100 °C inside a well-insulated container.

   a The specific latent heat of ice is 330 000 J/kg. How much energy is needed to melt the block of ice in the container?

   b This energy is supplied by the hot water. By how much will the temperature of the water decrease as a result? (Specific heat capacity of water = 4200 J/kg °C.)

   c The melted ice and the water are now thoroughly mixed. What will their final temperature be?

Saving energy, money – and the world!

Figure 1b.1 This thermogram shows how a house loses energy to its surroundings. A lot of energy is escaping through the windows, but the loft is well-insulated.

In the UK, we have to heat our houses for much of the year because the outside temperature is too low for comfort. Figure 1b.1 shows energy escaping from a house, simply because it is warmer than its surroundings. There are three good reasons for trying to slow down the rate at which energy escapes.

- Energy costs money – it would be good to reduce fuel bills.
- Most houses use fossil fuels as their energy supply, and these will eventually run out. If we burn them now, they will not be available for future generations.
- Burning fossil fuels releases carbon dioxide to the atmosphere, and this is leading to global warming and climate change.

So there are selfish and unselfish reasons for making sure our homes are well insulated.

SAQ

1 Which of the reasons listed could be described as unselfish?

Home insulation

A well-insulated house can avoid a lot of energy wastage during cold weather. Insulation can also help to prevent the house from becoming uncomfortably hot during warm weather. Figure 1b.2 shows a house which has been designed to reduce the amount of fuel needed to keep it warm. The windows on the sunny side are large, so that the rooms benefit from direct radiation from the Sun. The windows on the other side are small, so that little energy escapes through them.

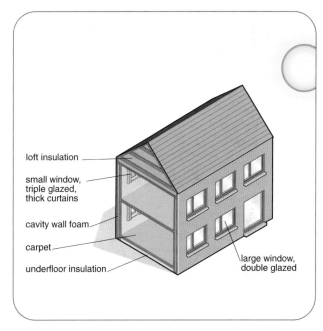

loft insulation

small window, triple glazed, thick curtains

cavity wall foam

carpet

underfloor insulation

large window, double glazed

Figure 1b.2 A well-designed, energy-efficient house.

Here are some ways of retaining energy in a house:
- fit thick curtains and draught excluders
- use loft **insulating materials** (fibreglass, mineral wool)
- fit carpets and underfloor insulating materials
- double glaze the windows
- make use of cavity walls
- fit reflective foil in the wall cavity
- install foam or rockwool in the wall cavity

Several of these make use of the fact that air is a very good insulator. For example, a **cavity wall** has two layers of bricks with an air gap between them. Energy escapes more slowly through a cavity wall than through a single layer of bricks, or even a double layer of bricks with no air gap.

The cavity may be filled with foam, a material whose volume is mostly holes filled with trapped air. Loft insulation works in the same way – it is a material which contains many tiny pockets of trapped air.

When new houses are built (or extensions added to older houses), the Building Regulations set out the amount of insulation that must be included. Buildings inspectors check that they have been correctly installed.

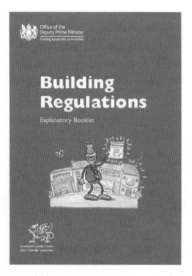

Figure 1b.3 Regulations set out the amount of insulation that must be included.

SAQ

2 It makes sense to build well-insulated houses. It is more expensive to fit insulation after the house has been built. Look at the list of ways of insulating a house (page 265).

  a Which methods of insulation could be added easily to an older house?

  b Which methods are best fitted when the house is first built?

Saving money

It costs money to insulate your home well, but is it worth it? That depends on two things:
- how much the insulation costs
- how much money you save.

It costs a lot to double glaze a window, and this doesn't save a lot of energy. Here are some figures.

A large double-glazed window costs £200 to install. In a year, it saves £10 on heating bills. From this we can work out the **payback time** – that is, how many years it will take to get back the cost of the window.

$$\frac{\text{payback}}{\text{time}} = \frac{\text{initial cost}}{\text{annual saving}} = \frac{£200}{£10 \text{ per year}} = 20 \text{ years}$$

Asbestos worries

If you ever pass through the Australian town of Wittenoom, beware! There are warning signs at each entrance to the town, advising you to close your car windows and drive through without stopping. The problem is that asbestos was mined at Wittenoom from 1937 to 1966, and there is still asbestos dust blowing around.

Asbestos is a fibrous mineral used for heat insulation. It is quite harmless for as long as it remains in a solid piece. However, processing the mineral releases large amounts of dust which can prove fatal to those who breathe it in. By the year 2000, more than 10% of the population of Wittenoom had died of lung disease. At one time, this was the largest town in the north of Western Australia – Rolf Harris worked there as a young man. Today, just 30 residents remain, reluctant to accept that they are putting themselves in danger.

Figure 1b.4 The Australian asbestos-mining town of Wittenoom has been largely abandoned.

continued on next page

Asbestos worries - *continued*

It took a long time to convince the world that asbestos was hazardous. The big mining companies denied that their product was responsible for illness amongst their employees (although lung disease was common among asbestos miners in South Africa in the early 1900s). Today, after many thousands of deaths, asbestos is treated with great caution. If it is found as part of the insulation of old buildings, it must be removed in such a way that there is no hazard to the public or to the people working with it.

Figure 1b.5 Safety clothing must be worn when removing old asbestos insulation.

This is a long time to wait, but most types of insulation pay for themselves much more quickly than this.

SAQ

3 It costs £120 to fit loft insulation in a terraced house and this cuts fuel bills by £40 per year. Calculate the payback time.

Energy efficiency

We want to waste as little energy as possible, to save money and to conserve resources and the environment. So it makes sense to buy appliances which use energy efficiently. In the showroom,

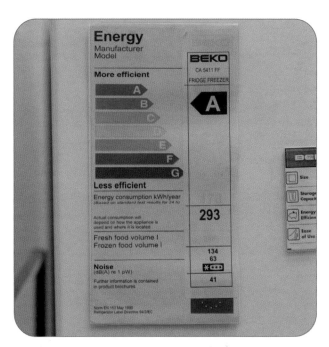

Figure 1b.6 An efficient fridge can save up to £100 each year in running costs.

washing machines, fridges etc. carry labels indicating how efficient they are in their use of energy (see Figure 1b.6).

Several different types of energy-efficient light bulb are available (see Figure 1b.7). These give more light for every joule of electrical energy they are supplied with than the less-efficient filament lamps.

Figure 1b.7 Energy-efficient lamps, on the left, will save you money in the long run, even though they are more expensive to buy.

Useful energy

To decide how efficient a light bulb is, we have to think about how much useful energy we get out of it. We want a light bulb to produce light energy but they also produce heat energy, which is not what we want. So the light energy is the useful energy output of the light bulb. We can show this using a Sankey diagram (Figure 1b.8).

We can use the information shown to calculate the **efficiency** of each type of light bulb:

$$\text{efficiency} = \frac{\text{useful energy output}}{\text{total energy input}}$$

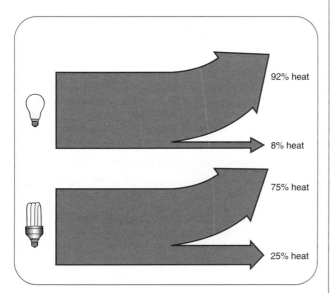

Figure 1b.8

For the filament lamp:

$$\text{efficiency} = \frac{8\,\text{J}}{100\,\text{J}} = 0.08 = 8\%$$

So 92% of the energy supplied to the lamp is wasted as heat energy.

SAQ

4 Calculate the efficiency of the energy-saving light bulb.

Summary

You should be able to:

◆ describe how insulation reduces the rate at which energy escapes from houses

◆ explain why insulating materials often include air

◆ solve problems using

$$\text{payback time} = \frac{\text{initial cost}}{\text{annual saving}}$$

◆ solve problems using

$$\text{energy efficiency} = \frac{\text{useful energy output}}{\text{total energy input}}$$

Questions

1 A builder recommends fitting cavity wall foam in an old house. The house owner wants to know what the payback time will be.

   a Explain what is meant by *payback time*.

   The builder explains that the cost will be £1200 and it will save £80 each year in fuel costs.

   b Calculate the payback time.

2 A boiler is used to heat water in a central heating system. For every 1000 J of energy supplied to the boiler, 300 J are wasted as heat, which escapes.

   a How much useful energy does the boiler supply?

   b What is its efficiency?

H 3 A new hot water boiler is advertised as having an efficiency of 80%. The burning fuel releases 2000 J of energy every second. How much heat energy is supplied to the house every second?

Retaining energy

Figure 1c.1 House at Tiengen, Germany.

It is hard to imagine that a house with transparent walls could be good at saving energy. The house in the photo is at Tiengen in Germany. Its walls are of glass with a transparent **insulating material** behind. Light passes through this and is absorbed by a dark, solid material behind, which gets warm. Heat energy finds it very difficult to escape outwards through the insulation.

Method of insulation	Why it works
thick curtains, draught excluders	prevents cold air from entering and warm air from leaving
loft insulating materials (fibreglass, mineral wool)	reduces conduction of heat through ceilings
carpets and underfloor insulating materials	reduces conduction of heat through floors
double glazing of windows	vacuum between glass panes cuts out losses by stopping conduction and convection
cavity walls	reduces heat losses by conduction
reflective foil in wall cavity	reduces heat losses by radiation
foam or rockwool in wall cavity	reduces heat losses by convection

Table 1c.1 Ways of retaining energy in a house.

To understand how different types of insulation work, we need to think about the three ways in which heat energy travels: **conduction**, **convection** and **radiation**. Table 1c.1 shows how these are affected by different methods of insulating a house.

Reducing conduction

First, let's look at conduction. Heat can conduct through the brick and glass of walls and windows. These materials are quite good insulators (you should recall that metals are the best conductors of heat, as well as electricity), but a lot of energy still escapes through walls and windows. Cavity walls provide better insulation than single brick walls because they are thicker, and there is a layer of air to reduce conduction.

Heat energy cannot conduct through a vacuum (empty space), and that is how double-glazing works. Two panes of glass have a vacuum between them so that no heat is lost by conduction.

SAQ

1 Some old windows have metal frames. Why is it a good idea to replace them? What materials should the new frames be made of?

Reducing convection

Now let's look at convection. Heat energy can be carried away by moving air. Hot air rises and is replaced by falling colder air – this is a convection current. We need some convection currents in a house – they are one way for warm air to spread out from heaters, radiators and fires. However, we don't want warm air to escape completely, so it is a good idea to fit draughtproofing around doors and windows.

Heat can also escape through a wall cavity by convection. The inner wall is warm; the outer wall is colder. The convection current transfers energy from inside to outside and so the house cools down. This is why it is useful to fit cavity foam. The foam has pockets of trapped air; this air cannot flow around, so there are no convection currents.

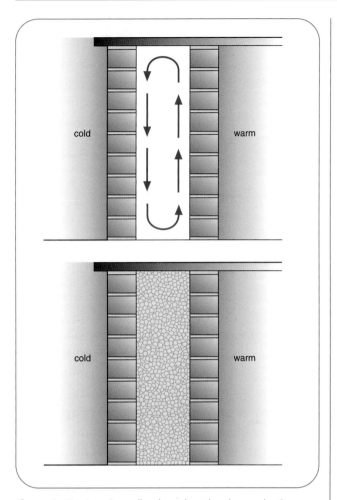

Figure 1c.2 a A cavity wall reduces heat loss by conduction, because the air in the cavity is a good insulator. However, a convection current can transfer energy from the inner wall to the outer. **b** Filling the cavity with foam or mineral (glass or rock) wool prevents convection currents from forming.

SAQ

2 When cavity wall foam is fitted to a house, the amount of energy lost by conduction increases. Explain why.

Reducing radiation

Lastly, let's consider radiation. Warm objects lose energy to their surroundings by **infra-red radiation**. If you sit in the sunshine, you will feel the warmth of this radiation on your skin. The thermogram of a radiator (Figure 1c.3) shows that radiators are warm objects, and so they give out infra-red radiation.

Figure 1c.3 This thermogram shows the effect of drying your underwear on a radiator.

Radiation in space

We experience infra-red radiation from the Sun – it is the heat which we feel on our skin on a sunny day. We are protected from the full strength of the Sun's radiation by the atmosphere. However, it is different for a spacecraft above the atmosphere.

In 2005, the European Space Agency sent its *Venus Express* spacecraft to visit Venus; because this planet is closer to the Sun, the intensity of infra-red radiation is twice what we experience here on Earth. That can be a problem, because the spacecraft absorbs infra-red radiation and there is a danger of overheating so that its instruments would malfunction.

Figure 1c.4 An artist's impression of *Venus Express*.

continued on next page

Radiation in space - *continued*

What can the engineers do about this? Firstly, the spacecraft is covered with a reflective coating, to reflect as much radiation as possible back into space. Secondly, there is a 23-layer covering beneath this coating, to limit conduction of heat as much as possible. Finally, the solar panels are used in a clever way. The solar panels face the Sun at all times (see Figure 1c.4), to generate as much electricity as possible. This means that the back of each panel faces away from the Sun; it has a black coating, which radiates heat out into space.

Because radiators are usually next to a wall, half of their radiation warms the wall, not the room. This is why reflective aluminium foil is sometimes fitted behind the radiator, to reflect the radiation back into the room.

Reflective foil is also fitted under the roof tiles when a new roof is built, to reflect radiation back into the loft.

SAQ

3 Look at the thermogram of the radiator. Explain how it shows that it is not a good idea to dry your washing in this way, if you want the room to be warm. Use the words *insulation* and *radiation* in your answer.

Explaining conduction, convection and radiation

best conductor	diamond	worst insulator
	silver, copper	
	aluminium	
	steel	
	lead	
	ice, marble, glass	
	polythene, nylon	
	rubber, wood	
	polystyrene	
worst conductor	glass wool, air	best insulator

Table1c.2 Comparing conductors of heat, from the best conductors to the worst.

A bad conductor is a good insulator. Table 1c.2 compares some solid materials as conductors and insulators. Almost all good conductors are metals; polymers (plastics) are towards the bottom of the list. Glass wool is an excellent insulator because it is mostly air.

Recall that it is a temperature difference which makes heat flow. What makes one material a better conductor of heat than another? To answer this, we have to think about the way in which heat energy conducts through a solid.

Explaining conduction in non-metals

Both metals and non-metals conduct heat. Metals are generally much better conductors than non-metals, and we need different explanations of conduction for these two types of material. We will start with non-metals.

Picture a long rod (Figure 1c.5a). One end is being heated, the other is cold. There is thus a temperature difference between the two ends, and heat flows down the rod. What is going on inside the rod?

We will picture the atoms that make up the rod as shown in Figure 1c.5b. (They are shown as being identical, and regularly arranged, although they may not really be like this.) At the hot end of the rod, the atoms are vibrating a lot; at the cold end, they are vibrating much less. As they vibrate, the atoms jostle their neighbours. This process results in each atom sharing its energy with its neighbouring atoms. Atoms with a lot of kinetic energy (KE) end up with less, those with a little KE end up with more. The jostling gradually transfers energy from the atoms at the hot end to those at the cold end. Energy is steadily

H transferred down the rod, from hot to cold. This is the mechanism by which poor conductors (such as glass, ice and plastic) conduct heat.

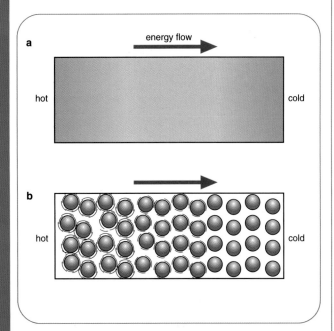

Figure 1c.5 Conduction of heat in non-metals. **a** A rod that is heated at one end and is cold at the other. Heat travels from the hot end to the cold end. **b** Energy is transferred because the vibrating atoms jostle one another. This shares energy between neighbouring atoms; the result is a flow of energy from the hot end to the cold end.

Explaining conduction in metals

Metals are good conductors of heat. If we look at the arrangement of atoms in a metal, we see the important difference between metals and non-metals (Figure 1c.6a). In a metal, there are many tiny particles called **electrons** between the atoms. These are sometimes known as **free electrons**, because they have escaped from the atoms of the metal and are free to move about inside the metal. As they move, they collide with one another, and with the vibrating atoms of the metal. Each collision results in a sharing of energy.

Now picture a metal rod which is hot at one end and cold at the other (Figure 1c.6b). The atoms at the hot end are vibrating more than those at the cold end, because they have more energy. The free electrons collide with these atoms and gain KE – they move slightly faster. When they collide with atoms at the cold end, they give energy to the atoms, which start to

H vibrate more – they have more KE. The overall result is that energy is transferred by the electrons from the hot end of the metal to the cold end.

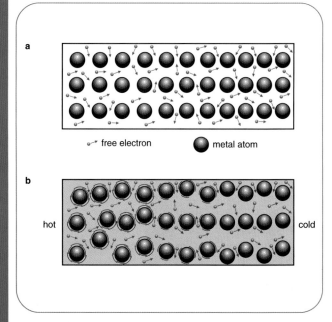

Figure 1c.6 Conduction of heat in metals. **a** Inside a metal, there are vibrating atoms and free electrons. The free electrons move randomly through the metal. **b** Heat is transferred through the metal by the free electrons. They collide with vibrating atoms at the hot end, gaining energy. They lose energy when they collide with the atoms at the cold end.

SAQ

4 Which non-metal is the best thermal conductor of all (see Table 1c.2)? Find out why this material is such a good conductor.

Explaining convection

'Hot air rises.' This is a popular saying, one of the few ideas from Physics that almost everyone who has studied a little science can remember.

When air is heated, its density decreases (its molecules move farther apart so that it expands). Because it is less dense than its surroundings, it then floats upwards, just as a cork floats upwards if you hold it under water and then release it.

When hot air rises, cold, denser air flows in to replace it. This is how a convection current is set up. Convection currents help to share energy between warm and cold places. If you are sitting in

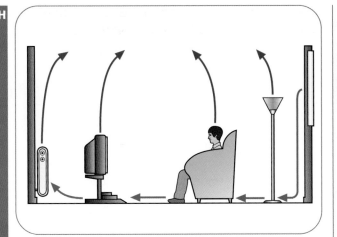

Figure 1c.7 Convection currents rise above the warm objects in a room. Cold air flows down from the window to replace it.

a room with an electric heater, energy will be moving around the room from the heater as a result of convection currents rising from the heater. You are likely to be the source of convection currents yourself, because your body is usually warmer than your surroundings (see Figure 1c.7).

Convection is a phenomenon which can be observed in any fluid (liquid or gas).

SAQ

5 Figure 1c.8 shows a domestic refrigerator. The freezing compartment is at the top. Explain how, with the door closed, a convection current is set up in the fridge, and how this ensures that the foods in the bottom compartment are kept cold.

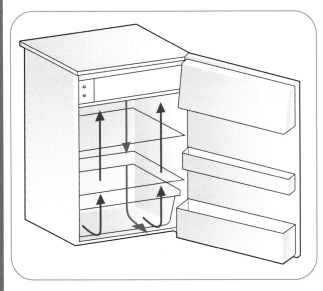

Figure 1c.8

Explaining radiation

All objects give out infra-red radiation. This is rather like light radiation but is invisible to our eyes. It can travel through empty space – we know this because it can reach us from the Sun through the vacuum of space.

On a hot, sunny day, car drivers may park their cars with a sunshield behind the windscreen (Figure 1c.9). Such a sunscreen is usually white or silver because this reflects away light and infra-red radiation which would make the car get uncomfortably hot. The black plastic parts of the car (such as the dashboard) are very good absorbers of infra-red, and they can become too hot to touch.

It is the surface which determines whether an object absorbs or reflects infra-red radiation:
● a surface which is a good reflector is a poor absorber
● a good absorber is also a good emitter
● shiny or white surfaces are the best reflectors (the worst absorbers)
● matt black surfaces are the best absorbers (the worst reflectors).

The radiators which are part of a central heating system are often painted white. This might seem to be a bad idea, because white surfaces are poor emitters of heat. However, special paint or enamel is used on such radiators; although it is a poor emitter of visible light, it is a good emitter of infra-red radiation.

Figure 1c.9 A sunscreen reflects away unwanted radiation which would otherwise make the car unbearably hot.

H *SAQ*

6 A modern double-glazed window has a vacuum between two panes of glass. One inner surface is coated with a thin film of a material which allows light to pass through but which reflects infra-red radiation. Draw a diagram to show how this window should be fitted to a house, and explain how its design reduces heat loss by conduction, convection and radiation.

7 Hot water tanks should be well insulated. It is usual to heat the water in the morning and the evening, when most hot water is likely to be needed. Some people argue that it is best to keep the water hot throughout the day, because allowing the water to cool is a waste of energy and more energy will be needed to warm it up again in the evening. Use what you know about how heat energy travels to evaluate this idea.

Summary

You should be able to:

◆ describe how energy escapes from houses by conduction, convection and radiation

◆ describe how convection currents carry heat energy around the house

◆ describe how a house can lose heat energy by infra-red radiation

H ◆ explain how heat energy conducts through metals and non-metals

◆ explain the process of convection, including the formation of convection currents

◆ describe how heat energy can be carried away by infra-red radiation, and relate this to the nature of the radiating or absorbing surface

Questions

1 In cold climates, it is important to keep a house well insulated. Listed below are three ways of insulating a house. For each, explain how it reduces heat loss. In your answers, refer to *conduction*, *convection* or *radiation*, as appropriate.

  a Heavy curtains, when closed, trap air next to a window.

  b Shiny metal foil is fitted in the loft, covering the inside of the roof.

  c Glass wool is used to fill the gap in the cavity walls.

2 When new houses are built, their brick walls have a cavity which is filled with insulating foam. One surface of the foam is covered with reflective foil. Explain how this design reduces heat losses through the walls:

  a by conduction   b by convection   c by radiation.

 3 Which methods of energy transfer are being described here?

  a A change of density causes a fluid to flow.

  b Kinetic energy is transferred from a vibrating particle to its neighbours.

  c Energy is carried by infra-red waves.

Ready, steady, cook with radiation

Figure 1d.1 How many electrical appliances can you see in this kitchen?

If you visit a department store or electrical showroom, you will discover many different and unusual electrical devices for the kitchen: electric egg boilers, sandwich toasters, vegetable steamers, chip fryers. Most people can manage without these. However, most kitchens do have a conventional oven and a microwave oven. In this chapter, we will look at the science of how these cook food.

In the oven

If you are about to cook something in a conventional oven – a baked potato, for example – it is usual to turn the oven on in advance so that it will heat up. The interior of the oven is heated by electric elements or by gas flames. Within a few minutes, the inside of the oven is at the right temperature. You have to be careful when putting the potato on the shelf, because the inside of the oven is very hot.

You can feel the radiation which is being **emitted** by the inner surface of the oven. This is infra-red radiation, and the walls of the oven are a dull black colour because this is the best emitter of infra-red. (Remember: a good absorber

is also a good emitter.) The hotter the oven, the more radiation it emits.

How does the potato get cooked? Some of the infra-red radiation which strikes the potato is **absorbed** by the surface of the potato. This makes the skin of the potato very hot, and heat then conducts into the interior of the potato and cooks it. It may take an hour for enough energy to penetrate to the centre of the potato – if you take it out too soon, you will find that the centre is still hard and uncooked.

Some people wrap their potatoes in foil, to stop them shrinking. This slows things down, because shiny surfaces reflect infra-red radiation. It would be better to use a black material, because black surfaces absorb infra-red radiation well.

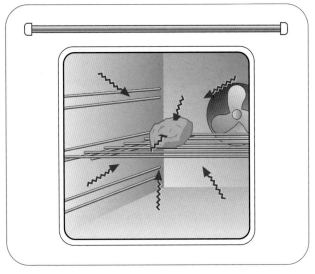

Figure 1d.2 Infra-red radiation is absorbed by the particles on the surface of the potato; this increases their kinetic energy.

SAQ

1 Describe a surface which is:
   a a good absorber of infra-red radiation
   b a good reflector of infra-red radiation.

H 2 As a potato is baked, energy is transferred to its centre.
   a Explain why convection plays no part in this energy transfer.
   b Suggest a food in which energy is transferred by both conduction and convection as it cooks in an oven.

Microwave oven

Microwave ovens are different from conventional ovens. You don't have to pre-heat them, and the inner walls don't get hot. They work in a different way, using a different (but closely related) type of radiation called **microwaves**. Both infra-red radiation and microwaves are examples of **electromagnetic radiation**; we say that they belong to the **electromagnetic spectrum**.

Figure 1d.3 A microwave oven.

If you look inside a microwave oven, you may be able to see the rectangular hole, a few centimetres across, through which the microwaves enter the oven. The way in which a potato cooks is rather different from what happens in a conventional oven.

Microwaves are reflected by metal surfaces. This means that they bounce around inside the oven until they hit the potato. They can penetrate a centimetre or so into the potato before they are absorbed by water in the potato. Water is a good absorber of microwaves and potatoes contain a lot of water. This heats the potato and the potato is cooked.

Some people say that a microwave oven 'cooks food from the inside out'. This isn't strictly true; if the food is very thick, the microwaves may be absorbed before they can penetrate right to the middle and so the middle may not be properly cooked when the outside is ready to eat.

Microwaves can pass straight through some materials, including glass, china and plastic. That is why a dish can remain cool in the microwave oven even though the food on it is piping hot.

SAQ

3 For each of the materials in the list, say whether microwaves are absorbed, reflected or pass straight through:
plastic   water   aluminium   glass

H Comparing infra-red and microwave radiation

Remember that infra-red radiation is absorbed by the surface of the potato. The energy of the radiation is absorbed by the particles of the potato and so they move faster – their kinetic energy (KE) has been increased. Then the energy conducts into the inside of the potato.

Microwaves are different; they can penetrate a centimetre or so into the potato, so they are more effective at getting into the potato than infra-red radiation. The energy of the microwaves heats the outer layers of the potato by increasing the KE of the water molecules, and energy conducts to the middle.

Infra-red and microwaves both travel as a type of vibration or wave, similar to light waves. The difference is that they have different **frequencies** – that is, different numbers of vibrations per second. Infra-red waves have higher frequencies than microwaves, and this means that they are more energetic and they can do more damage when absorbed by living tissue.

SAQ

4 Explain why it takes less energy to cook a potato in a microwave oven than in a conventional oven. (Think about where energy is wasted in a conventional oven.)

5 Explain why a microwave oven must be switched on for twice as long if you want to cook two potatoes.

Cooking your brain?

Mobile phones use microwaves to transmit their messages. When you use a mobile phone, the aerial is right next to your brain, and your brain contains a lot of water. So are you in danger of cooking your brain when you make a long phone call?

Figure 1d.4 The microwave radiation inside a microwave oven is about 100 000 times as intense as that from a mobile phone.

Figure 1d.5 This children's playground in southern Spain is dominated by nearby mobile phone aerials. The radiation they give out is low intensity – but is it safe?

No one has yet put a thermometer inside a phone user's head, but computer simulations have shown that the area of your brain close to the phone might be heated by around 0.1 °C. This is a small effect and we can't be sure whether it is harmful – it might even be good for you. Little is known about possible long-term effects.

The UK government has been concerned about this for some time and an official report has suggested that, although it is unlikely that mobile phones are causing much harm, it is inadvisable for young people (up to the age of 8) to make frequent calls because they are at an age when their brains are developing rapidly.

It is also important to think about the radiation which comes from base station aerials. These are the masts which communicate back and forth with nearby phones. They are constantly sending out microwave radiation, so people who live or work nearby may be exposed to this radiation for long periods of time. Is this a hazard? The masts are usually positioned on high buildings and they are designed so that their radiation spreads outwards and downwards. The more it spreads out, the weaker it is. Provided you are a metre or two away, the radiation is at what is thought to be a safe level.

In some places, it can be difficult to get a good signal on a phone. Think about the direct line from the phone's aerial to the mast – is it 'in line of sight'? Microwaves can pass through walls and windows but, if there is a big obstacle, such as a hill, they will be blocked.

SAQ

6   Explain why a mobile phone user is likely to lose a signal when their train goes into a tunnel.

7   Why does the radiation from a mobile phone mast get weaker as you move away from it?

8   Why is text messaging less likely to be harmful than making a normal call on a mobile phone?

Diffraction and interference effects

Mobile phone systems have to be designed with care. It would be very annoying for users if they found that signal reception was patchy, so that they had to keep moving to a point where the signal was strong. So engineers have to understand the factors which might reduce the strength of a signal.

There are two things which cause problems.

Diffraction is the spreading out of waves as they pass the edge of an object, or through the gap between two objects.

Interference happens when there are two signals with the same frequency (or similar frequencies), so that a phone cannot separate out the signal which was intended for it.

To overcome these problems, mobile phone companies set up a large number of masts fairly close together. This means that users will usually be close to a mast and the signal they receive will not have been affected by diffraction and interference effects. The masts are high up so that they broadcast downwards to users, avoiding obstacles. Neighbouring masts transmit at different frequencies, so that they do not interfere with each other.

(There is more about diffraction and interference effects in Item P1f *A wireless world*.)

Summary

You should be able to:

◆ state that infra-red radiation and microwaves are part of the electromagnetic spectrum

◆ explain how a conventional oven uses infra-red radiation to transfer energy to food

◆ explain how a microwave oven uses microwaves which penetrate food and which are absorbed by water in the food

◆ describe how the absorption of infra-red and microwave radiation changes the kinetic energy of the particles of food, and how this energy is then transferred through the food by conduction and convection

◆ describe how mobile phones use microwave radiation to transmit and receive messages

◆ discuss concerns about possible radiation hazards to mobile phone users and those living near phone masts

◆ describe the effects of diffraction and interference on microwave signals, and how these can be reduced

Questions

1 a Name the radiation that cooks food in a conventional oven.

 b To what family of radiations does this radiation belong?

 c What substance in most foods is good at absorbing microwaves?

2 A microwave oven has a safety catch on the door. When the door is opened, the microwaves are automatically switched off. Why is this an important safety feature?

3 Give a step-by-step explanation of the way in which a microwave oven cooks an item of food. In your explanation, say how energy is transferred to the particles on the surface of the food and in its interior.

Infra-red signals

When you press a button on the TV remote control, you are sending an infra-red signal to the TV set. You can't see the signal but it consists of a series of pulses of infra-red radiation. The pattern of on–off pulses tells the set which channel you want to switch to.

Some car owners have a similar device for opening the garage doors. Their electronic 'key' sends a signal to the door mechanism; only the correctly coded signal opens the doors.

Human beings send out infra-red 'signals', too. Because our bodies are warm, they give out heat radiation. This is what is detected by some burglar alarms and security lights, and by thermal-imaging cameras.

An **infra-red sensor** is used in each of these examples, to detect the infra-red radiation.

Figure 1e.1 This computer has a wireless keyboard and mouse. These transfer data to the computer using an infra-red signal, but they only work over a short distance.

SAQ

1   List as many examples as you can of uses of infra-red radiation in the home.

2   How could you show that the infra-red signal from a TV remote control is reflected when it strikes a glass windowpane?

TV remote control

Take a look at a TV remote control. At one end (the end you point towards the TV set) is a small 'bulb', an infra-red emitting diode. This is what produces the infra-red signal detected by the TV. On the front of the TV is a small panel, behind which is an infra-red sensor.

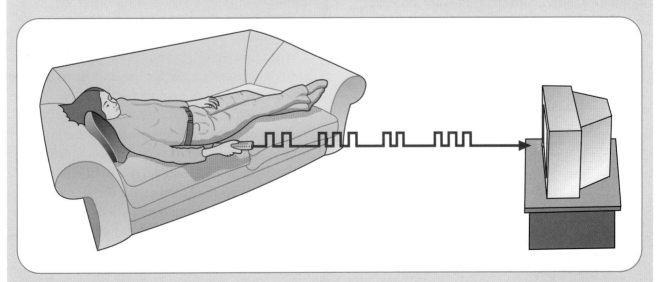

Figure 1e.2 Using a TV remote control.

continued on next page

TV remote control - *continued*

Address code		Command code	
0, 1	TV sets	0–9	Channels
5, 6	VCRs	12	Standby
17	Tuner	16, 17	Volume +, –
18	Tape	18, 19	Brightness +, –
20	CD player	55	Record

Table 1e.1 Codes for a TV remote control.

The signal is a series of on–off bursts or pulses of infra-red radiation, similar to Morse code. The signal obviously depends on which button you press, and it is decoded by an electronic circuit in the TV. The signal is in two parts.

First, the address code. This indicates whether the signal is intended for the TV set, DVD player, CD player or some other device.

Second, the command code. Numbers 0 to 9 are for the channels you are selecting. Higher numbers are for volume, brightness and other controls.

SAQ

3 Why is it necessary to include an address code in the remote control's signal?

Analogue and digital

Today's telephones use digital technology. Whether the signal is carried by microwaves through the air, by electricity along a wire or by light along an optical fibre, it consists of a series of on–off pulses, similar to the signal from a TV remote control. How does this work?

When you speak into a phone, you are producing sounds. We can represent these sounds as waves which vary up and down; the louder you speak, the more they vary. The microphone in the telephone handset converts these into matching electrical waves.

Both of these are examples of **analogue signals**. The signal has a small value when you are talking quietly and a large value when you are loud. It can have any value in between.

Inside the phone, a processor turns this into a **digital signal**, which is a series of electrical pulses. All of the pulses have the same voltage, either on or off. These are often represented by the digits 0 (off) and 1 (on).

Analogue versus digital

Digital signals have several advantages over analogue ones. An important advantage is to do with **noise**.

When any signal is transmitted – along a cable or fibre, or through the air – it is likely to become degraded. Instead of being a perfect replica of the original signal, it will have acquired some noise. It is also likely to have lost power, so that its amplitude is smaller. It needs to be **amplified**.

- When an analogue signal is amplified, the noise is amplified along with the signal.
- When a digital signal is amplified, the amplifier removes the noise and only amplifies the signal.

This is possible because the amplifier knows that the signal has a value of either 0 or 1 (on or off) – it can't be anything in between. With an analogue signal, there is no way of knowing what is the signal and what is the noise.

Figure 1e.3 A mobile phone converts analogue signals to digital.

Digital radio and TV broadcasts are usually clearer than analogue ones because the receiving set can clean up the signal, removing any noise from other stations which might interfere with the station you want to hear or see.

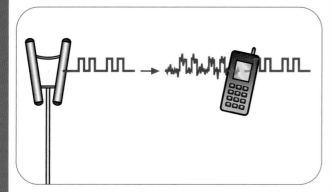

Figure 1e.4 Removing 'noise' from a transmitted signal.

Optical fibres

The telephone system is used for many purposes. It forms a network connecting billions of people together so that they can speak to one another. At the same time, a vast amount of computer data is travelling around the same system – that's the Internet.

Much of the system is connected together using **optical fibres**. These are very fine glass threads, thinner than a human hair. At one end of a fibre, a tiny laser sends out pulses of infra-red radiation, many millions every second. The radiation travels along the fibre, reflecting from side to side as it goes. At the other end, a small detector senses the radiation and converts it into an electrical signal.

Light travels very fast – in a long enough optical fibre, it could go several times around the world in a second. This means that optical fibres can transmit masses of digital data over long distances almost instantaneously.

Figure 1e.5 shows how the radiation follows the shape of a curved fibre. Each time it strikes the side of the fibre, it reflects off it (just as a ray of light reflects off a mirror).

SAQ

4 What can you say about the two angles shown in Figure 1e.5?

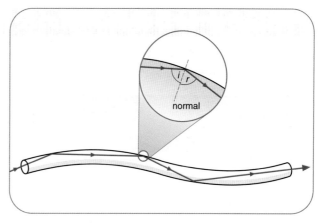

Figure 1e.5 A ray of light or infra-red radiation can travel along a curved optical fibre. The fibre must be made of very high purity glass, so that the ray is not absorbed as it goes along. '*i*' is the angle of incidence and '*r*' is the angle of reflection.

Total internal reflection

How does light reflect inside an optical fibre? It is as if the surface was a perfect mirror, keeping 100% of the light inside the fibre each time it is reflected.

The secret is in the angle at which the ray strikes the inner surface of the glass fibre. It only works if the ray strikes the surface at a sufficiently big angle of incidence (recall that this angle is measured between the ray and the normal). The angle must be greater than the **critical angle**, otherwise some of the light will be refracted out of the fibre (Figure 1e.6).

Because the ray is *totally reflected*, and it happens *inside* the glass, we call this process **total internal reflection (TIR)**. The material involved doesn't have to be glass; it also happens when a ray is travelling out of water, clear plastic or any other transparent material into air.

Doctors make use of optical fibres to see inside their patients. Figure 1e.7 shows an **endoscope** in

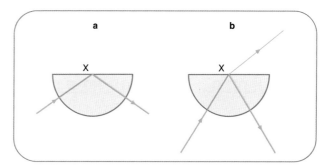

Figure 1e.6 a Angle of incidence greater than critical angle: 100% of light reflected inside glass. **b** Angle of incidence less than critical angle: some light lost by refraction out of the glass.

use. You can see where the long, flexible end of the endoscope enters the patient's mouth. Images of the patient's insides are then shown on a screen for all to see.

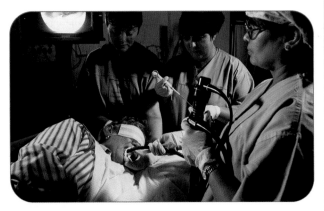

Figure 1e.7 This medical team is using an endoscope to look at a patient's stomach.

ⓗ *SAQ*

5 Figure 1e.8 shows what happens when a ray of light strikes the inner surface of a glass block (point X in the diagrams). Copy the diagrams and add the normal at point X on each. Indicate an angle which is equal to the critical angle. For each diagram, write a sentence explaining what happens to the ray of light when it strikes point X.

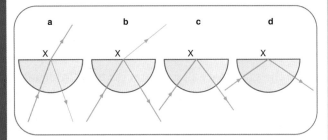

Figure 1e.8 Total internal reflection.

ⓗ Message after message

There is another advantage of digital signals rather than analogue. Think about sending a message by Morse code. It takes several seconds to send the code for a long word, much longer than it takes to say it. Digital signals in optical fibres take much less time to send a message than it takes to speak it. Here's why.

You are speaking into the phone. The sound waves of your voice have to be coded into digital on–off pulses. It takes *thousands* of pulses to code one second's worth of speech. However, a laser can send *millions* or even *billions* of pulses down an optical fibre every second. So a fibre can carry millions of calls at the same time. To do this, your message is divided up into a series of batches of pulses. The laser flashes the first batch of your message, then a batch of pulses from someone else's, then a batch from a third message. Eventually it gets back to your message and sends the next batch. At the other end of the fibre is processor which separates out the batches and recombines them into the correct sequence – you hope!

The interleaving of many digital signals as they are transmitted down a single data line is called **multiplexing**. This is what allows you to use a single domestic phone line for both phone calls and broadband internet access at the same time.

Another advantage of optical fibre communications is that it is very unlikely that one signal will interfere with another. When your phone call travels along a fibre, it is impossible for any other signal to get into the cable from the outside, and it is impossible for your signal to leak out and interfere with anybody else's signal.

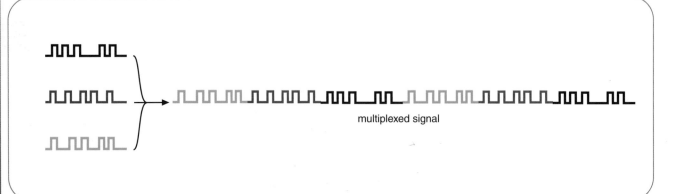

multiplexed signal

Figure 1e.9 Multiplexing.

SAQ _____

6 In the world of espionage, spies are always trying to intercept telephone signals. It is more difficult to intercept a signal which is being

transmitted along an optical fibre than one which is transmitted through the air or along an electrical cable. Explain why this is.

Summary

You should be able to:

◆ give examples of ways in which infra-red radiation is used for sending signals over short distances

◆ describe the differences between analogue signals (which have a continuously variable value) and digital signals (which are either on or off, 1 or 0)

◆ describe how total internal reflection (TIR) allows light and infra-red radiation to travel down optical fibres, carrying digital signals

◆ state that, for a ray of light to undergo TIR, it must meet an internal boundary (e.g. glass-to-air) at an angle of incidence greater than the critical angle

◆ describe the advantages of using digital signals, including recovery from noise, multiplexing and lack of interference from other signals

Questions

1 a Which type of signal (analogue or digital) can have any value over a range of values?

   b Write a similar description of the other type of signal.

2 Figure 1e.10 shows a fibre optic light. Explain where the lamp is and why light comes out of the tips of the fibres.

3 We usually think of light travelling in straight lines. However, a ray of light can travel along a curved optical fibre. Copy the diagram in Figure 1e.11 and complete it, to show how this happens.

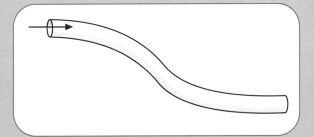

Figure 1e.11

Figure 1e.10 A fibre optic light.

 4 Outline the major advantages of digital signals over analogue ones.

Signals without wires

Figure 1f.1 Each of these young people has a mobile phone which can pick out the correct messages for its owner from the many thousands which fill the air around us.

There are signals all around us – signals for mobile phones, radio and TV transmissions, and wireless computer connections. We can't detect these with our bodies but that doesn't mean they aren't there. Tune in a radio and you will find that it can pick up many stations – perhaps dozens. All of those radio signals are available to you all of the time, coming from different transmitters in different places.

These signals are described as *wireless* signals, because they come to us through the air, rather than along wires. They make use of three types of radiation: radio waves, microwaves and infra-red radiation. You have already learned a bit about two of these, microwaves and infra-red radiation; all three belong to a family of radiation called **electromagnetic radiation**.

The diagram shows the different members of this family. You probably recognise their names; now you know that they all belong together. In this chapter, you will learn some things that all of these types of radiation have in common.

radio waves	microwaves	infra-red	visible light	ultraviolet	X-rays	gamma rays

Figure 1f.2 The electromagnetic spectrum.

Getting the message

Wireless communications are extremely convenient. You don't need to plug in anywhere, because there is no wiring involved. This means that you can walk around with a portable radio, mobile phone or wireless laptop computer. Some business centres, hotels and even trains are 'wireless hotspots' where anyone can access the internet without needing a wired connection.

You will have noticed that mobile phones work inside buildings. This shows that microwaves can pass through brick and glass. Similarly, a wireless radio link can work between computers in different rooms, showing that radio waves can also pass through some solid materials.

Figure 1f.3 High buildings block signals from the transmitter.

continued on next page

Getting the message - *continued*

However, there can be problems.

- Your mobile phone won't work if you are on a train that goes into a tunnel. The signals are absorbed by the thick layer of rock above you.
- A portable radio may not work inside a building which has a lot of steel in its structure. In this case, it is because radio waves are reflected by metals.
- Infra-red links between computers only work when the devices are in the same room, because infra-red radiation is easily absorbed by solids.

SAQ

1 Look at the diagram of the electromagnetic spectrum. Which type of electromagnetic radiation can we see with our eyes? Which two types are used for cooking?

2 When we speak, we send signals to each other. Are we using electromagnetic radiation?

Absorbing and reflecting

You can see that it can be difficult for a signal to get through if it is **absorbed** or **reflected** before it gets to its destination. This explains why, if you have a mobile phone, you will have noticed that it is easier to get a signal in some places than in others.

SAQ

3 A car radio usually has an aerial on the outside of the car. Suggest why the radio would not work if the aerial was inside the car.

Tuning in

With a radio or TV set, you can tune in to whichever station you want. What's going on when you do this?

The transmitter sends out radio waves in all directions – that's what 'broadcast' means. (Even TV programmes are broadcast using radio waves.) It can broadcast several stations at the same time. When you tune your set, it picks out only the signals which you want to hear or see.

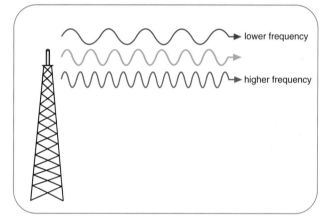

Figure 1f.4 Broadcasting signals.

How do these signals differ? In the drawing, you can see how we picture these signals. They travel through the air as **electromagnetic waves**. The wave for each station has a different **frequency**; that is, each station sends out a different number of waves each second.

The frequency of a wave is the number of waves in each second, measured in hertz (Hz).
1 Hz = 1 wave per second.

When you tune a radio set, you are selecting the frequency of the station which you want to hear. On many radio sets, the tuning dial shows the frequencies of the different stations.

There has to be a gap on the tuning dial between neighbouring stations, otherwise your radio will not be able to separate them. We say that, if the frequencies of two stations are too close together, they **interfere** with each other.

There is another way that this can happen. If you live on the south coast of England, for example, you pick up signals from a UK transmitter. However,

Figure 1f.5 a This radio has a tuning dial which shows the frequencies which it can tune into. **b** This is a digital radio; digital broadcasting can give better reception.

France is not very far away, and your radio and TV may pick up signals from there, too. French stations use the same frequencies as UK ones, so you may notice interference between them. The TV picture may have blurry shapes moving across it as a French soap opera interferes with a British one. There are international agreements to try to prevent this from happening.

Digital radio
If you listen to digital radio, you are likely to find that the quality of the signal is better than for an old-fashioned analogue radio. As we saw in Item P1e *Comunicating with infra-red radiation*, digital signals have the advantage that they can be electronically 'cleaned up' to remove any noise which could reduce their quality.

Digital radio signals are broadcast using radio waves. Each frequency can carry the signals for up to four channels; these are multiplexed in a similar way to phone calls along an optical fibre. This means that the radio waves carry a short burst of the signals for one channel, then for the next, and so on. After the fourth channel, it goes back to the first channel again.

Inside the digital radio set, there is an electronic circuit which separates out the four channels and strings the sections together before letting you hear the programme. This means that there is a short time interval (a few seconds) between when the programme is broadcast and when you hear it. If you listen to the same programme on an analogue set and a digital set at the same time, you will immediately notice this time gap.

Electromagnetic spectrum
Signals may be carried by radio waves or microwaves. What's the difference? The difference is in the frequencies. Microwaves are electromagnetic waves with higher frequencies than radio waves.

Radio waves are at one end of the **electromagnetic spectrum**, the end with the lowest frequencies. At the other end are gamma rays, which are electromagnetic waves with very high frequencies.

Radio stations broadcast in different 'bands'. The FM band uses higher frequencies than the AM bands. TV stations use even higher frequencies.

SAQ
4 Put these types of electromagnetic radiation in order, from lowest frequency to highest:

light, gamma rays, ultraviolet, radio waves

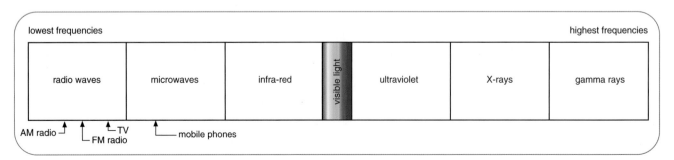

Figure 1f.6 The electromagnetic spectrum shows all types of electromagnetic radiation in order, from lowest frequencies to highest frequencies.

A change of direction

Sometimes, a radio set may pick up a station broadcast in a distant country, perhaps a thousand kilometres away. How does this happen?

Transmitters tend to send their radio wave signals out in all directions. As they spread out from the transmitter, they get weaker (simply because they are more spread out). Some of the waves go straight upwards and disappear out into space. Others are affected by the atmosphere as they pass through it, and they change direction. This is an example of **refraction** and it happens because the radio waves go faster where the atmosphere is thinner.

Figure 1f.7 shows one effect of this. Waves refract a lot when they reach a layer of the atmosphere called the **ionosphere**. Some bend back down towards the Earth and can be picked up thousands of kilometres from the transmitter. So someone in the UK may be able to listen to broadcasts from stations in Spain or Russia.

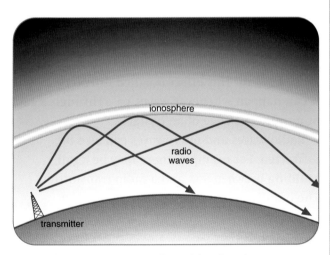

Figure 1f.7 Radio waves are refracted (bent) as they pass through the atmosphere. This is how radio waves can travel around the curved surface of the Earth.

SAQ

5 What do we call the bending of radio waves as they pass through the atmosphere?

6 Look at the diagram of radio waves being bent by the atmosphere. State one useful consequence of this, and one bad consequence.

H Long-distance communication

To understand why radio waves are refracted, we need to think about the atmosphere. The higher you go, the less dense it becomes – that's why mountaineers often carry oxygen supplies. Also, the atmosphere is formed of different layers. The ionosphere, which is around 100 km up, is a layer of electrically charged particles, and this affects radio waves greatly.

You should recall that refraction occurs when light or other radiation changes speed. Radio waves change direction as they enter the ionosphere and the effect is so great that lower-frequency waves are bent back down towards the Earth. This is an example of total internal reflection (TIR). Higher-frequency radio waves and microwaves are bent less, and travel out into space.

Satellite broadcasts use microwaves because these are least affected by refraction in the atmosphere. A microwave signal is sent up to the satellite. The satellite then retransmits it down to Earth. Anyone with a satellite dish pointing at the satellite can receive the signals, but you need a decoder to be able to see the transmissions.

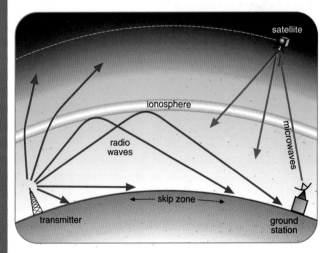

Figure 1f.8 The atmosphere affects radio waves more than microwaves.

H

SAQ

7 Which frequencies of waves are bent most by the ionosphere, low frequencies or high frequencies?

8 The 'skip zone' is an area where high-frequency signals cannot be picked up, although receivers farther away *can* pick them up (see Figure 1f.8). Explain why this happens.

9 Astronomers have to send messages up to spacecraft and to receive messages in return. Should they use microwaves or radio waves? Explain your answer.

Figure 1f.9 This large 'dish' aerial is used to communicate with spacecraft, high above the atmosphere.

Diffraction effects

There is another way in which radio waves and microwaves can be caused to change direction. It's easiest to start by thinking about sound waves. If someone is talking in the next room, the sound of their voice can reach you through an open door, even if you cannot see the person. This

H

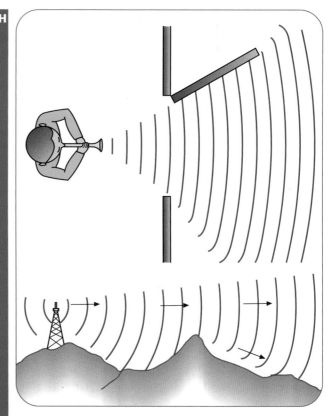

Figure 1f.10 Diffraction of waves.

happens because, as the sound waves pass through the doorway, they spread out into the space beyond. This effect is called **diffraction**, and it can happen whenever waves pass through a gap or around the edge of an obstacle.

Radio waves and microwaves can also be diffracted. For example, there may be a tall building between you and a radio transmitter. The building would block the signals from reaching you if it wasn't for the fact that they are diffracted as they pass it, spreading into the space behind.

Similarly, people in mountainous regions may have poor radio reception, because the mountains get in the way. However, the lowest frequency radio waves, which are diffracted the most, tend to bend down into the valleys, so there is a chance that they may be able to receive these.

Microwave signals are sent out from 'dish' aerials. Often, the sender wants to send a narrow beam of microwaves from one dish to another, perhaps as part of a mobile phone network. The microwaves are diffracted by the edges of the dish so that they spread out instead of forming a narrow beam. This results in a weakening of the signal.

Figure 1f.11 These dish aerials send beams of microwaves from one transmitter to another. The waves are diffracted as they leave the dish, so that the beam becomes spread out.

SAQ

10 What is the difference between *refraction* and *diffraction*?

Summary

You should be able to:

◆ give some uses of wireless signals

◆ explain how wireless signals can be carried by infra-red, radio waves and microwaves

◆ describe how electromagnetic waves can be reflected and refracted

H ◆ explain how refraction and diffraction can affect signals

◆ describe how long-distance communications make use of the reflection of radio waves by the ionosphere

◆ describe how satellite transmissions of broadcasts work

Questions

1 Copy the table and complete it to show three examples of the use of wireless technology. For each, state the type of radiation used.

Use of wireless technology	Type of radiation used
radio and TV broadcasts	

H 2 Radio waves may be *refracted* as they pass through the atmosphere. Explain what this means and why this happens. Why is this effect useful for transmitting signals around the curve of the Earth?

3 Electromagnetic waves may be *diffracted*. Explain what this means. Give one benefit which may come from diffraction of radio waves, and one disadvantage of the diffraction of microwaves.

4 Why must microwaves be used for communications between Earth and an orbiting spacecraft? Describe how a satellite TV broadcast travels from the ground station to the receiving set.

Describing waves

We say that electromagnetic radiation travels as waves – radio waves, microwaves and so on. These waves are not quite the same as waves on the sea or sound waves, but they do have a lot in common with them. Figure 1g.1 shows how scientists think of waves.

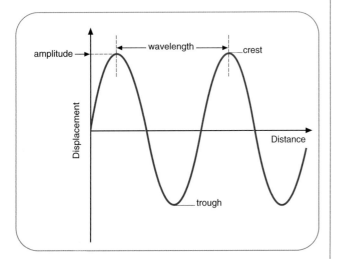

Figure 1g.1

- A wave is a regular sequence of equally spaced **crests** and **troughs**, travelling along.
- The distance from one crest to the next is called the **wavelength**. (This is the same as the distance from trough to trough.) Wavelength is measured in metres (m).
- The height of the wave above the centre line is its **amplitude**.

Take care with the amplitude! It is not measured from crest to trough. For water waves, think of it as the height of the wave above the level of the undisturbed water (when there is no wave there).

SAQ

1 Figure 1g.2 shows a water wave. How many crests are shown? What is the wavelength of the wave? What is its amplitude?

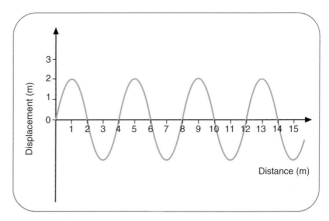

Figure 1g.2

Tsunami 2004

In December 2004, a giant tsunami devastated coastal regions in several countries around the shores of the Indian Ocean. Hundreds of thousands of people lost their lives as the huge tidal wave crashed onto thousands of kilometres of beaches.

What causes a tsunami? Although they are also known as tidal waves, they have nothing to do with the tides. In fact, they are set off by some energetic event in the Earth's crust – a volcano, an earthquake or a landslide. An impact from a comet or asteroid can have the same effect.

The tsunami of 2004 was caused by an underwater earthquake which ripped open a

Figure 1g.3 The tsunami wave in Thailand. This Dutch family were lucky to survive.

continued on next page

Tsunami 2004 – *continued*

crack in the seabed. Millions of tonnes of rock disappeared into the resulting fissure, triggering the great wave.

Although the effects of the wave were devastating, some small islands survived with little damage. Here is why. On the open sea, the tsunami travels fast. Its amplitude is small – perhaps a metre or two. It is scarcely affected by small islands but, when it reaches the mainland, it slows down as it drags on the seabed. This causes a pile-up of water, and the amplitude increases. Eventually, the wave is so high that it breaks on the beach.

It is a long time since a serious tsunami struck the British Isles, but there are concerns that one might do in the not-too-distant future. The island of La Palma in the Canaries is volcanic. One day, an eruption there is likely to cause half of the island to collapse into the sea. The resulting tsunami will travel across the Atlantic, drowning the eastern seaboard of North America; part of the wave will travel up the English Channel, causing serious damage along the south coast of England. However, this may not happen for thousands of years.

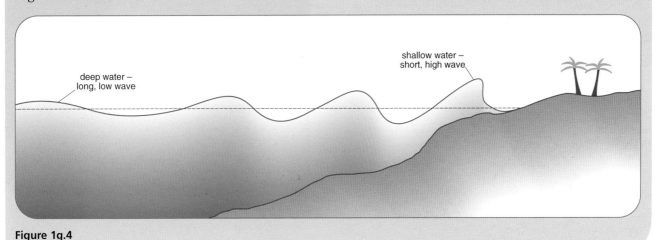

deep water –
long, low wave

shallow water –
short, high wave

Figure 1g.4

Transverse waves

Imagine floating in a wavy sea. The waves travel along horizontally, across the surface of the water, but you bob up and down. The wave is a disturbance of the water; although the wave may move quite fast, the water itself doesn't travel along; it merely moves up and down.

A wave like this, in which the movement of the material carrying the wave is at right angles to the movement of the wave itself, is called a **transverse wave**.

SAQ

2 Explain why, when you go to the seaside, you see seagulls bobbing up and down, rather than zooming along on the surface of the water, carried by the waves.

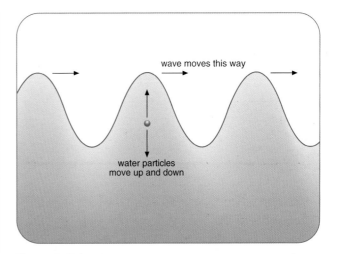

wave moves this way

water particles
move up and down

Figure 1g.5 A transverse wave.

Electromagnetic waves

Electromagnetic waves are waves of electric and magnetic force, speeding along. They move very fast; in fact, in empty space, they all move at the speed of light:

speed of light = 300 000 000 m/s

That is the fastest anything can go. If they leave empty space and travel through a different material, they go more slowly – for example, light travelling through glass, or radio waves travelling through air.

The diagram shows three different electromagnetic waves. You can see that they have different wavelengths and different frequencies. The wave with the longest wavelength has the fewest waves, so it has the lowest frequency. The wave with the shortest wavelength has the highest frequency.

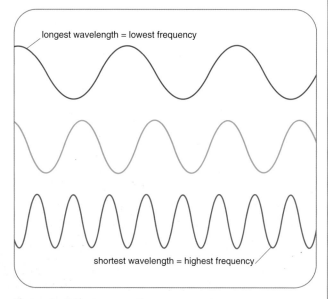

Figure 1g.6 Electromagnetic waves.

SAQ

3 Figure 1f.6 on page 286 shows the electromagnetic spectrum, from the lowest frequency to the highest frequency. Copy the diagram and add extra labels to show the waves with the shortest wavelengths and the waves with the longest wavelengths.

Laser light

A **laser** gives a narrow, intense beam of light. Compare this with a light bulb: the bulb may give out more light, but it spreads out in all directions.

There is another difference. The light waves from a laser are all of the same frequency, so its light is a pure colour. Also, all of its waves are in step with each other – we say that they are **in phase** with each other.

Figure 1g.7 The red laser beam on the right was used to guide tunnelling equipment during the construction of the Channel Tunnel. This ensured that the two teams working from opposite ends met in the middle with pinpoint accuracy.

The light waves from a light bulb are of many different frequencies, covering the complete spectrum from red to violet.

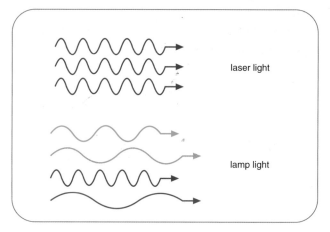

Figure 1g.8 Light from a laser compared with light from a bulb.

Morse code - communicating with light

Figure 1g.9 This operator is listening to messages in Morse code and then passing them on to the next exchange.

Morse code may seem like a simple idea, but it brought about a revolution in communications. There were two ways in which messages could be sent using the code:

- by flashing lights at night, from one hilltop to another
- as a series of electrical pulses, along wires.

This was particularly important in the USA when railroads were being built across the continent. Instead of using riders to carry messages on horseback, they could be sent almost instantaneously over large distances.

Today, Morse code is rarely used. Instead, people use SMS text messaging. When experienced Morse signallers have challenged SMS texters, it is the Morse coders who have proved to be the fastest at sending their messages – something of a surprise to all the

Letter	Code	Letter	Code	Number	Code
A	•—	N	—•	0	—————
B	—•••	O	———	1	•————
C	—•—•	P	•——•	2	••———
D	—••	Q	——•—	3	•••——
E	•	R	•—•	4	••••—
F	••—•	S	•••	5	•••••
G	——•	T	—	6	—••••
H	••••	U	••—	7	——•••
I	••	V	•••—	8	———••
J	•———	W	•——	9	————•
K	—•—	X	—••—	full stop	•—•—•—
L	•—••	Y	—•——	comma	——••——
M	——	Z	——••	query	••——••

Figure 1g.10 Morse code uses a system of dots and dashes to represent letters and numbers.

fans of the latest digital technology! And some mobile phone manufacturers are even thinking of adding a Morse function to their products.

When Morse code was invented, people rapidly invented quick ways of saying things (just like the abbreviations used in SMS messages). The best known of these is the distress call, SOS. Others included ADS for 'address', YL for 'young lady' and 88 for 'hugs and kisses'.

Speed, frequency, wavelength

How fast do waves travel across the surface of the sea? If you stand on the end of a pier, you may be able to answer this question. Suppose the pier is 60 m long and you notice that exactly five waves fit into this length (see Figure 1g.11). From this information, you can deduce that their wavelength is 12 m.

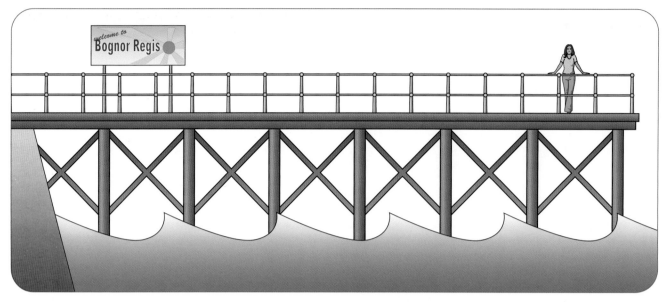

Figure 1g.11 By timing waves and measuring their wavelength, you can find the speed of waves.

Now you time the waves arriving. The interval between crests as they pass the end of the pier is 4 s. How fast are the waves moving? One wavelength (12 m) passes in 4 s. So the speed of the waves is

$$\frac{12\,\text{m}}{4\,\text{s}} = 3\,\text{m/s}.$$

Hence **wave speed**, **frequency** and **wavelength** are connected. We can write the connection in the form of an equation:

wave speed (m/s)
= frequency (Hz) × wavelength (m)

Another way to think of this is to say that the wave speed is the number of waves passing per second times the length of each wave. If 100 waves pass each second (frequency = 100 Hz) and each is 4 m long (wavelength = 4 m) then 400 m of waves pass each second.

SAQ

4 Some ripples are travelling across shallow water. Their wavelength is 2 cm and their frequency is 5 Hz. What is their speed? (Give your answer in cm/s.)

Worked example 1

An FM radio station broadcasts signals of wavelength 3.0 m and frequency 100 MHz. What is their speed?

Step 1: write down what you know and what you want to know.

frequency = 100 MHz = 100 000 000 Hz

wavelength = 3.0 m

speed = ?

Step 2: write down the equation for wave speed. Substitute values and calculate the answer.

wave speed = frequency × wavelength

wave speed = 100 000 000 Hz × 3.0 m
= 300 000 000 m/s

So the radio waves travel through the air at 300 000 000 m/s. You should recognise that this is the speed of light, the speed at which all electromagnetic waves travel through empty space.

H

Worked example 2

A pianist plays the note middle C, whose frequency is 264 Hz. What is the wavelength of the sound waves produced? (Speed of sound in air = 330 m/s)

Step 1: write down what you know and what you want to know.

frequency = 264 Hz

wave speed = 300 m/s

wavelength = ?

Step 2: write down the equation for wave speed; rearrange it to make wavelength the subject.

wave speed = frequency × wavelength

$$wavelength = \frac{wave\ speed}{frequency}$$

Step 3: substitute values and calculate the answer.

$$wavelength = \frac{330\ m/s}{264\ Hz} = 1.25\ m$$

So the wavelength of the note middle C in air is 1.25 m.

SAQ

5   What is the frequency of sound waves whose wavelength is 0.11 m and which are travelling at 330 m/s?

Drawing comparisons

You have learned about several different ways of sending signals – using light, infra-red radiation, radio waves and microwaves. (These are all forms of electromagnetic radiation.) Signals can also be carried by electric currents along wires – that is how traditional telephones work.

H

Each has its own advantages and disadvantages, as shown in the table.

Advantages	Disadvantages
Light waves	
High speed. Carry many signals on a single ray in an optical fibre. Difficult to intercept in optical fibre.	In air, absorbed by dust and water vapour.
Radio waves	
High speed. Wide range of frequencies allows many stations to broadcast. Reflected by ionosphere – broadcast around curve of Earth. Diffract behind obstacles.	Signals may interfere with each other. Signals reflected by buildings, blocked by mountains. Don't pass through metal.
Microwaves	
High speed. Travel through atmosphere with little refraction. Can carry mobile phone and satellite TV signals.	Diffraction by edges of transmission dishes causes spreading out of signals. Don't pass through metal.
Infra-red radiation	
Useful for local signals.	Need to have clear line of sight from source to detector.
Electrical signals	
Can be amplified easily. Can be sent to particular receivers, rather than being broadcast to everyone.	Need continuous wiring from source to receiver. Need low resistance cables. Tend to pick up noise.

Table 1g.1 Comparing different signals.

Summary

You should be able to:

♦ describe the main features of a transverse wave

♦ calculate the speed of a wave using the equation
  wave speed = frequency × wavelength

H ♦ use the above equation to calculate frequency and wavelength

♦ explain the advantages and disadvantages of different means of transmitting signals

Questions

1 a What quantity is the same for all electromagnetic waves in space?

   b How do microwaves differ from radio waves?

2 A laser produces an intense beam of light. State *two* ways in which this light differs from the light produced by a filament lamp.

3 Give the units of:
   a wavelength
   b frequency.

4 Draw a diagram to illustrate what is meant by the *wavelength* and the *amplitude* of a wave.

5 Sound waves of frequency 150 Hz have a wavelength of 2.0 m. Calculate their speed.

H 6 Radio waves of wavelength 1500 m have a frequency of 200 kHz (1 kHz = 1000 Hz). Calculate the speed of these waves.

Seismic waves

Figure 1h.1

Figure 1h.2 This seismograph is in an earthquake laboratory in Taipei, Taiwan. Its trace shows the tremors it detected when a major earthquake occurred on 14 June 2001.

An earthquake is a frightening thing to live through. Things that you always thought of as stable – the ground, trees, your home – start to shake. Buildings may collapse, roadways split open. What's going on?

An earthquake is a violent rearrangement of rocks close to the Earth's surface. **Shock waves** called **seismic waves** spread out from the centre of an earthquake. (The **epicentre** is the point on the Earth's surface, directly above the centre.) Earthquake recording laboratories are stationed all around the globe, and these use instruments called **seismometers** (or **seismographs**) to pick up details of the tremors produced by an earthquake which may be on the other side of the Earth. The photograph shows an example of a seismograph and the trace it recorded of an earthquake.

When an earthquake occurs, it is recorded by most seismographs. The further they are from the focus, the longer it takes the seismic waves to arrive. Geologists can then deduce how long the waves have been travelling. From this information, they can work out where and when the earthquake took place. They can also work out something about the materials through which the waves have been moving.

Two types of seismic waves can be identified, **primary** or **P waves**, and **secondary** or **S waves**. P waves travel faster than S waves, so they arrive first at a detector – hence the words 'primary' and 'secondary'. The differences between these are summarised in Table 1h.1. Figure 1h.3 shows representations of both types.

Name of wave	Nature of wave	Speed	Travels through ...
primary (P wave)	longitudinal	faster (5 km/s)	solids and liquids
secondary (S wave)	transverse	slower (3 km/s)	only solids

Table 1h.1 The properties of P and S seismic waves.

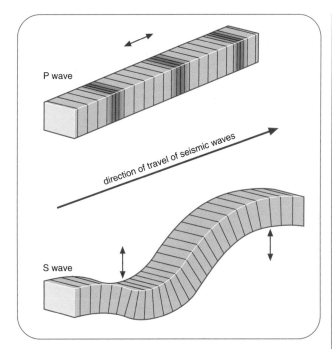

Figure 1h.3 The nature of P and S seismic waves. P waves are longitudinal, while S waves are transverse.

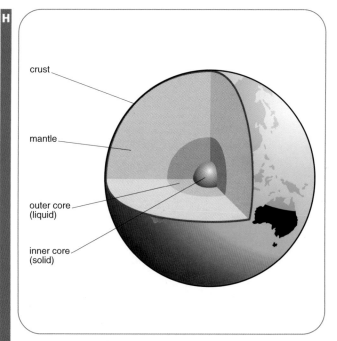

Figure 1h.4 The internal structure of the Earth. The outer core is liquid.

P waves are described as **longitudinal** waves. This means that the rock through which they are travelling is pushed back and forth along the direction in which the wave is travelling. (This is similar to the vibrations which make up sound waves.) S waves are described as **transverse** waves; the vibrations of the rock are at right angles to the direction in which the wave is travelling (like up-and-down waves on the surface of water).

From the table, you can see an important difference between P waves and S waves: P waves can travel through both solids and liquids but S waves cannot pass through liquids.

SAQ

1 Why are P waves detected sooner after an earthquake than S waves?

2 The Earth's core is partly liquid. Which type of seismic wave *cannot* pass through the core?

3 What instrument is used to detect seismic waves resulting from an earthquake?

Refracted waves

Figure 1h.4 shows the internal structure of the Earth but how do we know that it is like this? In particular, how do we know that the outer core is

liquid? Seismic waves caused by earthquakes have helped geologists to deduce the Earth's hidden structure.

As P and S waves travel outwards from the focus of an earthquake, their directions change. This is because they are refracted by the material through which they are moving. Figure 1h.5 shows typical paths of such waves. You will notice that the biggest changes of direction occur when P waves enter or leave the core. This is because their speed changes dramatically here. The waves also curve slightly within the mantle and core, because the material does not have a uniform density so the waves gradually change speed.

The S waves travel only within the mantle; they are absorbed by the liquid core.

From Figure 1h.5, you can see that there are regions of the Earth's surface where no waves arrive. These are known as **shadow zones**. These result from the fact that P waves are refracted away by the core, while S waves are absorbed. This is the most striking evidence we have for the existence of the Earth's liquid core.

When geologists are prospecting for oil, they often use a seismic technique like an earthquake in miniature. They set off an explosion on the Earth's surface and detect the waves which are reflected and refracted by layers of rock

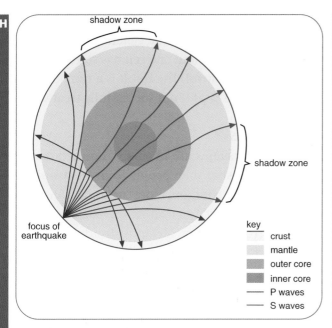

Figure 1h.5 Geologists have learnt about the interior of the Earth by studying the movement of seismic waves (P and S waves) through it. The waves are refracted as their speed changes. Refraction by the core results in shadow zones, where no waves arrive at the surface.

underground. In a similar way, explosions have been set off on the surface of the Moon; space probes on the surface have detected the seismic waves which result and this data is used to find out about the interior of the Moon.

Light from the Sun

The Sun is our principal source of energy. Its radiation has been falling on the Earth for 4.5 billion years. We see the Sun as a bright yellow disc in the sky. Figure 1h.6 shows a satellite view – you can see that there is quite a lot going on up there. (The Sun is far too bright for us to look at it directly; we have a natural reflex to look away from such a dangerously bright object.)

The surface temperature of the Sun is about 6000 °C. Such a hot object emits all types of electromagnetic radiation, but mostly **infra-red** (heat), **visible light** and **ultraviolet**. These three types of radiation lie side by side in the electromagnetic spectrum.

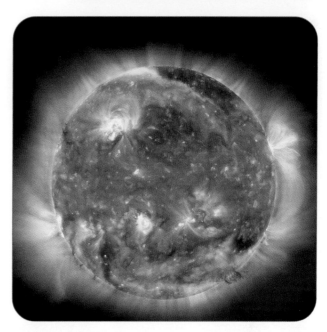

Figure 1h.6 This image of the Sun was produced by the SOHO satellite using a camera that detects the ultraviolet radiation given off by the Sun. You can see some detail of the Sun's surface, including giant prominences looping out into space. The different colours indicate variations in the temperature across the Sun's surface.

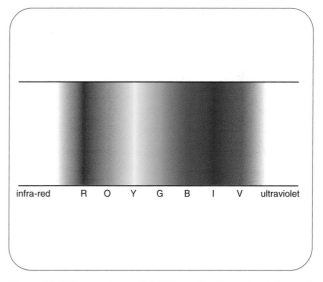

Figure 1h.7 The spectrum of light from the Sun extends beyond the visible region, from infra-red to ultraviolet.

Skin damage

Ultraviolet (UV) radiation has higher frequencies than visible light or infra-red, and that makes it more dangerous. Here is what exposure to UV radiation can do.

● A weak dose over a period of days can cause a **suntan** in people with fair skin. This happens because exposure to UV produces a black pigment called melanin in the skin.

- A more intense dose can cause **sunburn**. UV radiation has been absorbed by cells of the skin, and the energy of the radiation damages the cells.

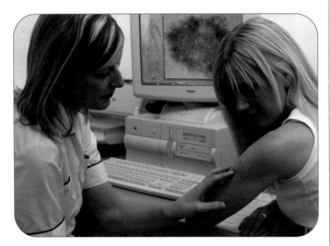

Figure 1h.8 This young woman is being screened for skin cancer, which can be caused by even low doses of UV radiation. The nurse checks moles on the skin; you can see an abnormal mole on the screen behind. Skin cancer can prove fatal unless detected and treated at an early stage.

- Any dose of UV can cause **skin cancer**. The radiation has damaged the genetic material (DNA) in the cells, and one cell has started to divide out of control. The result is a tumour, and this can prove fatal if not successfully treated.

Worth the risk?

People whose origins are in sunny countries tend to have darker skin than people from less tropical areas. Dark skin contains more melanin, which gives greater protection against the hazards of radiation from the Sun. Melanin absorbs UV radiation, so that less reaches the sensitive underlying tissues.

However, people often try to change the colour of their skin for reasons of fashion. Dark-skinned people may use skin-lightening cosmetics. Fair-skinned people may use cosmetics to darken their skin, or they may try to get a tan. By exposing themselves to a lot of UV radiation, they could be putting themselves in danger.

Staying safe

People with fair skins are particularly advised to avoid exposing themselves to too much strong sunlight. People with darker skins have more of the protective pigment melanin, which absorbs UV radiation before it can reach the sensitive inner layers of the skin.

UV radiation can also damage the eyes, increasing the possibility of cataracts in the lens. This leaves the lens opaque, so that a new one may have to be fitted. That is why welders wear masks when they are working – their torches produce strong UV radiation from which they must be protected. The mask has a glass panel to see through; glass is a good absorber of UV radiation, which is why people don't get tanned by sunlight coming through windows.

Winter sports holidays and beach holidays can both be risky. On snowy mountains, you are high up and there is less atmosphere above you to absorb dangerous UV radiation. Sunburn can happen quickly, so suntan lotion and protective

clothing are advisable. Goggles protect eyes against the harmful rays.

Figure 1h.9 People with fair skins in particular should avoid excessive exposure to UV radiation. Today, many people from temperate climes have the opportunity to visit hot countries for holidays. At the same time, the damage to the ozone layer means that we are exposed to more intense UV radiation. As a consequence, the incidence of skin cancer is increasing rapidly.

Figure 1h.10 Look for the sun protection factor (SPF) on any bottle of sun block.

Sun block is a substance designed to reduce the hazards of UV radiation. It is a mixture of chemicals; the most important one is titanium dioxide (the colouring used in white paint). This is in the form of microscopic 'nanoparticles'. If you apply sun block to your skin, these are smeared to form a very thin layer all over your body, where they absorb UV radiation. This means that you can safely stay out in the sunshine for longer – the risks are reduced.

The sun protection factor (SPF) tells you how much longer you can stay out. For example, an SPF of 30 suggests that you can spend 30 times as long in full sunlight, although it's not recommended that you do.

SAQ

4 On a bright day, fair skin may burn in 15 minutes. How long could a fair-skinned person stay in the sun if they applied sun block with SPF 20?

H Ozone layer

Space is a dangerous place. Above the Earth's atmosphere, there is intense UV radiation from the Sun. Astronauts require absorbent glass or plastic in the windows of their spacecraft and spacesuits. Fortunately for the rest of us, down below, we are protected by the atmosphere, which absorbs most of the UV radiation from the Sun.

Most absorption occurs in the **ozone layer**. This is a layer of the atmosphere, between about 20 and

H 60 km up, which contains ozone (O_3), a form of oxygen. Ozone is good at absorbing UV radiation.

Early on in the life of the Earth, the ozone layer had not formed and deadly UV radiation reached the Earth's surface. The earliest lifeforms had to withstand this; nowadays, there is still enough UV radiation to cause a significant number of genetic mutations, and this may be important for evolution to continue.

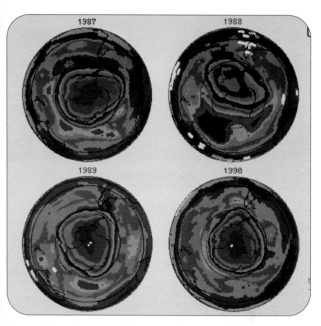

Figure 1h.11 These satellite images show the hole (pink) in the ozone layer above Antarctica, in four consecutive years. The size of the hole varies from year to year, depending on the weather. Recent measurements suggest that we still cannot be sure that the hole is shrinking.

In 1980, scientists first observed that the ozone layer was being damaged. Holes were appearing in it, particularly above the Earth's polar regions. (These 'holes' are in fact regions where the ozone layer has thinned down to less than half of its normal thickness.) It was soon realised that man-made chemicals, released into the atmosphere, were responsible for this. The main culprits are CFCs (chlorofluorocarbons), chemicals used in refrigerators. When old refrigerators were dismantled, their CFCs were released; they reached the ozone layer, where they reacted with the ozone, thereby depleting the layer.

Today there is an international agreement, the Montreal Protocol, to protect the ozone layer. Alternative chemicals are replacing CFCs and the

H

Figure 1h.12 This disused quarry in East Sussex is used to store old refrigerators before the CFCs can be extracted from them.

CFCs from old refrigerators are disposed of safely. The ozone layer may be recovering, but it will take decades before it returns to its original condition.

But there are new worries. In future, we may make more use of hydrogen as a fuel for cars. Hydrogen is another substance which reacts with ozone, and any escapes from filling stations may lead to new problems for the ozone layer.

A warmer world

We rely on the Earth's atmosphere to make the Earth a comfortable place to live. It is not just that we need air to breathe. The atmosphere also helps to keep the Earth at a comfortable temperature. Without the influence of the atmosphere, the Earth would be 20 or 30 degrees colder, like the Moon.

Carbon dioxide (CO_2) makes up only a small fraction of the atmosphere (less than 400 parts per million) but it is an important contributor to the **greenhouse effect**, which makes the Earth warmer than it would otherwise be.

Human influences

We tend to think of our climate as 'natural'. However, now there are worries that human activities are having an impact on the climate. The graph shows that the amount of CO_2 in the atmosphere is gradually increasing, which means that the greenhouse effect is stronger and the Earth's temperature appears to be rising. This is known as **global warming**.

The graph shows that the level of CO_2 has gradually increased; you can also see that the level goes up and down during the year. This is because trees and other plants absorb a lot of CO_2 while they are growing rapidly; in the winter, they grow much more slowly. At the same time, power stations are busier in the winter, so they produce more CO_2 then as well.

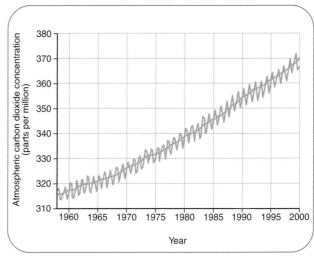

Figure 1h.13 The level of carbon dioxide (CO_2) in the atmosphere has been monitored over a long period of time. Data courtesy of the National Oceanic and Atmospheric Administration (NOAA).

Natural causes

Natural phenomena can also change our climate. In 1815, a volcano called Tambora erupted in Indonesia. This is thought to have been the biggest eruption in recorded history. It is thought to have put up to 200 cubic kilometres of dust into the atmosphere, with dramatic effects. Dust reflects sunlight, so less energy reached the Earth from the Sun. The result was that temperatures dropped around the world. The following year, crops failed and the European winter was the bitterest in living memory. 1816 became known as 'the year with no summer'. (Mary Shelley, who was 19 at the time, went on holiday in the Alps. The weather was appalling, so she stayed in and wrote a novel, *Frankenstein*.)

In a similar way, an increase in cloud cover can lead to cooling, as can an increase in the area of the polar ice caps.

Figure 1h.14 This photo, taken from the International Space Station, shows a volcano erupting on the Caribbean island of Montserrat. You can see the cloud of dust which it emits. Notice also how the clouds reflect sunlight back into space.

Cities tend to be warmer than the surrounding countryside. This is partly because industry, traffic and so on release a pall of dust into the atmosphere. So heat which escapes from buildings is reflected back downwards, preventing it from disappearing into space.

Carbon dioxide levels in the atmosphere are increasing for two main reasons.

- We are burning carbon-containing **fossil fuels**, and this produces CO_2. Previously, this carbon was locked up, underground. Now it is in the atmosphere.
- Forests are being cut down, particularly in tropical regions. (Temperate areas were deforested centuries ago.) Trees are stores of carbon and, when they are cut down, they are burned or they rot, releasing carbon dioxide.

SAQ

5 Why would an expansion of the polar ice caps lead to global cooling?

H 6 An increase of a degree or two in the Earth's average temperature could have a big impact on our climate. However, it is quite hard to tell if the Earth's *average* temperature is increasing. Suggest some reasons why this is.

Summary

You should be able to:

♦ describe the two types of seismic waves, P waves and S waves

H ♦ explain how analysis of seismic waves allows us to deduce the inner structure of the Earth

♦ describe the effects of ultraviolet radiation on the human body

H ♦ state that we are protected from most of the Sun's ultraviolet radiation by the ozone layer

♦ calculate safe exposures to ultraviolet radiation, given values of sun protection factor (SPF)

♦ explain how human activity and natural phenomena can affect our climate

Questions

1   There are two types of seismic wave: P waves and S waves. Which of these:

   a   travels faster through the Earth?

   b   is a transverse wave?

   c   can travel through both solids and liquids?

2   State *two* hazards that can arise from exposing the skin to ultraviolet radiation from the Sun.

3   A bottle of sun block is labelled 'Sun Protection Factor 30'.

   a   Explain what this means.

   b   If a person with fair skin can safely expose their bare skin to the sun for 10 minutes, how much longer can they spend in the sunshine when wearing this sun block?

4   Draw diagrams to show why dust from volcanoes leads to global cooling, while dust from a factory leads to its surroundings being warmer.

5   a   Why do seismic waves change direction as they pass through the interior of the Earth?

   b   Explain how shadow zones are formed at the surface of the Earth during an earthquake.

6   If we use fossil fuels more slowly, will this reverse the effects of global warming? Explain your answer.

Solar cell technology

Solar cells, also known as **photocells**, have been around for a long time. They have been used in solar-powered calculators and in camera light meters. You may have used them in Technology projects. However, it is only in the past decade or two that they have gained many new uses. Large arrays of solar cells are often to be seen at the roadside, powering emergency phones, for example. Some buildings are equipped with many solar cells to generate the occupants' electricity supply.

Figure 2a.1 An emergency phone powered by the array of solar cells shown.

In some ways, a solar cell is similar to a battery. It has a positive connection and a negative connection, and the **voltage** across the cell makes a current flow when it is part of a complete circuit. The current flows from positive to negative. Current like this, which always flows in the same direction, is known as **direct current (DC)**.

So, what is the difference between a solar cell and a battery? A battery contains chemicals which store energy to produce the current; when the chemicals are used up, the battery is dead ('flat').

A solar cell uses the energy of sunlight to make the current flow. As long as sunlight shines on the solar cell, it will go on producing electricity. This means that the solar cell never goes flat.

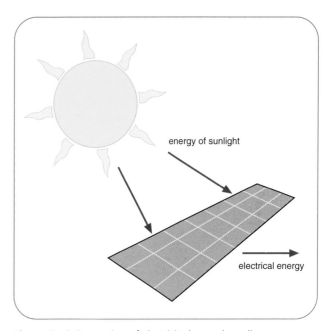

energy of sunlight

electrical energy

Figure 2a.2 Generation of electricity by a solar cell.

Why use sunlight?

The Sun's energy comes to us free of charge. The Sun's radiation travels through space to reach the Earth. It consists of all types of electromagnetic radiation, particularly infra-red (heat), visible light and ultraviolet. This radiation reaches us at a very steady rate, so we know we can rely on it. (Compare that with the wind, which is very variable and often doesn't blow at all.) Because we can't use up the Sun's radiation, we say that it is a **renewable source of energy**.

There are problems, of course:

- the Sun doesn't shine at night
- clouds may block the Sun's rays
- sunlight is weaker the farther you are from the equator.

SAQ

1 Solar cells are often connected to batteries, which store the energy collected by the solar cells. Which problems does this help to overcome? How does it help?

Remote living

There are many people – about one-third of the world's population – who live in parts of the world without a reliable electricity supply. They may be able to do without washing machines, electric lights and central heating, but there are some benefits of electricity that can make their lives much better.

Think of a medical service. A clinic in a remote part of the world can be of great help to local people. However, today's medicines need to be looked after properly. Vaccines, for example, can save lives, but they need to be stored in a cool place. A solar-powered fridge can be just the thing if there is no mains electricity.

BP is an international company best known for selling petrol, but it has put a great deal of effort into developing improved solar cell technology and putting it to use in remote parts of the world. Here are some examples of BP-promoted schemes.

- The Shanti Bhavan boarding school in Tamil Nadu, India. Solar cells power the school's lighting, water pump and educational equipment.

- A telecommunications link between the two capital cities of Vietnam, Ho Chi Minh City and Hanoi. It is difficult to provide electricity to optical fibre stations in mountainous regions.

- 200 health centres in rural Zambia now have solar-powered vaccine fridges and solar lighting.

Solar cells are excellent in these circumstances because, once they are installed, they need very little maintenance and there are no fuel costs to worry about.

Figure 2a.3 Solar cells are useful in remote locations like this village in Mali. This TV is powered by a car battery that is charged from solar cells on the roof.

SAQ

2 Figure 2a.1 shows an emergency phone in the French Alps. It is powered by photocells. You can also see the aerial which transmits the phone call to a nearby receiver. Explain why photocells are a good choice for powering a phone in a situation like this.

Electricity for ever?

Photocells are a great invention, but can they be the answer to our future electricity needs? Here are some advantages of photocells:

- they have no moving parts, so they need very little maintenance
- they are rugged (difficult to damage)
- they need no fuel and their energy source is renewable
- they do not require miles of electric cables to connect up to a distant power station
- they do not produce any waste gases or other polluting substances
- they are long-lived – they may last 20 years or more.

All of these make photocells an attractive way of generating electricity. However, there are some other features which are not so good:

- they cannot generate electricity all of the time, for example, at night
- they take up a lot of space
- they are expensive to manufacture
- some polluting waste is produced during the manufacture of photocells and the batteries which they charge up.

Why do photocells have to be so big? The reason is that the Sun's rays are not a very concentrated source of energy. For an average person in the UK, it takes several square metres of photocells to provide their daily electricity supply at home. The photo shows what this means in practice – most of the roof of a house has to be covered in photocells to provide for a family of four.

The rate at which photocells (or any other electricity supply) transfer energy is called their **power**. The greater the area of the photocells, the more energy they collect from the Sun's rays and the greater the power (energy per second) they can then supply.

Figure 2a.4 This Swiss house gets most of its electricity supply from the photocells fitted on its roof.

SAQ

3   Which one word means 'the rate at which energy is transferred'?

4   Photocells could be described as 'robust'. Look at the list of advantages of photocells on page 306. Find a word which means the same as 'robust', and explain why robustness is an advantage.

5   Discuss whether photocells would be a good choice to supply the electricity needs of people living in a tall block of flats.

6   Many different uses of photocells have been mentioned in this chapter. List them and add some more of your own.

How photocells work

Photocells generally look black or dark blue. This shows that they are absorbing sunlight, rather than reflecting it. How do they do this?

At the heart of a solar cell is a material called **silicon**. Some of the electrons of silicon atoms are only weakly held to the atoms; when sunlight falls on silicon, these electrons capture the energy and break free. Then they can move around to form an electric current.

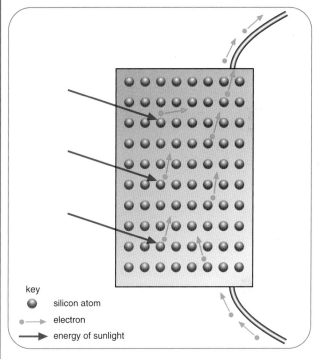

key

●   silicon atom

• →   electron

→   energy of sunlight

Figure 2a.5 Generating electric current in a photocell.

A photocell is thin, but it has a large area to catch as much sunlight as possible. Each cell produces only a small voltage (less than 1 V). A panel consists of many cells; several are connected in series to give a bigger voltage and then several of these sets are connected in parallel to make a bigger current flow. This is why a greater area produces more electricity.

A cell produces electricity more quickly (its power is greater) when the sunlight is more intense (more energy is falling on the cell each second). Usually, photocells are tilted so that they are at right angles to the Sun's rays, to maximise the amount of energy they collect (Figure 2a.6). To make the most of the Sun's energy, the photocells should also turn so that they track the Sun as it moves across the sky in the course of a day.

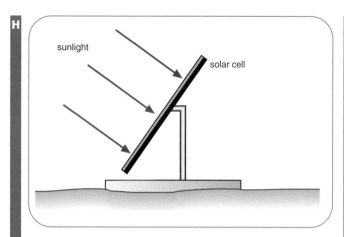

Figure 2a.6

Solar heating

Figure 2a.7 These university residences in Glasgow make use of passive solar heating.

It is useful to be able to turn the energy of sunlight into electricity, because electricity has so many uses. However, there are simpler ways to make use of sunlight.

● Buildings can be designed to capture as much sunlight as possible, reducing the need for heating. This is called **passive solar heating**. Figure 2a.7 shows a university hall of residence designed in this way.

● Water can be heated by installing a **solar water heater** on the roof. This works like a radiator in reverse; sunlight is absorbed by the black panel, and its energy is transferred to water inside. The hot water is then piped into the house.

SAQ

7 Why is the interior of the solar water heater in Figure 2a.8 painted black?

Figure 2a.8 This house in Barbados has a free hot water supply thanks to the solar water heater on its roof.

How it works

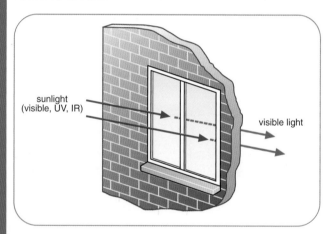

Figure 2a.9

Sunlight enters a room through the windows. Glass lets through visible light, but most of the infra-red and ultraviolet radiation in the Sun's rays are absorbed by the glass. The visible light is absorbed by objects in the room, making them warm.

Any warm object emits infra-red radiation – the hotter it is, the faster it emits radiation. This radiation is trapped in the room, because it cannot escape through the windows or walls. So the room gradually warms up.

Modern double-glazed windows use glass which has a special coating designed to reflect infra-red radiation. This means that more of the energy which comes into a room is trapped inside. In

A solar power station

Figure 2a.10 The Schelhino power station in the Ukraine – you can see many of the mirrors, and the central tower towards which they reflect the Sun's rays.

In the Ukraine, close to the Black Sea, is an unusual power station. It consists of many curved mirrors that reflect the Sun's light towards the top of a tall tower. The mirrors are arranged in curved rows around the tower. At the top of the tower, things get very hot! The sunlight reflected by the mirrors heats water in a boiler to a temperature of hundreds of degrees Celsius. The resulting steam turns a turbine and generator, converting the energy into electrical

energy. The station produces enough electricity for several thousand homes.

The mirrors are curved because this helps to bring the Sun's rays to a focus, concentrating them at a single point. As the Sun moves across the sky, the mirrors are turned so that the light they reflect is always focused on the top of the tower.

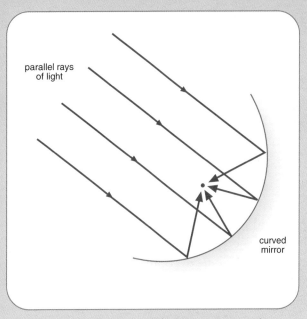

Figure 2a.11 A curved mirror focuses the Sun's rays.

addition, on a very sunny day, infra-red radiation is reflected away by the windows, so the room does not overheat.

SAQ

8 Why does the glass of a window get hot on a sunny day?

Wind power

Here's another way we make use of energy from the Sun – with a **wind turbine**. The wind is an example of a **convection current**, caused when the Sun shines. Air is warmed by the Sun, and rises. Cooler air sinks and flows in to replace it.

Moving air has **kinetic energy (KE)**, and it is this energy which we make use of with a wind

turbine. The turbine is the set of blades which are forced to turn by the wind. This is connected to a generator which turns to produce electricity.

Figure 2a.12 The wind farm at Cemmaes, Powys, Wales.

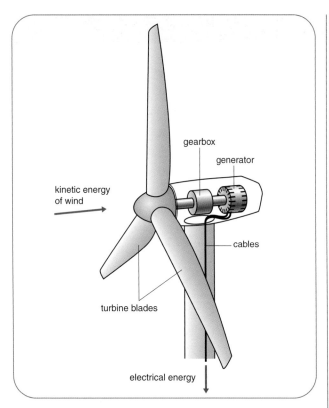

Figure 2a.13 Energy conversion by a wind turbine.

A wind turbine is usually high above the ground, where the wind tends to be steadier. Its mast is likely to be over 50 m tall, and each blade may be as long as 30 m. A big turbine can provide enough electricity for several thousand homes.

SAQ

Figure 2a.14 In the desert.

9 Figure 2a.14 shows a hot desert with cooler mountains in the distance. Draw a similar diagram and on it show warm air rising above the desert and cold air flowing down from the mountains to replace it. Show how a wind turbine could be positioned to take advantage of this.

H Pros and cons

Wind turbines have the great advantage that they do not produce polluting waste substances. (Fossil fuel power stations produce polluting gases and nuclear power stations produce radioactive waste.) The wind is a renewable resource and modern wind turbines can be built to last for 20 years or more. There are concerns about **visual pollution** (see Figure 2a.12).

The greater the wind speed, the more energy it carries and the more electricity can be generated. Wind turbines are designed to be rugged, because they have to cope with the worst weather conditions. However, they can only operate over a range of wind speeds. If the wind is too slow, they will not turn. If it is too fast, they may be damaged, so then they are usually turned sideways to the wind, to avoid its impact.

Wind turbines have to be large and wind farms must cover a large area because the wind is not a very concentrated source of energy. To replace a large coal-fired power station requires a wind farm of perhaps 100 turbines covering several square kilometres of countryside. Of course, the land can still be used for farming or other purposes.

SAQ

10 Wind power (electricity generated from the wind) is a renewable resource. Explain why.

11 What is meant by 'visual pollution'?

Environmental concerns

The UK is a windy place compared with the rest of Europe. This makes wind power a useful resource for generating electricity. However, not everyone thinks we should rely on it too much. Country Guardian is an organisation opposed to the development of new wind farms in the UK. What reasons do they have? Here is what they say:

- a wind farm, consisting of hundreds of turbines, can be a blot on the landscape
- wind turbines can be noisy
- the electricity they produce is expensive
- they are unreliable – they only generate for a fraction of the time
- birds and bats may be killed as they fly past.

  Country Guardian has been successful in helping local groups to oppose new wind farms.

One response from the wind energy industry has been to build new wind farms out at sea, where they do not spoil anyone's view. However, that makes them difficult to service, and undersea cables are needed to bring the electricity ashore.

Figure 2a.15 Opposition to wind farms.

Summary

You should be able to:

- ◆ state that the Sun is a stable source of energy (light and heat)
- ◆ state that photocells use the energy of sunlight to produce direct current (DC) electricity
- ◆ state that the greater the area of the photocells, the greater the electrical power they produce
- **[H]** ◆ explain that the energy of sunlight is absorbed by electrons in silicon, so that they become free to move
- ◆ explain that, the greater the intensity of light and the greater the area of the photocell directed towards the light, the greater the rate at which energy is transferred to electricity
- ◆ describe how passive solar heating makes use of light which enters through glass; the light is absorbed, causing warming
- **[H]** ◆ state that glass transmits light but absorbs or reflects infra-red radiation
- ◆ describe how wind power makes use of convection currents in the air which drive round turbines
- ◆ describe how the kinetic energy of the wind is transferred to electricity
- **[H]** ◆ state that wind turbines use a renewable energy resource and produce no pollution
- ◆ explain how wind turbines can produce more electricity when the wind is stronger
- ◆ state that wind turbines are large and may cause visual pollution

Questions

1 Electricity can be generated from several different sources. Which of the following are *renewable* energy sources?

   wind; coal; nuclear fuel; sunlight.

2 In the UK, most houses receive their electricity from power stations which are situated many kilometres away. A few houses have photocells on their roofs to provide electricity. Give three advantages of using photocells.

3 A wind turbine uses the energy of the wind to generate electricity.

   a What form of energy does the wind have?

   b Copy the boxes below in the correct order, to show how the electricity from a wind turbine comes from the Sun.

This convection current is what we call the wind.	The turbine causes the generator to turn.	The wind turns the turbine blades.	Sunlight warms the air, causing it to rise. Cooler air flows in to replace the rising air.

4 Figure 2a.16 shows sunlight falling on a window. The visible light enters the room.

   a Name two forms of electromagnetic radiation which do not pass through the glass.

   b The visible light falls on a book with a black cover. Explain why the book becomes warm.

   c What type of radiation does the book emit?

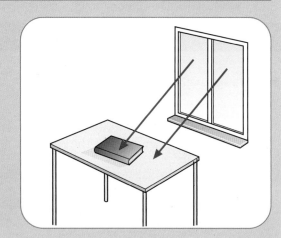

Figure 2a.16

5 George lives in a caravan (Figure 2a.17). He is not connected to the mains electricity supply. He has a photocell fixed flat on the roof of the caravan to provide electricity.

   a Explain why the power supplied by the photocell is less in the evening than at mid-day.

   b George decides that it would be better if the photocell was tilted towards the Sun. Explain why this would give more electricity.

   c The photocell does not provide electricity at night, when George needs it most. Explain how George could adapt this system so that he can use his computer at night.

Figure 2a.17

Seeing the supply

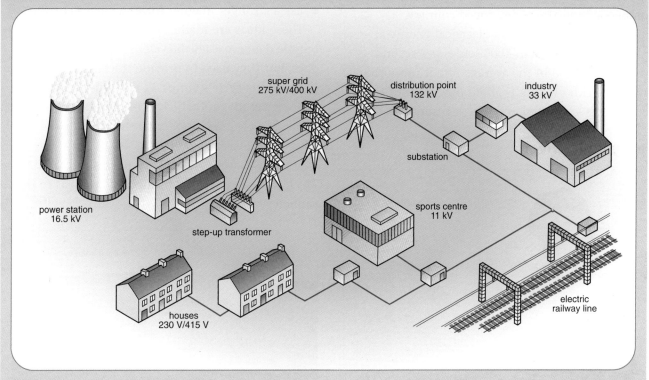

Figure 2b.1 Electricity is usually generated at a distance from where it is used. If you look on a map, you may be able to trace the power lines which bring electrical power to your neighbourhood.

Fortunately, we usually don't have to think about the electricity we use. We plug in a computer or switch on a light and they work. Often, we have no idea where the electricity we use is generated or how it gets to us. You may have noticed some clues as you travel around. Figure 2b.1 shows what to look out for.

Power stations may be 100 km or more from the places where the electricity they generate is used.

Cooling towers often have clouds of water vapour pouring out into the sky.

Lines of **pylons** stride across the countryside, supporting the high-voltage **power lines**, heading for the urban and industrial areas which need the power. These are the cables of the **National Grid**.

From the **substation**, electricity is distributed to neighbouring houses; larger buildings such as shopping centres, hospitals or sports centres may have their own substation.

Figure 2b.2 Look out for electricity substations in your neighbourhood. Here the voltage of the supply is reduced and power is distributed to all the nearby buildings.

Supplying electricity

Houses, offices, factories, shops and schools – all must be supplied with electricity if we are to maintain the lifestyle which we are used to as members of a modern, developed country. Typically, we each consume electrical energy at an average rate of about 1000 joules per second, although in the USA the average rate is more than twice this. This chapter is about how most of this electricity is generated and how it reaches us.

The dynamo effect

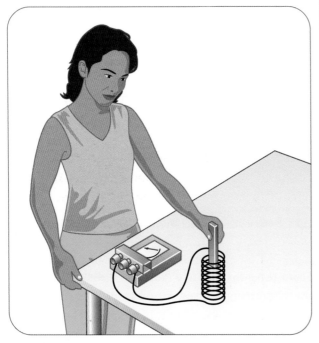

Figure 2b.3 Generating electricity.

Figure 2b.3 shows how you can demonstrate the principle of electricity generation. Move a magnet in and out of a coil of wire; the meter shows that a current flows in the wire. (A current only flows if the coil is part of a complete circuit.) Alternatively, you can move the coil back and forth near the magnet; it doesn't matter which is being moved. This is called the **dynamo effect**.

Here are five further observations:

● move the magnet or the coil more slowly and the current which flows is smaller

● hold the magnet and coil stationary and no current is generated

● with the magnet further from the coil, the current is smaller

● if the coil of wire has more turns, the current is bigger

● use a stronger magnet to get a bigger current.

These ideas are made use of in a traditional bicycle dynamo – few bicycles have one these days, but they used to provide the electricity for lighting the cycle's lamps. In the illustration, you can see the magnet which is made to spin round as the bicycle goes along, and the coil of wire in which the current flows.

You may have seen wind-up radios and torches. These have a **generator** inside which works in the same way.

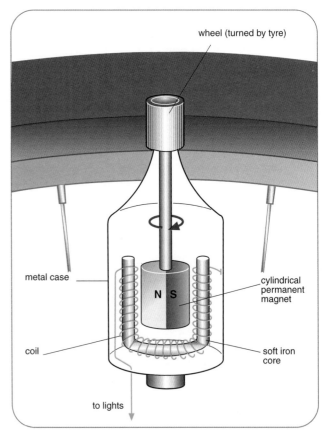

Figure 2b.4 A bicycle dynamo generates an alternating current that works the bicycle's lights.

SAQ

1 List four ways in which the current produced by the dynamo effect can be increased.

AC generators

It is not very convenient to generate electricity by moving a magnet in and out of a coil of wire. Instead, most generators have a coil of wire which is made to rotate in a magnetic field. Figure 2b.5 shows the principle of this.

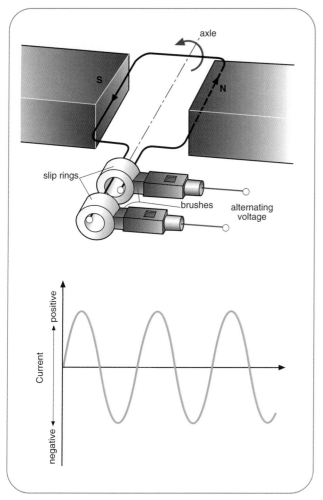

Figure 2b.5 A simple alternating-current generator. The slip rings and brushes are used to connect the current to the external circuit. The graph shows how the current varies.

This **generator** makes use of the dynamo effect, as follows:
● there is a coil of wire and a pair of magnets
● the coil is fitted closely in between the magnets
● the coil is made to move relative to the magnets.
    The generators in a modern power station are giant versions of a generator like this.

Figure 2b.6 The turbines and generators in the generating hall of a modern power station.

Alternating current

The current from a generator flows back and forth in the wires. (You can see this in the graph in Figure 2b.5 – the value of the current keeps changing from positive to negative.) This is called **alternating current**, or AC for short.

    AC is different from the **direct current** (**DC**) produced by batteries and photocells. DC flows steadily in the same direction all the time.

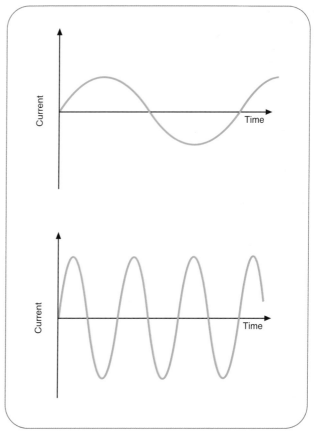

Figure 2b.7 The lower trace is produced when the generator is turned more rapidly.

An oscilloscope can be used to show alternating current. If an AC generator is connected to the input of the oscilloscope, its screen shows a varying trace. Each complete rotation of the generator produces a single cycle of AC. Rotating the generator faster has two effects:

- the trace on the screen goes up and down more, showing that a bigger current is flowing
- the trace goes up and down more frequently, showing that the current is changing direction more times each second.

Measuring peak voltage

From an oscilloscope trace, we can work out the maximum **voltage** of the AC coming from the generator.

Worked example 1

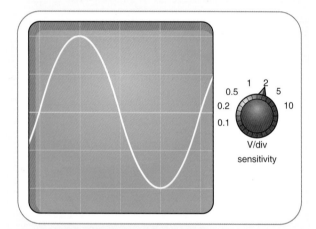

Figure 2b.8

Step 1: measure the height of the trace in Figure 2b.8, using the grid on the screen. Remember to measure from the central line, not from top to bottom.

Height = 2 divisions

Step 2: find the sensitivity dial, which tells you how many volts each division represents.

Sensitivity = 2 V/div

Step 3: use these two numbers to calculate the maximum voltage.

Maximum voltage = 2 div × 2 V/div = 4 V

So the maximum voltage is 4 V.

SAQ

2 When you cycle faster, the bicycle's dynamo (Figure 2b.4) turns faster. In what two ways will the alternating current produced by the dynamo change?

Measuring frequency

The frequency of an alternating current is the number of cycles per second, measured in hertz (Hz). In Europe, including the UK, the mains frequency is 50 Hz. In other parts of the world, it may be 25 Hz or 60 Hz.

You can work out the frequency from the oscilloscope trace.

Worked example 2

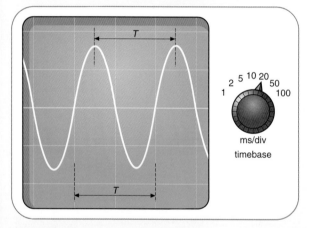

Figure 2b.9

Step 1: measure the time for one cycle (marked T on the diagram in Figure 2b.9) in divisions across the oscilloscope screen.

One cycle occupies 2 divisions

Step 2: note down the timebase setting. This tells you the time represented by each division across the screen.

Timebase setting = 20 ms/div
(This means 20 milliseconds per division.)

Step 3: use these two numbers to calculate the period of the AC.

continued on next page

H

Worked example 2 - *continued*

Period $T = 2\,\mathrm{div} \times 20\,\mathrm{ms/div} = 40\,\mathrm{ms}$
(Each division represents 20 ms, and one cycle occupies 2 divisions or 40 ms.)

Step 4: calculate the frequency of the AC (this is easier if we write 40 ms as 40×10^{-3} s).

Frequency $f = \dfrac{1}{T} = \dfrac{1}{40 \times 10^{-3}\,\mathrm{s}} = 25\,\mathrm{Hz}$

So the AC frequency is 25 Hz (25 vibrations per second).

SAQ

3 Suppose that the frequency of the AC in Worked example 2 was increased to 50 Hz. How would the oscilloscope trace change?

Power stations

In the UK, most of our electricity is generated in power stations which use fossil fuels. They burn coal, oil or gas, and the energy released is used to boil water in a boiler. This gives high-pressure steam, which turns a **turbine**. A rotating shaft connects the turbine to a generator, which produces high-voltage electricity.

Unfortunately, only a fraction of the energy provided by the fuel ends up being carried away by the electricity.

- The most modern gas-fired power stations waste about half of the energy of the gas.
- Older coal-fired stations may waste three-quarters of the energy of the coal.

Energy escapes in a number of ways.

- When the fuel burns, hot gases escape up the flue (just as hot gases go up the chimney from a domestic fire).
- The hot steam must be condensed using cooling water, and this takes away a lot of energy. (You can see clouds of warm, moist air rising up from cooling towers.)
- Some energy is wasted in the generator, which gets hot.

You can see from this list that the waste energy escapes from the power station as heat. It ends up warming up the surroundings. Engineers are always trying to improve the design of power stations so that they waste less energy, but generating electricity by burning fuels will always result in a lot of energy being wasted.

SAQ

4 Name these parts of a power station:

  a forced to rotate by high-pressure steam

  b produces electricity

  c contains water which is heated by burning fuel.

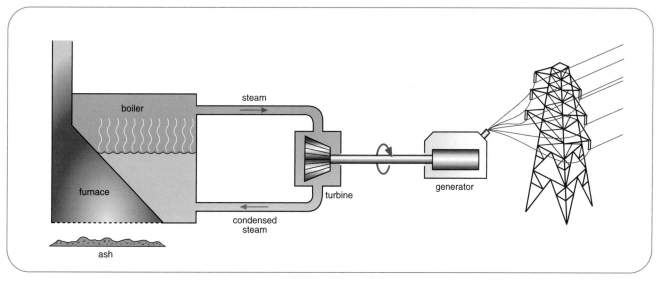

Figure 2b.10 Power station.

Calculating efficiency

We say that a power station which wastes half of the energy supplied to it has an **efficiency** of 50%. Think of it like this.

The **input** to the power station is the energy stored in the fuel.

The useful **output** is the energy carried by the electricity.

The remaining energy output is the waste energy.

Fuel energy input = electrical energy output
+ waste energy output

(You should recall that energy can't simply disappear, so the total output must equal the total input.)

The efficiency of a power station tells us what fraction of the energy input appears as useful output:

$$\text{efficiency} = \frac{\text{electrical energy output}}{\text{fuel energy input}}$$

Efficiency is often calculated as a percentage, like this:

$$\text{efficiency} = \frac{\text{electrical energy output}}{\text{fuel energy input}} \times 100\%$$

Worked example 3

Figure 2b.11 shows the energy that flows through a typical coal-fired power station in 1 s. Calculate its energy efficiency.

Step 1: identify the energy input. This is the energy provided by the coal, so:

energy input = 1200 MJ

Step 2: identify the useful energy output. This is the energy carried away by the electric current in the wires, so:

useful energy output = 420 MJ

Step 3: calculate the energy efficiency:

$$\text{energy efficiency} = \frac{\text{useful energy output}}{\text{energy input}} \times 100\%$$

$$= \frac{420\,\text{MJ}}{1200\,\text{MJ}} \times 100\% = 35\%$$

From this, you can see that the power station wastes almost two-thirds of the energy supplied to it. (Notice that the energy units, MJ, cancel out and we are left with a number without units.)

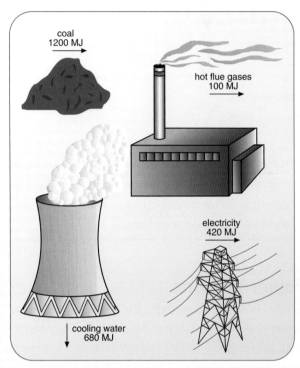

Figure 2b.11 Energy flows through a large coal-fired power station, per second.

SAQ

5   A power station produces 47 J of electrical energy for every 100 J of energy in its fuel supply. What is its efficiency?

6   A power station is 40% efficient. How much energy must its fuel provide for every 100 J of electrical energy it produces?

Grid of power

In the UK, most power stations are sited in industrial areas, although nuclear power stations and wind farms tend to be in rural locations. The National Grid is the network of cables which carries the electricity from the power stations to the millions of users in all parts of the country. There are over 22 000 pylons holding up the power lines (cables) of the grid.

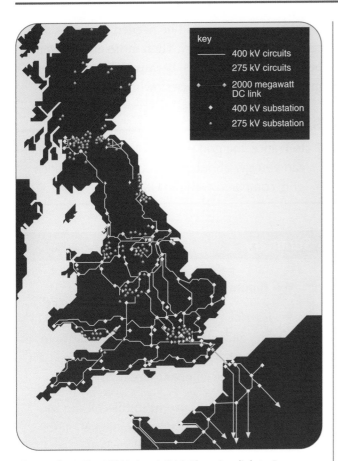

Figure 2b.12 The UK National Grid. There is a link to France so that electricity can be imported or exported when necessary.

The cables carry electricity at very high voltages – 275 000 V or 400 000 V. Operating at such high voltages is desirable because less energy is wasted as the electric current flows through the cables. It is always good to avoid wasting energy, and it saves money, too!

Voltage up and down

The voltage of electricity is changed as it is distributed by the grid.

Power stations generate electricity at 25 000 V (25 kV).

- It is distributed at 275 kV or 400 kV.
- It is reduced to 240 V for domestic users.
- Other users may have a 415 V or higher supply.

These changes of voltage are achieved using devices called **transformers** (see Figure 2b.16). A transformer consists of two coils of wire (two electromagnets) linked by an iron core. The electricity supply is connected to the primary coil; the new voltage (higher or lower) is obtained by connecting across the secondary coil.

Working on the grid

The National Grid is a major piece of engineering. We rely on the grid to supply us with electricity at all times, and people become very concerned if there is ever an interruption to their supply.

The engineers who work on the grid are very safety conscious – who wouldn't be, when working with such high voltages? Sometimes it is possible to switch off a section of the grid when repairs are needed, but it is also possible for engineers to work on cables which are at very high voltages. They can safely touch the cables so long as they are not in contact with the pylon or the ground, so that current cannot flow through them. For the same reason, it is safe for birds to sit on power lines.

Figure 2b.13 These engineers are fitting new insulators to a pylon.

Figure 2b.14 The power cables are attached to the lower end of the insulators, so that current cannot flow into the pylon.

Figure 2b.15 An engineer checks switching equipment at the point where electricity leaves a power station.

The transformer in the picture has more secondary turns than primary, so it increases the voltage. It is described as **a step-up transformer**. If the primary coil had more turns than the secondary, it would reduce the voltage. It would be a **step-down transformer**.

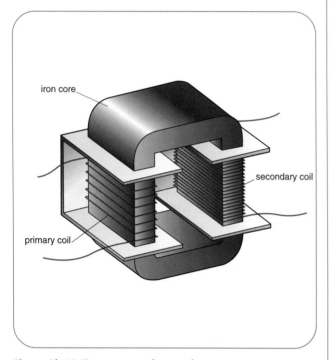

Figure 2b.16 The structure of a transformer.

Transformers are a vital part of the electricity supply system, but you will also find much smaller ones in more familiar situations. For example, a mobile phone charger contains a transformer which reduces the mains voltage (240 V) to the correct voltage (perhaps 6 V) needed to charge the phone's battery.

More volts, less amps
We use electricity because it is an excellent way of transferring energy from place to place. You should be familiar with these ideas.
- Electric current flows in cables. As it flows, it loses some of the energy it is carrying because of the resistance of the cables. The cables get warm.
- Voltage tells us about the energy carried by the current. The greater the voltage, the more energy it is carrying.

These ideas can help to explain why the grid works at very high voltages. The grid is transferring energy. If it worked at a lower voltage,

a greater current would be needed to carry the same amount of energy. Then more energy would be wasted in the cables, because a bigger current means more energy losses owing to resistance.

SAQ
7 In France, the grid works at 1 million volts. What can you say about energy losses in the French grid, compared with the UK grid?

Summary

You should be able to:

- describe the dynamo effect: when a coil of wire and a magnet are moved relative to each other, a current flows in the coil (if it is part of a complete circuit)

- state that the current in the coil can be increased with a stronger magnet, more turns of wire on the coil, or faster movement

- state that a generator consists of a coil of wire turning in a magnetic field; it produces alternating current (AC)

- state that the frequency of AC is the number of cycles per second

- describe how, in a conventional power station, fuel is burned to heat water, producing steam; this turns a turbine, which turns a generator

- state that a fraction of the energy of the fuel is wasted, ending up as heat in the surroundings

- efficiency $= \dfrac{\text{electrical energy output}}{\text{fuel energy input}}$

- state that electricity is distributed in the grid at high voltage; this reduces energy waste and costs

- state that distributing electricity at high voltage requires a smaller current, so there is less heating of the cables

Questions

1   If you hold a coil of wire next to a magnet, no current will flow in the wire. What else is needed to make a current flow?

2   Name a device which can be used to increase or decrease the voltage of an electricity supply.

3   In an AC generator, a coil of wire is turned in a magnetic field. A current is made to flow in the wire.

   a   How could the coil be altered to make a bigger current flow?

   b   State two other ways in which a bigger current could be made to flow.

4   A laptop computer works from a 9 V battery. Alternatively, it can be operated from the mains supply. Explain why a transformer is necessary for this.

5   This question is about a coal-fired power station. Each second, it uses coal whose energy content is 500 MJ.

   a   If its efficiency is 44%, how much electrical energy does it supply each second?

   b   How much energy is lost to the environment each second?

   c   What effect does this waste energy have on the environment?

6   The diagram in Figure 2b.17 shows two traces on an oscilloscope screen. These traces represent two voltages.

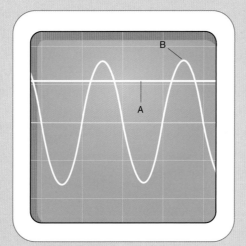

Figure 2b.17

   a   Which trace, A or B, represents the voltage of an AC generator?

   b   Each vertical division of the grid represents 100 V. What is the maximum voltage of the AC?

   c   Each horizontal division of the grid represents 0.05 s. What is the frequency of the AC?

Fuels for power

Choosing fuels

Almost all of the electricity used in the UK comes from thermal power stations. In these stations, water is boiled to make steam, which turns the turbines to spin the generators. Several different fuels are used to boil the water.

- **Fossil fuels** – coal, oil and natural gas – account for about 75% of our electricity. Burning them releases energy as heat.
- **Nuclear fuels** – **uranium** and **plutonium** – account for about 20% of our electricity. These are not burned; instead, they are purified and made into fuel rods which release heat when they are packed together in a reactor.
- **Biomass fuels** – including wood, straw and manure – can also be burned to release heat.

Fossil fuels and nuclear fuels are **non-renewable resources**, but biomass fuels are **renewable**.

Figure 2c.1 A thermal power station.

SAQ

1 Explain why biomass is a renewable energy resource.

Biomass fuels

In the past, we made much greater use of biomass fuels as an energy source. Before the days of electricity and petrol, the UK was home to millions of working horses. They needed oats and hay, and these crops accounted for about 20% of all agricultural land.

Today, there are sporadic attempts to generate more of our electricity from biomass; this is desirable because biomass is a renewable energy resource. A farmer in Suffolk started a small power station which uses chicken manure as its fuel. Other stations have been built to burn wood, which can be supplied by farmers who plant rapidly growing willow and poplar trees.

Sometimes biomass is burned directly. However, an alternative approach is to ferment it. This can produce methane (natural gas) or alcohol. The alcohol can be mixed with petrol and used as fuel in cars, when it is known as gasohol.

Some industries produce biomass as a waste product and it makes sense to use this to generate electricity. Figure 2c.2 shows a power station in Hawaii which burns the remains of sugar cane, after the sugary sap has been extracted. It is capable of burning other types of biomass, too.

Figure 2c.2 A sugar-cane-burning power station.

H What's best?

What is the best choice of energy source for generating electricity? There are many different factors to take into account when evaluating the different possibilities.

- How concentrated is the energy source?
- Is the resource readily available, and can you be sure of the supply in the future?
- How expensive is it? And how expensive is the power station which uses it?
- What are the running costs? How much will it cost to decommission (dismantle) the power station at the end of its life?
- What waste materials are produced? Will they damage the environment?

SAQ

2 Give examples of energy sources which match the following descriptions – you may be able to think of more than one for each:

   a not very concentrated

   b likely to run out in the near future

   c produces radioactive waste materials

   d burns biomass which is readily available in countries other than the UK.

Electrical power

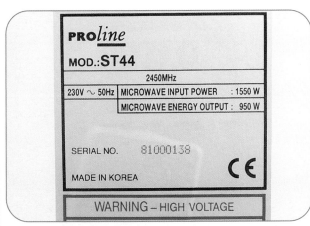

Figure 2c.3 This label is fixed to the back of a microwave oven. The power rating indicates the maximum power it draws from the mains supply when the oven is operating at full power.

Most electrical appliances have a label which shows their power rating. An example is shown in Figure 2c.3. Power ratings are given in watts (W)

H

or kilowatts (kW). Typical values for some domestic appliances are shown in Table 2c.1. Items which produce a lot of heat, such as heaters and washing machines, have a high power rating. Small electrical appliances such as personal stereos and calculators, which can run from batteries, have much lower power ratings. (These are only rough values; power ratings may vary quite a lot from one model to another.)

Appliance	Power rating
washing machine	3 kW
electric heaters	1 kW, 2 kW, 3 kW
light bulbs	40 W, 60 W, 100 W, 150 W
television set	150 W
stereo system	100 W
refrigerator	100 W
portable radio	5 W
personal stereo	1 W
calculator	0.0004 W

Table 2c.1 Typical power ratings for some domestic electrical appliances.

From Table 2c.1, you can see that light bulbs come in several different power ratings. A 40 W bulb is much less bright than a 150 W bulb (although each runs from the same 240 V mains supply). In other words, it transforms energy at a lower rate. The power rating of an appliance shows the rate at which it transforms energy, so the higher the rating, the faster the energy is being transformed.

Power is the rate at which energy is transferred (from place to place) or transformed (from one form to another):

$$\text{power (W)} = \frac{\text{energy transformed (J)}}{\text{time (s)}}$$

The unit of power is the watt:

1 watt = 1 joule per second (1 W = 1 J/s)

1 kilowatt = 1000 watts (1 kW = 1000 W)

SAQ

3 A light bulb is rated at 100 W. How many joules of energy does it transform each second?

H Calculating power

To find out the power of an appliance, you need to know two quantities:

● the voltage of its supply
● the current flowing through it.
    Here is how you calculate its power:

power = voltage × current

Worked example 1

What is the power of a heater which has a current of 5 A flowing through it when connected to the 240 V mains supply?

Step 1: write down the equation for power:

power = voltage × current

Step 2: substitute values for current and voltage, and calculate the answer:

power = 240 V × 5 A = 1200 W

Notice that the answer comes out in watts (W). We could also write this in kilowatts (kW): 1 kW = 1000 W:

power = 1.20 kW

SAQ

4   What is the power of a car headlamp that has a current of 3 A flowing through it when connected to the car's 12 V battery?

5   What current flows through a 6 W torch bulb when connected to a 3 V battery?

Paying for energy

When you pay for electricity, you pay for the energy you have used. The higher the power rating of an appliance, the more energy it transforms each second and the more you will have to pay. Also, the longer an appliance is used for, the more energy it transforms and the more you will have to pay. An electricity meter is fitted near where the mains supply enters the house, and this records the total consumption (Figure 2c.4). An electricity meter records consumption in units called **kilowatt-hours (kWh)**. This is a unit of energy, but it is not a standard SI unit. If you run a 2 kW heater for 3 hours, the energy transferred is 6 kWh. So

energy transferred (kWh) = power (kW) × time (h)

The reason for using the kilowatt-hour rather than the joule is that the joule is a very small unit of energy; cook yourself a full English breakfast and you will have used many thousands of joules of energy. The kilowatt-hour is more practical. Take care! It is a kilowatt-hour, not a kilowatt per hour. This should remind you that you multiply the kilowatts by the hours, rather than dividing:

● 1 kilowatt-hour is the energy transferred when an appliance transfers energy at the rate of 1 kW for 1 hour.

From the meter reading, the electricity supply company knows how many kilowatt-hours of electricity have been used. Given the price of an individual unit, they can then work out the total cost, like this:

cost = number of kilowatt-hours × price per kilowatt-hour

Worked example 2 shows how to make such calculations.

Figure 2c.4 A domestic electricity meter. The readout shows the number of kilowatt-hours (sometimes simply called 'Units') of electricity that have been used. If you can see inside the meter, you may notice a metal disc spinning steadily round. This spins faster the more power you are using, and the figures on the readout then change faster.

Worked example 2

A student uses two 150 W lamps for 6 hours. If the price per kWh of electricity is 10p (10 pence), what is the cost of this?

Step 1: calculate the power being used, in kW:

power = 2 × 150 W = 300 W = 0.3 kW

Step 2: calculate the energy transferred, in kWh:

energy transferred = power × time
$$= 0.3 \text{ kW} \times 6 \text{ h} = 1.8 \text{ kWh}$$

Step 3: calculate the cost:

cost = number of kilowatt-hours × price per unit
$$= 1.8 \text{ kWh} \times 10p = 18p$$

Notice that you could combine steps 2 and 3 into a single calculation.

SAQ

6   A 2 kW fire is used for 4 hours. How many kWh of energy are supplied? If each kWh costs 8p, what is the cost of using the fire?

7   An electric heater is rated at 2 kW. It has a thermostat which switches it on as the temperature falls and off again when the temperature rises. The heater is on for 25 minutes in every hour, on average. If the heater is used for 12 hours, how many kWh of electricity are used? If each kWh costs 8p, what is the cost of using the heater?

Off-peak power

Some houses have a second electricity supply. This comes through a different meter, coloured white, and is only available during off-peak hours (typically from 11 p.m. until 6 a.m.). This electricity is cheaper to use; people may choose to run their washing machine at night, using the off-peak supply. There is an extra charge for having a white meter, so the customer must weigh this up against the benefits.

Why do electricity supply companies operate this system? The reason is that consumers use more electricity at some times of the day than others. The peaks in demand are usually around breakfast time and again in the early evening. Electricity generating companies must be able to meet this peak demand, which means that some of their power stations are only used at the busiest times; they are idle for the rest of the time, particularly at night. It makes sense to encourage consumers to use less electricity during peak times and to shift their use to the quieter times.

Night storage heaters often use the cheaper off-peak supply. They heat up at night and then gradually cool down during the day as they release their store of heat. Some users find that they have cooled down too much by the evening.

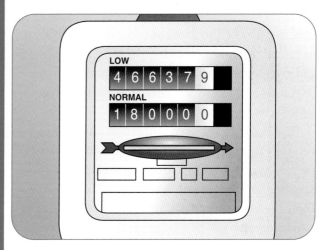

Figure 2c.5 Using electricity in off-peak times can be cheaper.

Nuclear power

Nuclear power stations use uranium or plutonium as their fuel. Figure 2c.6 shows the basic principles of nuclear electricity generation. Uranium is a metal, mined from the Earth. There are good reserves at present but one day it may all be used up.

SAQ

8   Is uranium a renewable or non-renewable resource?

Radioactive waste

Any method of generating electricity is bound to damage the environment. Burning fossil fuels produces carbon dioxide, a gas which contributes to global warming. Nuclear power stations do not

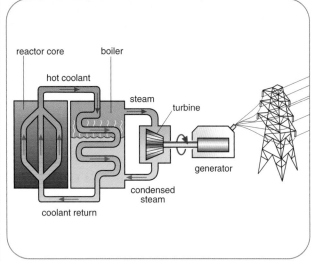

Figure 2c.6 Nuclear electricity generation.

Figure 2c.7 These containers hold radioactive waste. The security guard has a badge on his belt to monitor his exposure to ionising radiation.

produce any of the greenhouse gases which lead to global warming, but they do produce **radioactive waste** – this is what remains of the fuel rods after as much energy as possible has been extracted from them.

This is the major environmental concern in relation to nuclear power. How should the waste be handled? Should it be dumped underground, or stored above ground where its condition can be monitored for decades into the future?

Radioactive waste is hazardous because of the radiation it produces. This radiation, described as **ionising radiation**, can penetrate the body and damage cells, leading to cancer.

One of the substances in radioactive waste is **plutonium**, a radioactive element similar to uranium. This can be extracted and used as fuel in power stations specially designed to use it, but it can also be used to make nuclear bombs. This means that radioactive waste must be stored carefully, where it cannot fall into the wrong hands.

SAQ

9 Name two nuclear fuels.

Weighing up the costs

Nuclear power stations make use of advanced technology. In some ways, we are still learning to make this technology work well, so that it provides electricity both cheaply and safely.

Nuclear power has the advantage that it can help to make the country more independent of fossil fuel supplies. It is relatively easy to stockpile large quantities of nuclear fuel.

In assessing the value of nuclear power, it is essential to consider the complete life cycle of a power station and its fuel.

- All power stations are expensive to build, and the operator gets no return on investment until electricity starts to flow. This may be as long as 10 years from the start of planning and construction.
- Fuel must be processed from uranium ore, and this results in large volumes of radioactive waste substances.
- Spent fuel is another form of radioactive waste, and dealing with it is expensive.
- Operating costs are generally high, because any maintenance work is difficult to carry out and must be done to a very high standard, to avoid any leaks.
- Finally, at the end of its life, a nuclear power station must be dismantled and made safe (Figure 2c.10). This is called **decommissioning**. The site must be looked after for up to 100 years.

Nuclear costs and benefits

The nuclear power industry has had a troubled history. Nuclear power stations produce electricity which is the same as the electricity from any other power station. However, rather than burning a fuel such as coal or gas, it uses a completely different mechanism to produce the heat: nuclear fission, the splitting of uranium atoms. In its early days, half a century ago, nuclear power was seen as cheap, clean and safe way of providing electricity on a large scale. Today, a significant proportion of the world's energy resources come from nuclear power – roughly 20% of the UK's electricity, more than 75% of France's (Figure 2c.8).

Figure 2c.8 Most of France's electricity is generated in nuclear power stations like this one at Penly, near Dieppe.

The first countries to develop nuclear power were those which had developed nuclear weapons during the Second World War – the USA, the UK and the Soviet Union (now Russia, Ukraine etc.). Some people opposed this development because early reactors were often designed to produce plutonium for use in nuclear weapons as well as electricity. There is still concern that, when nuclear power technology is sold to other countries, they are being given the means to produce nuclear weapons.

A nuclear power station is very unlikely to explode like a bomb, but there have been accidents in which things have gone terribly wrong. The Chernobyl disaster (Figure 2c.9) may eventually result in thousands of deaths, and other leaks and accidents add to the death toll. However, this is nothing compared with the numbers who have died or suffered long-term health problems in coal-mining or in extracting oil and gas from hostile environments around the world.

Figure 2c.9 In 1986, the power station at Chernobyl (Ukraine) exploded after the operating engineers ignored instructions in the safety manual and the reactor core caught fire. Large amounts of radioactive material were spread over a wide area.

At the same time, we should be conscious of the benefits that have come from the nuclear industry. The radioisotopes which are used in medicine (for detecting and treating diseases) are all made in nuclear reactors, as are the radioisotopes used in the other industrial applications discussed in Item P2d *Nuclear radiations*.

Figure 2c.10 Berkeley, in the west of England, was one of the UK's first nuclear power stations. At the end of its working life, much of it was dismantled. Here you can see the large rusty-red boilers being laid down on the ground while radioactive steel girders are being cut up before removal.

All this adds to the cost, so that the original idea of electricity 'so cheap that it would not be worth the expense of metering' proved an over-optimistic dream.

SAQ

10 Which of the following statements are true, and which are false?

   a A nuclear power station produces greenhouse gases.

   b Fuel processing results in radioactive waste.

   c Nuclear power stations have low maintenance costs.

   d Nuclear power can replace fossil-fuel power stations.

Summary

You should be able to:

◆ state that fossil and biomass fuels are burned in power stations to generate electricity

◆ state that nuclear power stations use uranium and plutonium, which release heat energy, as their fuels

◆ state that the unit of power is the watt (W), equal to 1 joule per second

◆ state that high power ratings are given in kilowatts (kW); 1 kW = 1000 W

◆ state that appliances with a high power rating consume energy faster and that they consume more energy the longer they are switched on for

◆ state that power = voltage × current

◆ state that the unit of electrical energy is the kilowatt-hour (kWh)

◆ state that electrical energy supplied (kWh) = power (kW) × time (h)

H ◆ state that off-peak electricity makes use of electricity when demand is low and the electricity is cheaper

◆ explain why nuclear power stations do not contribute to global warming

◆ state that radioactive waste from nuclear power stations produces ionising radiation, which can cause cancer

◆ state that plutonium can be extracted from radioactive waste and used to make nuclear bombs

H ◆ state that nuclear power stations do not produce greenhouse gases but that there are many other costs in their life cycles

Questions

1   Calculate the power rating of a washing machine that has a current through it of 15 A when connected to a 120 V supply.

2   a   Name *three* forms of biomass.

   b   What gas is produced when biomass is fermented?

3   Uranium is a fuel used in nuclear power stations. What form of energy does it release?

4   Plutonium is a substance which does not occur naturally.

   a   Where is it formed?

   b   Give *one* use for plutonium.

5   A 4 kW heater is used for 10 hours.

   a   How many kilowatt-hours of electrical energy are consumed in this time?

   b   If the price of electricity is 8p per kWh, what is the cost of using the heater?

6   a   What distinguishes the radiation produced by radioactive waste from ultraviolet or microwave radiation?

   b   Why is this hazardous to people (and other living creatures)?

7   A car sidelight has a power rating of 15 W. It runs from the car's 12 V battery. Calculate the current which flows through the light.

8   A 6 kW electric motor is run for 15 minutes. How much electrical energy does it consume? Give your answer in kilowatt-hours.

9   A heater is run for 5 hours. In this time, it consumes 4 kWh of electrical energy. What is the power rating of the heater? Give your answer in watts (W).

10   Give *two* advantages and *two* disadvantages of a nuclear power station compared with a coal-fired power station.

Radioactivity all around

We need to distinguish between two things: **radioactive substances** and the **nuclear radiation** which they give out. Many naturally occurring substances are radioactive; usually, these are not very concentrated so that they don't cause a problem. If nuclear radiation hits our bodies, we say that we have received a dose of radiation – we have been **irradiated**. As we saw in Item P2c *Fuels for power*, exposure to radiation can result in cancer.

In fact, we are exposed to low levels of radiation all the time; this is known as **background radiation**. In addition, we may be exposed to radiation from artificial sources, such as the radiation we receive if we have a medical X-ray.

Figure 2d.1 shows the different sources which contribute to the average dose of radiation received by people in the UK. It is divided into natural background radiation (about 87%) and radiation from artificial sources (about 13%). We will look at these different sources in turn.

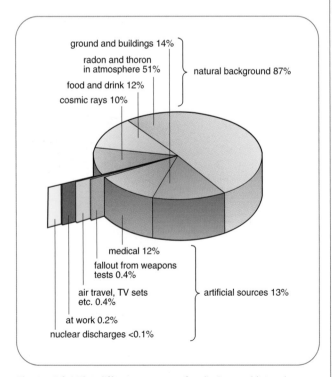

Figure 2d.1 The different sources of radiation and how they contribute to the average dose of radiation received each year by someone living in the UK. The main division is between natural background radiation and radiation from artificial sources.

Sources of background radiation

- Air is radioactive. It contains a radioactive gas called **radon** which seeps up to the Earth's surface from radioactive rocks underground. Because we breathe in air all the time, we are exposed to radiation from this substance. This contributes about half of our annual exposure. (This varies widely from one part of the country to another, depending on how much uranium there is in the underlying rocks.)
- The ground contains radioactive substances; we use materials from the ground to build our houses, so we are exposed to radiation from these.
- Our food and drink are also slightly radioactive. Living things grow by taking in materials from the air and the ground, so they are bound to be radioactive. Inside our bodies, our food then exposes us to radiation.
- Radiation reaches us from space in the form of **cosmic rays**. Some of this comes from the Sun, some from further out in space. Most cosmic rays are stopped by the Earth's atmosphere; if you live up a mountain, you will be exposed to more radiation from this source.
- Most radiation from artificial sources comes from medical sources. This includes the use of X-rays and gamma rays for seeing inside the body, and the use of radiation for destroying cancer cells. There is always a danger that

Figure 2d.2 Using a Geiger counter to monitor radiation levels close to a nuclear power station. Regular checks are made on samples of air, soil and water for 20 kilometres around.

exposure to such radiation may trigger cancer; medical physicists are always working to reduce the levels of radiation used in medical procedures. Overall, many more lives are saved than lost through this beneficial use of radiation.

SAQ

1 What is the biggest contributor to background radiation?

2 Why are people who live high above sea level likely to be exposed to higher levels of background radiation?

3 What fraction of our annual average dose of radiation is from artificial sources?

4 List three sources of exposure to artificial radiation.

Three types of radiation

There are three types of radiation emitted by radioactive substances. These are named after the first three letters of the Greek alphabet, **alpha** (α), **beta** (β) and **gamma** (γ).

There are two ways in which these are described:

- **nuclear radiation** – because they come from the nucleus of a radioactive atom
- **ionising radiation** – because they ionise atoms when they collide with them.

Recall that ionised atoms are particles which have an electric charge.

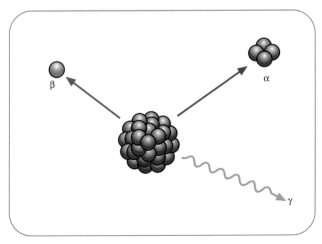

Figure 2d.3 Radiation comes from the nucleus of a radioactive atom.

(X-rays are ionising radiation, because they cause **ionisation**. However, they are not nuclear radiation, because they come from the electrons *outside* the nucleus and they can come from materials which are not radioactive.)

Penetrating power

When physicists were trying to understand the nature of radioactivity, they noticed that nuclear radiation can pass through solid materials. Different types of radiation can **penetrate** different thicknesses of materials. This is shown by placing the source of radiation on one side of the material and a detector (such as a Geiger counter) on the other.

- **Alpha particles** are the most easily absorbed. They can travel about 5 cm in air before they are absorbed; they are absorbed by a thin sheet of paper.
- **Beta particles** can travel fairly easily through air or paper, but are absorbed by a few millimetres of metal, such as aluminium.
- **Gamma radiation** is the most penetrating. It takes several centimetres of a dense metal like lead, or several metres of concrete, to absorb most of it.

Figure 2d.4 shows these different **penetrating powers**.

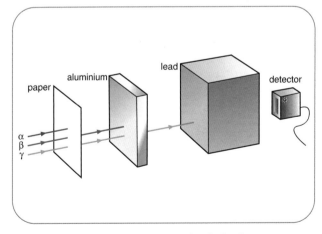

Figure 2d.4 The penetrating power of radiation is greatest for gamma radiation and least for alpha radiation.

SAQ

5 Put the three types of nuclear radiation (alpha, beta, gamma) in order, from most penetrating to least penetrating.

A radiation film badge

People who work with radiation have to monitor the amount of radiation they are exposed to (their radiation dose). Two types of detector are shown in Figures 2d.5a and b – a solid state detector, which is plugged into a machine to give a readout, and a film badge dosimeter. Figure 2d.5c shows the construction of a film badge.

The radiation is detected by a piece of photographic film inside the badge. The badge's cover has various 'windows', holes filled with different materials. The section of film behind the open window detects all of the radiation which enters through the window. (This is the only window which allows alpha radiation to reach the film.) The other windows allow through radiation of different energies; only the most energetic radiation can penetrate the lead window. In this way, when the film is developed, it is possible to say whether someone had a high or low dose of radiation, and how much of it was alpha, beta and gamma rays of different energies.

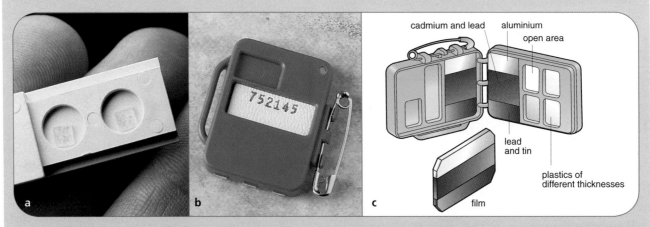

Figure 2d.5 a,b Two types of personal radiation dosimeter, a solid state detector and a film badge. **c** The windows of the film badge allow different energy ranges to penetrate to the film. Once the film is developed, a technician can work out how much radiation of each type the wearer has been exposed to.

Ionisation

Picture an alpha particle passing through air. As it collides with a molecule of the air, it may knock an electron from it so that the air molecule becomes charged. We say that the air molecule has become **ionised**. The alpha particle loses a little of its energy; it must ionise thousands of molecules before it loses all of its energy and comes to a halt.

There are two ways in which a particle (atom or molecule) can become ionised:
- it may lose an electron (and become a positively charged ion)
- it may gain an electron (and become a negatively charged ion).

Alpha radiation is the least penetrating of the three types of radiation; gamma is the most penetrating. We can make use of this to discover which type of radiation is coming from a radioactive source.
- If the radiation is absorbed by a thin sheet of paper, it must be alpha radiation.
- If the radiation can pass through paper but is absorbed by a few millimetres of aluminium, it must be beta radiation.
- If the radiation can pass through both paper and a few millimetres of aluminium, it must be gamma radiation.

SAQ

6 Explain how a radiation badge (see Figure 2d.5c) makes use of the different penetrating powers to identify alpha, beta and gamma radiations.

Safe working

Today, we know more about radiation and the safe handling of radioactive materials than ever before. Knowing how to reduce the hazards of radiation means that we can learn to live safely with it.

People who work with 'open' sources of radiation, such as powders or liquids, wear protective clothing to avoid contaminating their own clothing or their skin. You won't have to dress up like this for lessons on radioactivity, because school sources are 'closed' and cannot escape from their holders – although their radiation can get out.

A storage box used for keeping closed radioactive sources in a school laboratory is shown in Figure 2d.7. Each source is kept in its own lead-lined compartment, and the whole box should be stored in a metal cabinet with a hazard warning sign.

Students aged under 16 are not allowed to handle radioactive sources themselves. The teacher has to do this. Sources like those in the photograph must be handled with tongs. The teacher and students should stand well back from the source while it is in use, and it should be out of its storage box for as little time as possible.

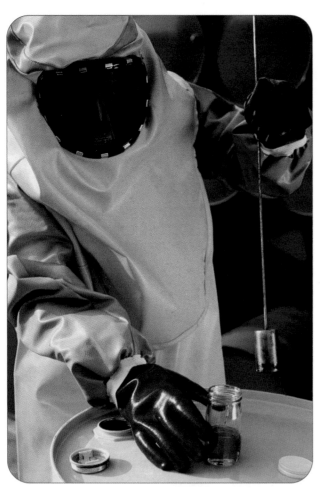

Figure 2d.6 Radiation worker wearing protective clothing.

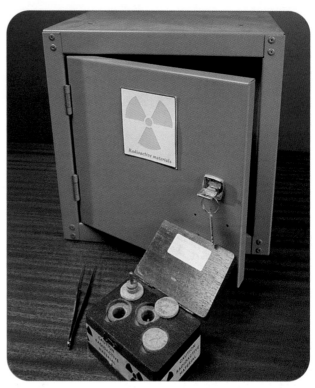

Figure 2d.7 A storage box for laboratory radioactive sources.

SAQ

7 Use what you know about radiation to explain the following:

   a the radioactive source storage box is lead lined

   b the sources are handled with tongs

   c the sources are replaced in their boxes as soon as possible.

Radioactive substances working for us

Smoke detectors are often found in domestic kitchens, and in public buildings such as offices and hotels. If you open a smoke detector to replace the battery, you may see a yellow and black radiation hazard warning sign (Figure 2d.8). The radioactive material used is americium-241, a source of alpha radiation. Here is how it works.

Radiation from the source falls on a detector, and the alarm is silent. When smoke enters the gap between the source and the detector, it absorbs the radiation. The detector notices the difference and sounds the alarm.

In this application, a source of alpha radiation is chosen because alpha radiation is easily absorbed.

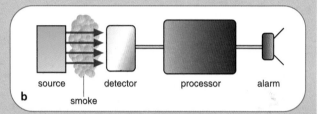

Figure 2d.8 a A smoke detector. The source of radiation is a small amount of americium-241. **b** The block diagram shows the circuit which sounds the alarm when smoke absorbs the alpha radiation.

Thickness measurements in industry often use beta radiation. Manufacturers of paper and plastic sheeting need to be sure that their product is of a uniform thickness. To do this, beta radiation is directed through the paper as it comes off the production line. A detector measures the amount of radiation getting through. If the paper is too thick, the radiation level will be low and an automatic control system adjusts the thickness. The same technique is used in the manufacture of plastic sheeting.

Beta radiation is used in this application because alpha radiation would be entirely absorbed by the paper or plastic; gamma radiation would hardly be affected, because it is the most penetrating.

Figure 2d.9 The beta source sends radiation down through the paper to the detector underneath.

Sterilisation of medical products also works by using radiation to kill cells. Syringes, scalpels and other instruments are sealed in plastic bags and then exposed to gamma radiation. Any microbes present are killed so that, when the packaging is opened, the item can be guaranteed to be sterile. The same technique is used to sterilise sanitary towels and tampons.

Figure 2d.10 Radiation was used to sterilise this syringe after it was packaged.

Fault detection in manufactured goods sometimes makes use of gamma rays. Figure 2d.11 shows an example in which an engineer is looking for any faults in some pipework. The radioactive source is strapped to the outside of the pipe and a photographic film is placed on the inside. When the film is developed, it looks like an X-ray picture and shows any faults in the welding. This is an example of non-destructive testing.

Figure 2d.11 Checking for faults in a metal pipe. The gamma source is stored in the box, but is pushed through the flexible tube to reach the pipe.

Radiation therapy. The patient shown in Figure 2d.12 is receiving radiation treatment as part of a cure for cancer. A source of gamma rays (or X-rays) is directed at the tumour which is to be destroyed. The source moves around the patient, always aiming at the tumour. In this way, other tissues receive only a small dose of radiation. Radiation therapy is often combined with chemotherapy, using chemical drugs to target and kill the cancerous cells.

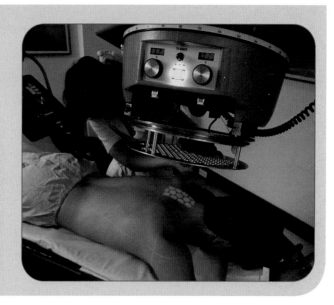

Figure 2d.12 A cancer patient about to be treated with gamma radiation.

Radioactive tracing. Every time you hear a Geiger counter click, it has detected the radioactive decay of a single atom. This means that we can use radiation to detect tiny quantities of substances, far smaller than can be detected by chemical means.

Engineers may want to trace underground water flows. For example, they may be constructing a new waste dump and need to be sure that poisonous water from the dump will not flow into the local water supply. Under high pressure, they inject water containing a radioactive chemical into a hole in the ground. Then they monitor how it moves through underground cracks using gamma detectors at ground level (Figure 2d.13).

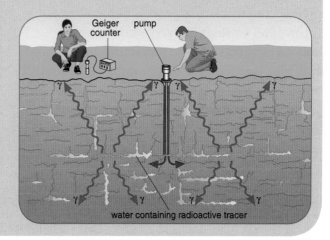

Figure 2d.13 Detecting the movement of underground water. Engineers need to know how water will move underground. This can affect the stability of buildings on the site. Water containing a source of gamma radiation is pumped underground and its passage through cracks is monitored at ground level.

DNA fingerprinting

Biochemists use radioactively labelled chemicals to monitor chemical reactions. The chemicals bond to particular parts of the molecules of interest, so that they can be tracked throughout a complicated sequence of reactions. The same technique is used to show up the pattern of a genetic fingerprint (Figure 2d.14).

Figure 2d.14 A DNA (genetic) fingerprint appears as a series of bands. Each band comes from a fragment of DNA labelled with a radioactive chemical. They show up on a photographic film.

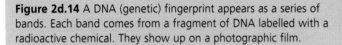

8 Why would beta radiation not be suitable for use in a smoke detector?

9 Why must gamma radiation be used for inspecting a welded pipe?

10 When medical equipment is to be sterilised, it is first sealed in a plastic wrapper. Why does this not absorb the radiation used?

11 Why must the engineers shown in Figure 2d.13 use a source of gamma radiation?

Dealing with radioactive waste

Nuclear power stations produce radioactive waste. Understanding nuclear radiation has helped engineers to think up safe ways of handling it.

Low-level waste is not very radioactive. It consists of things like old protective clothing and discarded wrappings. It can be dumped in landfill sites and covered with soil.

High-level waste is the most hazardous. It consists of the radioactive substances produced when nuclear fuel is used in a power station.

These are concentrated at a **reprocessing** plant. One way of disposing of this waste is to form it into solid glass blocks and store them underground.

Safe for ever?

Radioactive waste from the nuclear industry will remain hazardous for a very long time – perhaps thousands of years. It would be good if we could dump the waste and forget about it. One way to do this would be to dump the waste underground, perhaps in old mine workings. However, there is a problem with this. The radioactive substances might leak out of their packaging and contaminate groundwater, and from there enter our water supplies.

If the waste is to be stored above ground, there are other dangers. Terrorists might attack a storage site and steal the waste, which they could use in a 'dirty bomb' to contaminate large areas of the country.

Summary

You should be able to:

- name the three types of nuclear radiation given out by radioactive substances and describe their relative penetrating power

H - describe how alpha, beta and gamma radiations can be identified by their penetrating power

- state that nuclear radiations cause ionisation and can damage living cells

H - explain what is meant by *ionisation*

- state the causes of background radiation

- describe some beneficial uses of radiation

- describe how radioactive materials can be handled safely

- describe how radioactive waste can be disposed of

H - explain the problems of dealing with radioactive waste

Questions

1 Table 2d.1 lists the main causes of natural background radiation and shows their contributions to the average person's dose of radiation in the UK.

   a What is meant by *background radiation*?

   b Draw a bar chart to show the different contributions to background radiation.

   c Give *one* reason why a person might have a higher-than-average dose of background radiation.

Source of background radiation	Contribution to average background dose in the UK
radon and thoron in the atmosphere	59%
ground and buildings	16%
food and drink	14%
cosmic rays	11%

Table 2d.1

2 On pages 334–336, you can read about some different ways in which nuclear radiation is used.

   a List any examples which make use of the fact that radiation can pass through solid materials.

   b List any examples which make use of the fact that radiation is easily detected.

   c List any examples which make use of the fact that radiation damages living cells.

H 3 A school has two radioactive sources for use in physics experiments. One is a source of alpha radiation, the other a source of beta radiation. They have lost their labels and the teacher wants to check which is which. Use your knowledge of the different penetrating powers of these radiations to suggest how this might be done.

Magnetic Earth

Magnetism of the Earth

The Earth has a **magnetic field**. Without it, we would not be able to find our way about using magnetic compasses.

Figure 2e.1 We can find our way about using magnetic compasses.

The Earth has two magnetic poles, one in the north and the other in the south. When you use a compass, its needle (which is a small magnet) is attracted round by the Earth's magnetic poles, so that it lines up to point north–south. The effect of the Earth's magnetic field is the same as you see when you place a small plotting compass near a bar magnet.

We draw **magnetic field lines** to represent the Earth's magnetic field. The field comes out of the South Pole, travels around the Earth, and re-enters at the North Pole. It is as if there was a giant bar magnet hidden inside the Earth – but of course there isn't. (In fact, there are giant electric currents inside the Earth which create the field – see page 339.) The magnetic field extends far out into space; it can be detected by spacecraft thousands of kilometres away.

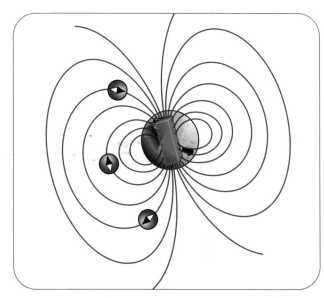

Figure 2e.2 Earth's magnetic field.

Magnetic navigation

Some animals (including several species of bird) are known to use the Earth's magnetic field to find their way around. Racing pigeons, for example, use the magnetic field as one clue to help them find their way home.

How do we know this? Investigators noticed that pigeons seemed to face towards home as they set off and suspected that the magnetic field was helping the birds. They attached tiny magnets to the birds' heads to cancel out the Earth's field and found that the birds lost the ability to set off in the right direction. This showed that the birds have tiny magnetic detectors inside their heads.

Don't worry! The birds found their way home in the end.

Figure 2e.3 Migrating birds may fly thousands of kilometres following the Earth's magnetic field.

Currents in the Earth

To understand *why* the Earth is magnetic, we need to think about what is inside the Earth. The important part is the **core**. When the Earth formed, it was much hotter than it is today. There was a lot of iron and nickel in the Earth, and these heavy metals were able to sink to the centre of the Earth, forming the core (see Items C2c *Does the Earth move?* and P1h *Beginnings and endings*). Today, the inner core is solid metal but the outer core is liquid, mostly iron. (The interior of the Earth is kept hot by the decay of radioactive materials in the core.)

Iron and nickel are magnetic materials and they also conduct electricity. Giant electric currents flow in the outer core and these produce the Earth's magnetic field.

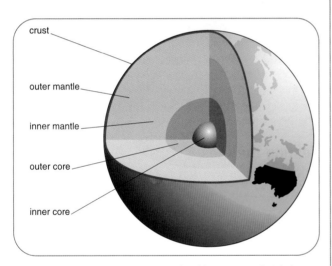

Figure 2e.4 The Earth's core is made of hot iron and nickel.

This may remind you of another way in which an electric current produces a magnetic field – in an electromagnet. An electromagnet is a coil of wire. When a current flows through the coil, it becomes magnetic. The pattern of the magnetic field is similar to the field of a bar magnet, and of the Earth.

SAQ

1 We show magnetic field lines coming out of a north magnetic pole and going into a south magnetic pole. Look at the drawing of the electromagnet.

   a Which end is its north pole?

b Where is the field strongest? How can you tell this from the drawing?

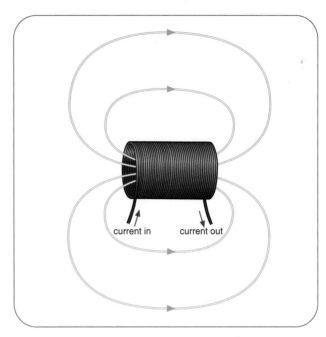

Figure 2e.5 An electromagnet has a magnetic field when a current flows through it.

The non-magnetic Moon

The Moon has no magnetic field. Astronauts have visited its surface and spacecraft have orbited it, and they have found no evidence of a field like the Earth's. Using seismic waves (see Item P1h), it has been shown that the Moon has only a tiny core of solid iron which is not big enough or hot enough to produce a magnetic field.

Figure 2e.6 shows how the structure of the Moon differs from the structure of the Earth. It is nearly all rocky mantle, with a core that is only 4% of its volume.

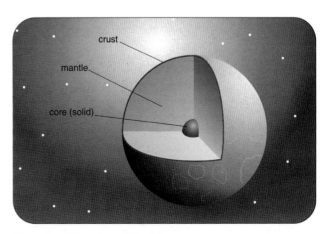

Figure 2e.6 Structure of the Moon.

Four theories

At the moment, astronomers cannot be sure how the Moon formed. There are several theories, each with its own supporters.

- The Earth may have been subjected to a violent impact from another large object – a small planet or a large comet or asteroid. Material from the damaged Earth collected together to form the Moon. This theory has the widest support among astronomers.
- The Moon may have formed as a separate, small planet. Later, it was attracted out of its orbit by the gravitational pull of the Earth.
- The Moon and the Earth may both have formed at the same time, close to one another, along with the rest of the solar system. They have continued to travel together ever since.
- The Earth may have initially spun much faster than today. The Moon may have arisen from matter thrown off by the rapidly spinning Earth.

Scientists don't always agree about scientific ideas. Each of these four theories has to be tested against the available evidence, and new evidence must be collected to help decide which (if any) is correct.

Figure 2e.7 One theory of how the Moon formed. An object the size of Mars is thought to have collided with the Earth. The Moon formed from the debris thrown into space.

How the Moon formed

Here is the most strongly supported theory of how the Moon formed. It formed when the solar system was young. At that time, it had not become the orderly system we know today. There were many small **planets**, **comets** and rocky asteroids whose paths crossed each other, so collisions were quite common. Astronomers think that a fairly large planet (at least the size of Mars) probably collided with the Earth.

- Lighter material (rocks from the Earth's crust and mantle) would have been thrown out into space, to form the Moon.
- Denser material (iron and nickel) would have been left behind.

SAQ

2 Look at Figure 2e.7, which shows the Earth in collision with another planet. This shows the Earth as it was over 4 billion years ago. How does it differ from the Earth today?

Testing theories

Scientists have two ways of testing this theory of how the Moon formed.

- They use computer models to see if this idea might have worked – and it does seem to be possible.
- They compare Moon rocks with rocks from the Earth, to see how similar they are.

So far, it seems that the theory is supported by the evidence. Moon rocks are similar to Earth rocks, so the Moon cannot have formed in another region of the solar system, where the composition of the planets is different. However, more samples will need to be collected to allow geologists to compare rocks from the Earth and Moon, to confirm this.

SAQ

3 Would you expect rocks from the Moon and Earth to be exactly the same? Why might they differ?

Radiation from space

The Earth is under constant bombardment from space. In Item P2d *Nuclear radiations*, we saw that **cosmic rays** are an important part of the background radiation which we are exposed to all the time. Cosmic rays are a form of ionising radiation from space.

At the same time, the Sun is giving out massive amounts of ionising radiation. The Sun may appear to shine at a steady rate but, from time to time, giant **solar flares** burst out from its surface – Figure 2e.8 shows four of these.

A flare is an outburst of very hot material with a temperature of millions of degrees. It shoots fast-moving charged particles into space, and they can reach the Earth in under an hour. These contribute to cosmic rays.

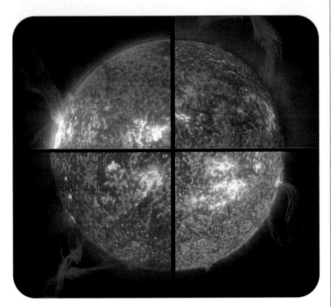

Figure 2e.8 A composite photo of the Sun, showing four solar flares, which occurred at different times.

Space hazard

Astronauts in spacecraft orbiting the Earth need to be warned of any solar flares. They are high above the Earth's atmosphere, so they have no protection against the radiation from a flare. They may turn their spacecraft so that its protective shield will block the incoming radiation. Some astronauts have reported seeing a flash in their eye when radiation passes through their eyeball.

Spacecraft themselves may be damaged by radiation from solar flares. There are many

hundreds of spacecraft in orbit around the Earth. They are known as **artificial satellites**. Among their uses are:

- telecommunications – transmitting phone and TV signals
- environmental monitoring – including weather forecasting
- military uses – including spying and missile control
- navigation systems – such as the GPS and Galileo systems.

Fast-moving charged particles from the Sun may damage these spacecraft. This radiation also produces strong magnetic fields which may interfere with satellite control and communication systems.

Figure 2e.9 The Astra TV satellite above the Earth.

SAQ

4 Spacecraft which orbit the Earth are sometimes known as artificial satellites.

   a Explain this meaning of the word 'artificial'.

   b What is the Earth's natural satellite?

Radiation and magnetism

Cosmic rays are a bit of an astronomical mystery. No-one is sure of their origin – they are very energetic, and astronomers are not certain how they might originate. Some seem to come from the very heart of our galaxy. Others may come from supernovae, which are explosions which mark the end of massive stars. Some weak cosmic rays come from the Sun as a result of solar flares.

Cosmic rays are mostly fast-moving charged particles called protons – particles which are usually found in the nuclei of atoms. When they hit the Earth's atmosphere, they collide with molecules of the air, producing a shower of gamma rays and other radiation which rains down on the Earth's surface. (It is this which contributes to background radiation.)

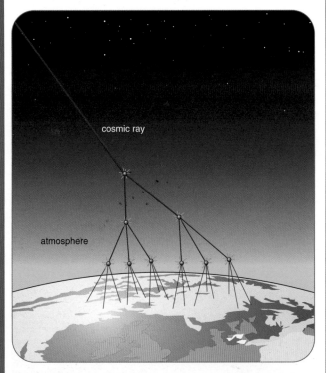

Figure 2e.10 A shower of cosmic rays is started when a fast-moving charged particle strikes a molecule of the air high in the atmosphere.

Because the fast-moving particles which travel through space have an electric charge (mostly positive), we can think of them as an electric current. An electric current is any flow of charge, and it creates a magnetic field of its own. (Think of the magnetic field produced by an electric current in a coil.) So now we have two magnetic fields:

- the Earth's magnetic field
- the field produced by the charged particles.

These two fields interact with each other, and the result is that the charged particles tend to spiral around the Earth towards its poles. This is

good for us, because it means that the particles have their greatest effect near the poles, where few people live. One consequence is the Aurora Borealis (the northern lights), seen in parts of the northern hemisphere towards the North Pole. This arises when charged particles lose their energy to molecules of oxygen and nitrogen in the atmosphere, causing these gases to glow. (There is a similar Aurora Australis, seen in regions close to the South Pole.)

Figure 2e.11 A dramatic view of the Aurora Borealis, photographed in Alaska. A strong aurora is often seen shortly after a major solar flare.

Space dynamo

If a solar flare is particularly strong and directed towards the Earth, it can have serious effects. Its magnetic field can affect the communications systems of satellites, and it can even damage electricity distribution grids.

- Think of the magnetic field of the solar flare, approaching the Earth. It's like a moving magnet.
- Think of an electric circuit – perhaps part of the National Grid. It's like a long length of wire.

As we saw in Item P2b, a moving magnet can make an electric current flow in a wire. That is the dynamo effect. So a large current may start to flow in the grid, and this can cause safety switches to trip off. It can even burn out electricity substations and generators. In March 1989, 6 million people in Quebec, Canada, woke up to find that their power supply had been cut by a magnetic storm from space, leaving them in the dark and the cold for the rest of the day.

Summary

You should be able to:

- describe the shape of the Earth's magnetic field, and the field of a coil carrying a current

- state that charged particles are deflected by a magnetic field

(H) - state that moving charged particles produce a magnetic field

- describe how the Moon is thought to have formed

(H) - discuss the evidence for this theory

- state the uses of artificial satellites

- describe the Sun as a source of ionising radiation, including the nature of solar flares

(H) - describe the effects on the Earth of solar flares

Questions

1 The Earth has a magnetic field.

   a Sketch the shape of the field.

   b In which part of the Earth does its field originate – core, mantle or crust?

   c What material is this part of the Earth made of?

2 List *four* uses of artificial satellites.

3 'Solar flares' are an unfashionable garment from the 1960s. True or false?

4 Figure 2e.12 shows the stages in the formation of the Moon, according to one possible theory. Describe what is thought to have happened.

Figure 2e.12 Formation of the Moon.

 5 Look back at question 4 above. What evidence do astronomers use to support this theory of the Moon's formation?

Exploring the solar system

Around the Sun

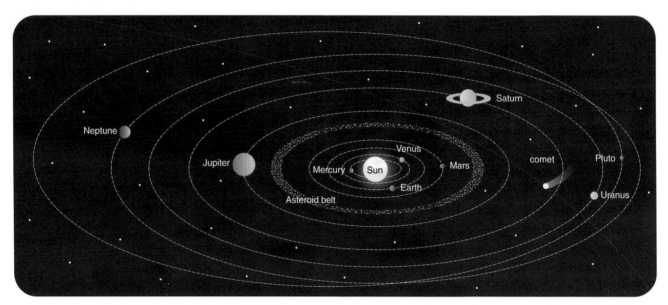

Figure 2f.1 The solar system.

We are used to the idea that the Earth orbits around our star, the Sun. We picture the Sun at the centre of the solar system, with the nine **planets** travelling around it in roughly circular orbits, held by the Sun's gravitational pull. What else is there in the solar system?

● The **asteroids** are lumps of rock and dust, mostly in orbit between Mars and Jupiter. A few have orbits which can bring them close to Earth – astronomers keep a close eye on these.

● **Comets** also orbit the Sun, but they spend most of their time far out in space. They are lumps of ice and dust. As they approach the Sun, they start to evaporate and we see their glowing tails.

● Behind them, comets leave trails of dust particles; if the Earth's orbit takes it through one of these trails, we see **meteors** ('shooting stars') in the night sky. These are dust particles burning up in the atmosphere.

SAQ

1 Write a mnemonic to help you remember the order of the planets, as shown in the illustration.

Our galaxy and beyond

The Sun is just one of billions of **stars** which make up our **galaxy**, the Milky Way. A galaxy is a large cluster of stars held together by gravity. The Milky Way is just one of billions of galaxies in the **Universe**.

We can see stars because they are hot, and so they give out their own light. Astronomers are only just beginning to learn about some of the other things in the Universe which we cannot see, because they are not hot.

One type of dark object is a **black hole**. These are objects which are so dense that their gravity is extremely strong; nothing can escape from the inside of a black hole, not even light – that is why it is described as black. Some black holes are quite small, with a similar mass to the Sun but much, much smaller. Others, supermassive black holes, are enormous and tend to gobble up any passing stars. There may be one lurking at the centre of our galaxy.

SAQ

2 Put these objects in order, from smallest to biggest:

star, planet, meteor, galaxy, comet.

Distant planets

In the past decade, scientists have discovered planets orbiting around stars other than the Sun. The planets are too small to see with a telescope, so how do they do it?

Using high-powered telescopes, the scientists watch an individual star. If they see it shifting regularly from side to side, they can guess that this is caused by a large planet orbiting it. The planet's gravity pulls on the star so that the star moves back and forth as the planet goes around it.

That is one technique for spotting a distant planet. Another is to measure the brightness of the light coming from a star. If it goes dim and then bright with a regular pattern, it may be that a large planet is passing in front of it, periodically blocking some of its light.

By 2006, over 150 planets had been discovered in this way, and the total keeps on rising. As instruments improve, astronomers can look for smaller planets – perhaps one day they will find another Earth.

Gravity at work

The planets orbit the Sun. They do not travel in straight lines; they follow roughly circular paths. An unbalanced force is necessary to keep them travelling along their curved paths. This force is the pull of the Sun's **gravity**. Similarly, the Moon is held in its orbit around the Earth by the pull of the Earth's gravity.

Any object travelling in a circle needs an unbalanced force to keep it in the circle. For example, if you whirl an apple around your head on the end of a piece of string, you must keep

pulling on the string. If you let go of the string or the string breaks, the apple will fly off.

The unbalanced force must pull the object towards the centre of the circle. Such a force is described as a **centripetal force**.

SAQ

3 a Draw a diagram showing the Earth in its orbit around the Sun. Add a labelled arrow to show the centripetal force which keeps it in its orbit.

   b Name the type of force which provides the centripetal force here.

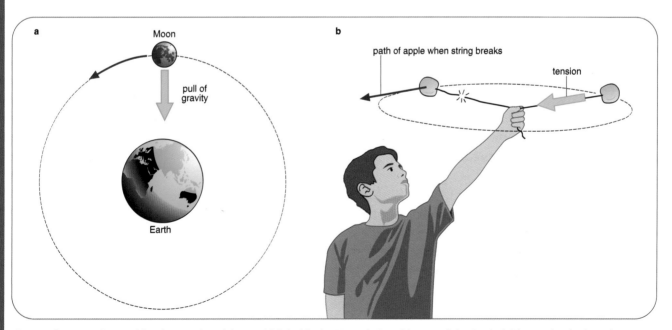

Figure 2f.2 a Gravity provides the centripetal force which holds the Moon in its orbit around the Earth. **b** The tension in the string provides the centripetal force to keep the apple moving in a circle.

Mission to Mars

Astronauts have orbited the Earth and visited the Moon but they have never gone further, to visit another planet. Perhaps one day they will, but it is a daunting task.

The first target planet must be Mars – Venus is too hot. What problems would astronauts experience?

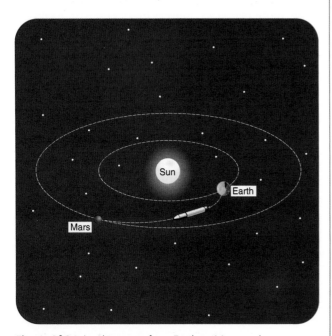

Figure 2f.3 It is a long way from Earth to Mars, and astronauts would want to get home again.

Mars orbits the Sun at a greater distance than the Earth. When the two planets are closest, their separation is about 75 million kilometres; when they are on opposite sides of the Sun, they are 380 million kilometres apart. That's a thousand times the distance to the Moon!

Astronauts travelling to Mars would need to communicate with Earth. Radio signals would take up to 45 minutes to travel from Mars to Earth and back again – that's a long time to wait before you get a reply!

Spacecraft travel quickly, but it would still take several months to reach Mars. A complete mission would take a couple of years, and astronauts would have to take all their supplies of food and water. (There are no chocolate bars on Mars.) They would also need to be able to maintain a stable atmosphere in their craft, and this would require supplies of oxygen and a means of removing carbon dioxide.

Because Mars is farther from the Sun, it is colder, so the astronauts would need fuel to keep them warm. They would have to wear heated spacesuits while exploring the planet.

Fuel supplies

It takes energy to launch a spacecraft and to get it moving fast on its journey. The rockets that do this are much bigger than the craft itself, because so much fuel is needed.

Fuel is also needed during the journey, because small rockets are used to adjust the speed and direction of the spacecraft.

At the same time, it takes fuel to slow down the spacecraft when it reaches its target. Retrorockets are fired to slow the craft for a safe landing; that takes more fuel. Then the spacecraft must lift off from Mars and return to Earth – yet more fuel is needed.

Figure 2f.4 An artist's impression of a spacecraft firing its retrorockets as it lands on Mars.

Keeping fit

Here on Earth, we are used to living with the Earth's gravity. In a spacecraft, when its rocket motors are switched off, gravity has no effect and astronauts feel weightless. That can be a problem.

Whenever we move around, we are exercising our muscles and keeping them toned up. Our bones, too, are built up when we exercise. (That's one reason why it's a bad idea to be a couch potato.) Astronauts, like those who spend months on board the International Space Station, have to make sure that they get regular exercise so that their muscles and bones do not start to waste away. They use a variety of exercise machines in which springs provide a force as a substitute for gravity.

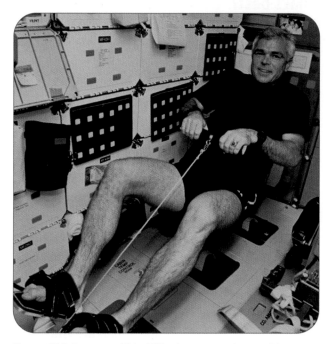

Figure 2f.5 Astronaut Rich Clifford uses a rowing machine on board a space shuttle.

Another problem with space travel is cosmic rays (as we saw in Item 2e). In space, there is no protection from the atmosphere. Although spacecraft have shielding around them to absorb cosmic rays, astronauts still receive a higher dose of radiation than if they had stayed on Earth.

SAQ
4 List all of the supplies which would be needed for a manned trip to Mars. Include any essential equipment for use in the spacecraft.

H Light-years
Spacecraft communicate with Earth using radio waves and microwaves. These are forms of electromagnetic radiation, so they travel through space at the speed of light – 300 000 km/s. Although that is fast, it still takes a significant time for signals to travel through the solar system.

H
- It takes 8 minutes for light to travel from the Sun to the Earth.
- It would take 5.5 hours for signals to reach Earth from a spacecraft passing Pluto.
- It takes 4 years for light to reach us from the nearest star after the Sun.
- It takes 14 billion years for light to reach us from the most distant galaxies.

We can use light as a kind of 'ruler' for measuring the Universe. The more distant an object, the longer it takes for its light to reach us. We measure these great distances in units called **light-years**.
- One light-year is the distance travelled by light in one year.
- The next nearest star is 4 light-years away.
- The most distant galaxies are 14 billion light-years away.

That's easier than saying the galaxies are at a distance of 130 000 000 000 000 000 000 000 km.

SAQ
5 A light-year is a measure of distance, time – or what? Write a sentence using the term correctly.
6 Light travels 300 000 km in 1 second. Using this information, calculate how many kilometres there are in a light-year.

Robot spacecraft
An alternative to sending people into space is to send spacecraft, perhaps carrying robot machines which can be landed on another planet to explore it. Spacecraft have already visited every planet in the solar system, and have landed on some of them. Mars has been a popular target because it is not too far away and because its solid surface is suitable for vehicles to explore. Figure 2f.6 shows a lander on Mars, with a robotic arm for taking soil samples. There is also a small rover vehicle which can move around on the surface.

Unmanned spacecraft have no need to carry food, water or oxygen supplies. However, they still need fuel and they must be designed so that they do not get too hot or too cold – otherwise their instruments might stop working.

Figure 2f.6 An artist's impression of the NASA lander *Spirit* on Mars.

Even if a spacecraft does not land on the planet it visits, it can carry instruments which allow it to tell us a lot about the planet. It can gather information about:

- temperatures
- magnetic field
- radiation
- gravity
- atmosphere.

Using radar, it can map the planet's surface.

SAQ

7 Explain why an unmanned spacecraft can withstand conditions which would be dangerous – even lethal – for humans.

What's best?

One day soon, astronauts may set out to visit Mars. But will it be worth it? Might it be better to send more unmanned craft?

- It would be cheaper to send an unmanned craft, because all of the systems needed to support and protect people wouldn't be necessary.
- It would obviously be safer not to send people, and it wouldn't be such a disaster if things went wrong.

However, there are advantages in sending people.

- Human astronauts can carry out repairs on the spacecraft or lander, if anything minor goes wrong.
- Humans are better at judging things which are of interest, and carrying out investigations. For example, they could decide which are the most interesting rock samples to bring back to Earth. This could be very important on a visit to another planet.

One day, people may decide to colonise other planets, or even to travel outside the solar system. This could be the first step along the way.

Summary

You should be able to:

- ◆ name the different types of object which make up the Universe
- ◆ state the relative positions of the Sun, Earth and other planets in the solar system
- ◆ state the relative positions of objects in the Universe
- ◆ state that gravity controls the motion of objects in the solar system
- ◆ state that gravity provides the centripetal force needed for orbital motion
- ◆ state that radio signals take a long time to travel through the solar system
- ◆ define the light-year (a measure of distance)

continued on next page

Summary - *continued*

♦ describe some of the difficulties of manned space flight

♦ describe some of the information which spacecraft can send back to Earth

H ♦ compare manned and unmanned spacecraft for exploring the solar system

Questions

1 An astronaut in a future mission to Mars could communicate with Earth by radio. He or she could speak to Earth, but there would be a gap of several minutes before the reply was heard. Explain why.

2 Which of the following objects may be found in the solar system?

   comet, galaxy, star, planet, black hole

3 When astronauts are in space, they may experience low gravity for a long time. What can be the consequence of this?

4 List *five* difficulties that would arise in a manned mission to Mars.

H 5 A TV satellite travels around the Earth in a circular orbit.

   a What force holds it in its orbit?

   b What name is given to any force which holds an object in a circular path?

   c How would the satellite move if the force stopped acting on it?

6 What astronomical objects are described here?

   a Its gravity is so strong that light cannot escape from it.

   b A cluster of billions of stars.

   c Dust from the trail of a comet that burns up in the Earth's atmosphere.

7 Both manned and unmanned spacecraft have been used to explore the solar system, although manned craft have only travelled as far as the Moon. Give *one* advantage and *one* disadvantage of using manned spacecraft for this.

Asteroid impact

The Moon's surface looks different from the Earth's. Its face is covered with craters, a sign that it has come under bombardment by rocks from space – but what rocks could these be? When did this happen? And why does the Earth look so different?

Figure 2g.1 shows that it isn't only our Moon which has a surface pockmarked with craters – many of the rocky objects in the solar system suffer from astronomical acne, too.

Figure 2g.1 This is Phoebe, one of Saturn's moons.

The Moon's craters were mostly created early in its life, when the solar system was young. At that time, the solar system was a much more hazardous place, with planets forming and colliding (recall how it is thought the Moon formed – Item P2e). Since then, things have calmed down a lot, but there are still dangerous objects moving in the solar system. There are two main types:

- **asteroids** – the rocky objects which normally orbit between Mars and Jupiter
- **comets** – the ice-and-dust objects which occasionally appear as they orbit the Sun.

When an asteroid or comet crashes into a planet or moon, it throws up a large mass of rock and dust, and a crater is formed. In July 1994, a comet was observed crashing into the planet Jupiter, so we know that these collisions still occur from time to time.

The craterless Earth?

There are craters to be found on Earth, but they are few and far between. Figure 2g.2 shows one, Meteor Crater in the US state of Arizona. It is several hundred metres deep, which gives an idea of the force of the impact when an asteroid, 50 metres or so in diameter, crash-landed 49 000 years ago.

Figure 2g.2 Meteor Crater was formed by the impact of an asteroid thousands of years ago.

In fact, the Earth must once have been as cratered as the Moon. However, the Earth has volcanoes and earthquakes, erosion by seas and rainfall, and the movement of tectonic plates; all of these tend to erase such features from the Earth's surface.

High impact

What happens when an asteroid or comet strikes the Earth?

- Hot rocks are flung upwards and outwards, forming a crater with a surrounding lip.
- Wildfires spread through neighbouring vegetation.
- Dust and smoke fill the sky, and may block sunlight for months or years.

Disappearing dinosaurs

Dinosaurs were once the dominant large animals on Earth, but they became extinct 65 million years ago. For a long time, no one knew why, but now the most popular theory links their extinction to an impact by an asteroid (Figure 2g.3).

The first evidence came when geologists noticed that rocks dated at 65 million years ago contained a thin layer of iridium. This is a metal which is common in asteroids but rare on Earth; the telltale layer was found all round the Earth.

Then a crater of the right age was found at the northern end of the Yucatán peninsula in Mexico. The asteroid that created this giant (but now eroded) crater must have been 10–15 kilometres across, much bigger than the one that created Meteor Crater.

Figure 2g.3 An artist's impression of the asteroid impact, which must have resulted in years of darkened skies and lowered temperatures, and drove the dinosaurs to extinction.

SAQ

1 What processes could have eroded the dinosaur-destroying asteroid crater?

- In extreme cases, the resulting cooling of the climate may result in the extinctions of many species of animals and plants.

We know this, because scientists have found supporting evidence by looking at known craters.

- They find the craters themselves, and their shape tells us they formed from an impact.
- They may find deposits of unusual elements such as iron and nickel for some distance around the crater. These have come from the asteroid which caused the impact.
- They may discover that there is a sudden change in the fossil species in adjacent layers of the surrounding rocks, showing that some creatures were wiped out.

Why asteroids?

To understand why the asteroids exist, we need to go back to the formation of the solar system.

Roughly 4.5 billion years ago, a giant, swirling cloud of dust and gas condensed to form the solar system where we live today. Gravity was the force which led to this condensation. Each particle in the cloud attracted every other particle, and they gradually came together. At the centre of the

cloud, a mass of hydrogen gas collected. It got hotter and hotter, glowing more and more brightly – the Sun had formed. Further out, dust and ice particles stuck together to form the planets (Figure 2g.4).

Figure 2g.4 The solar system is thought to have formed from a swirling cloud of dust and gas. The Sun and the planets formed at the same time. In this impression, the new Sun is beginning to glow while the surrounding cloud of dust and ice is condensing to form the planets. The force of gravity pulls all this material together.

In between Mars and Jupiter, one planet failed to form. Lumps of rock started sticking together, but each time their orbit took them close to Jupiter, its strong gravity pulled them apart again. This is still going on today; those lumps of 'would-be planet' are the asteroids.

SAQ

2 What force attracts particles of matter together to form a planet?

Comets

Several comets are likely to be detected each year, but only occasionally can they be seen

Figure 2g.5 Comet Hale–Bopp was a prominent feature in the night sky in April 1997. Its double tail consists of dust particles and glowing gas.

without a telescope. A comet is a dramatic object in the sky (Figure 2g.5); it moves across the background of the fixed stars, at a rate much greater than that of a planet. For a long time, it was believed that they were visitors from outside the solar system and that they foretold significant events. (Indeed, it seems likely that the 'Star of Bethlehem' was a comet which was visible from the Middle East in 5 BC.)

Now we know that comets are objects from the outer edges of the solar system. They are lumps of rock and ice which follow highly elliptical orbits around the Sun. (An ellipse is a shape like a circle which has been stretched out in one direction.)

The fact that their orbits are much more elliptical than those of planets means that comets behave in a different way to planets. A planet travels at more-or-less steady speed around its orbit. A comet moves most slowly when it is furthest from the Sun. This means that it creeps along in the furthest depths of the solar system, perhaps for hundreds of years, before it plunges back down towards the Sun, going faster and faster until it swings past the Sun, then more and more slowly as it travels back out into space.

A comet develops its tail as it approaches the Sun. The comet warms up, and frozen material starts to evaporate from it. Electrically charged particles streaming outwards from the Sun cause the tail to glow.

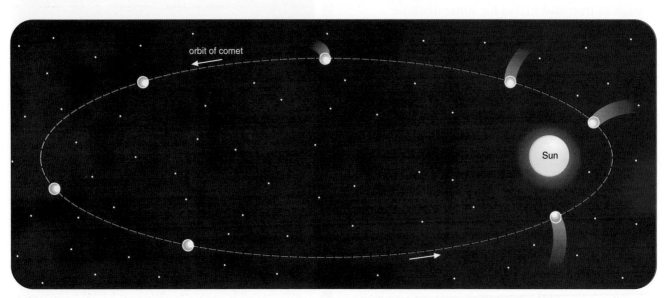

Figure 2g.6 A comet travels along an elliptical path around the Sun. It speeds up as it plunges in towards the Sun, then slows down again as it climbs back out into the depths of the solar system.

SAQ

3 Copy the drawing of the comet's orbit around the Sun from Figure 2g.6. Mark the point where it is moving most quickly and the point where it is moving most slowly.

H Changing speed

Why does a comet's speed vary as it travels around the Sun? We need to think about how the pull of the Sun's gravity changes as the comet moves along its orbit.

- When the comet is far from the Sun, the pull of the Sun's gravity is weak.
- When it is close to the Sun, the pull of gravity is stronger.

Picture the comet far from the Sun. It moves slowly, and the unbalanced force on it is small. As it gets closer to the Sun, the pull gets greater and it speeds up more and more. The comet 'falls' towards the Sun and it speeds up as it does so. It passes the Sun and then, as it moves away, it is slowed down by the Sun's gravity and so it moves more and more slowly.

SAQ

4 Complete your diagram (SAQ 3) by drawing arrows representing the force of the Sun's gravity pulling on the comet at different points in its orbit. Make sure that the arrows point in the correct directions. Draw longer arrows at points where the force is stronger.

Beware – near-Earth object

Why would an asteroid ever leave its orbit, 300 million kilometres away, and tumble in towards the Sun? The answer is that, if one asteroid should nudge another just a little, either of them might be pushed out of its orbit and start moving towards us.

Astronomers keep an eye out for unusual objects in the night sky. They are looking for comets, or even new planets. Every now and then, they announce that they have spotted something (an asteroid or a comet) whose orbit may take it close to Earth. This takes a large number of observations

to determine; they are looking at a bright speck of light in the night sky and watching how it moves against the background of stars from night to night. A computer works out the object's orbit and predicts its future path. If it is possible that it might strike the Earth at some time in the future, it is called a **near-Earth object (NEO)** – and we have a problem!

Figure 2g.7 Large telescopes like this are used to look for near-Earth objects.

SAQ

5 What two types of object might a near-Earth object be?

H Dealing with NEOs

How might we reduce the threat from NEOs?

Firstly, we need to know where they are. Astronomers need to survey the night sky with telescopes, looking for objects whose trajectories (paths) might bring them within the Earth's orbit.

Possible NEOs can be monitored by satellites. There have already been missions to take a close look at comets, so that we have more of an idea of their structure.

And what if we spot a NEO on a collision course with Earth? One approach would be to send up a

An expensive search

In the 1990s, many astronomers became concerned that a near-Earth object (NEO) might collide with our planet, with devastating consequences. If it landed in the sea (which is most likely), a giant tsunami could devastate the coasts of many countries. The energy released would boil off vast amounts of water, causing severe rainstorms. If it struck land, earthquakes and firestorms could result.

The Spaceguard Foundation was set up, with its headquarters in Rome, to bring together everyone with an interest in looking for NEOs. The Foundation encourages national governments to take the problem seriously and to think about how to deal with a possible future NEO problem.

It is tricky to get governments to spend money on this; the chances of an impact are small, but the consequences would be devastating. Should we spend money on a problem that may never arise?

Figure 2g.8 Crest of the Spaceguard Foundation.

H missile to blow it up – but that might just result in a lot of fragments hitting the Earth rather than one large object.

Figure 2g.9 shows another way of deflecting an approaching asteroid or comet. A large, curved mirror is positioned in space, close to the NEO. It reflects sunlight on to the object, heating it and causing ice to evaporate. This results in a gentle push on the object, changing its direction and guiding it away from the Earth. A nice trick if it works!

Figure 2g.9 Using sunlight to deflect a near-Earth object.

SAQ

6 An alternative method to deflect a NEO would be to explode a nuclear bomb next to it. Draw a diagram to show the path of a NEO approaching the Earth. Show where you would explode a bomb and how this would alter its path.

Summary

You should be able to:

◆ describe the origin and orbit of the asteroids

◆ describe the effects of an asteroid colliding with the Earth and evidence that this has happened in the past

H ◆ explain why the asteroids have never joined up to form a planet

◆ describe the composition of comets

◆ describe a comet's orbit and how its speed varies along its orbit

H ◆ explain why a comet's speed varies along its orbit

◆ state that a comet or asteroid whose orbit may bring it close to the Earth is called a near-Earth object

H ◆ suggest ways of reducing the threat of near-Earth objects

Questions

1 a Between which two planets is the orbit of the asteroids?

   b What material are asteroids composed of?

   c What evidence is there that the Earth has been struck by asteroids in the past?

2 a What materials are comets made of?

   b Describe the shape of a comet's orbit. Draw a sketch to illustrate your answer.

   c At what point in its orbit is a comet moving most quickly?

3 What name is given to a comet or asteroid whose orbit brings it close to the orbit of the Earth?

H 4 Explain why many small lumps of rock (the asteroids) orbit between Mars and Jupiter, rather than a solid planet.

5 A comet changes speed as it travels along its orbit.

   a What force holds it in its orbit?

   b How does this force change as the comet orbits?

   c How does this force affect the comet's speed?

6 Suggest a way in which a near-Earth object could be deflected from a collision course with Earth.

Beginnings and endings

Death of a star

One day – in about 4 billion years – the Sun will die. It will swell up to become many times its present size before collapsing inwards. How do we know this?

Telescopes allow us to see millions of stars. By looking at the differences between them, we can work out how **stars** form, live and die.

A star forms from a huge cloud of gas. This process can be seen in some parts of our own galaxy. However, most of the stars we see are in the steady, equilibrium state, producing energy at a constant rate. Some are at the end of their lives. How they die depends on their masses – see Figure 2h.1.

The Sun is a fairly typical, middle-aged, medium-mass star. In a few billion years, it will start to run out of the hydrogen fuel and it will no longer shine so brightly. As this happens, the Sun will blow up to form an enormous **red giant** star, engulfing the nearest planets, probably including the Earth. The cool outer layers will then blow off into space, forming a cloud of material around the remains of the star; this cloud is called a **planetary nebula** (*nebula* means *cloud*). Eventually, the inner material will collapse inwards (owing to gravity) and form a small **white dwarf** star (see Figure 2h.2). Finally, this will simply fade away as it cools to become a cold, dark lump of matter in space.

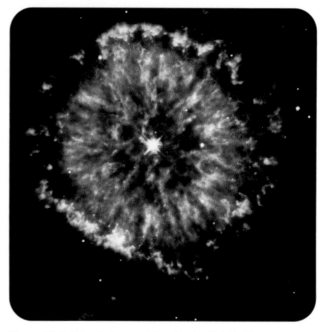

Figure 2h.2 This photo was taken by the Hubble Space Telescope. It shows a planetary nebula around a star which is collapsing to form a white dwarf. (A planetary nebula has nothing to do with planets.)

Massive stars

If a star is more massive than the Sun, it dies in a more spectacular way. It still forms a red giant but with a heavy core which collapses more and more. In a sudden searing flash, it bursts apart in a gigantic explosion. This is a **supernova**. Each year, astronomers detect a few of these catastrophic events out there in the Universe.

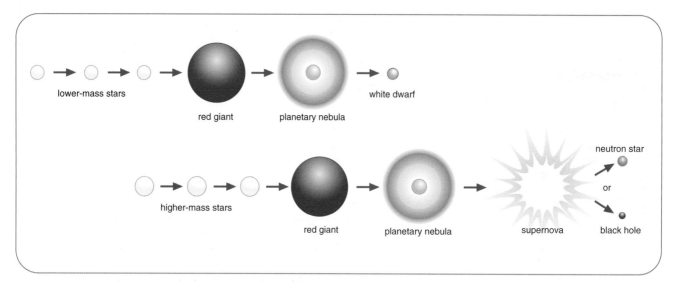

lower-mass stars

red giant

planetary nebula

white dwarf

higher-mass stars

red giant

planetary nebula

supernova

neutron star

or

black hole

Figure 2h.1 The way in which a star dies depends on its mass. It may end up as a white dwarf, a neutron star or a black hole.

The remnant of a supernova may be either a **neutron star**, which is a planet-sized object made mostly of neutrons, or, if the star was very heavy, a **black hole**. A black hole consists of matter which has collapsed inwards under its own gravity; its density is so great and its gravitational pull is so strong that nothing can escape from it, not even light.

SAQ

1 Put the following in order, to represent the life cycle of one type of star:

    star, black hole, supernova, red giant

2 What can you say about the mass of the star whose life cycle is given in SAQ 1?

How stars form

A star forms when a cloud of dust and gas pulls itself together by its own gravitational attraction; each particle is attracted by each of the others so that each feels itself pulled towards the centre of the mass. As the cloud collapses inwards, it gets hotter and hotter. (Gravitational potential energy is being transformed to kinetic energy.) The material at the centre of the cloud becomes a **protostar** and then a fully fledged star as nuclear reactions become established; planets may form around it. Figure 2h.3 shows a region of our own galaxy where new stars are forming in a region of space which is rich with gas and dust.

Figure 2h.3 New stars are seen here in the Eagle Nebula, forming from clouds of dust and gas.

The temperature in the central core of a star is millions of degrees Celsius. The particles of which it is made are rushing around very energetically. They press outwards, preventing the star from collapsing any further under its own gravity (see Figure 2h.4). Thus the outward force provided by the pressure of the moving particles balances the inward force of gravity. The two forces are in equilibrium.

Stars (including the Sun) produce light by the process of **thermonuclear fusion**. In fusion, the nuclei of light elements join together to form the nuclei of heavier elements. For example, four hydrogen nuclei may fuse to form a helium nucleus. In the process, energy is released, and this is the source of the star's radiation. For fusion to occur, very high temperatures are needed. The temperature inside a star reaches millions of degrees Celsius, so fusion is possible.

Inside the star, where it is hottest, fusion is going on at a steady rate. Energy is being released and this energy eventually escapes from the surface of the star.

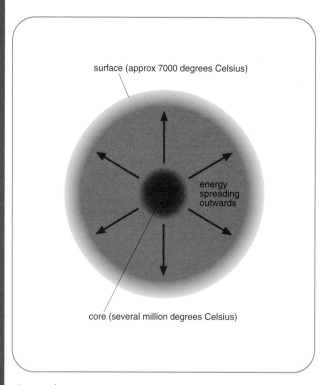

surface (approx 7000 degrees Celsius)

energy spreading outwards

core (several million degrees Celsius)

Figure 2h.4 Inside a star. The surface temperature of a star is a few thousand degrees Celsius. Inside, in its dense core, it is much hotter. This is where the process of nuclear fusion is going on, releasing energy. The inward force of gravity, tending to make the star collapse inwards, is balanced by the outward pressure of the moving particles.

A star can remain like this for millions or billions of years. The Sun has been glowing for over 4.5 billion years and is probably only halfway through its life. Stars like this are known as **main sequence** stars.

SAQ

3 What process is going on inside a star to release energy?

4 Why does a star not collapse completely under its own gravitational pull?

Black holes

Black holes are fascinating objects. They have a large mass compressed into a small volume, so their density is extremely high. Any nearby object *outside* the black hole is attracted by its gravity, which pulls it in so that it becomes part of the black hole. Any object *inside* the black hole cannot get out, no matter how fast it is moving. Even light (and nothing can move faster than light) cannot escape.

You might think that a black hole would be very difficult to see, and you would be right. One way in which their presence can be detected is by the X-rays which are produced when a black hole rips material from a nearby star (see Figure 2h.5).

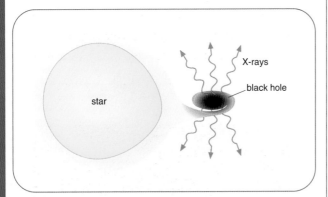

Figure 2h.5 A black hole is rather like a plughole in space. Any object which comes near it is likely to be sucked in. Here, material is being dragged in from a nearby star. As it accelerates into the black hole, it loses energy in the form of X-rays. Detecting X-rays coming from the site of a supernova hints at the existence of a black hole where a star used to be.

The life of the Universe

We live on a small planet circling an average star, towards the edge of a spiral galaxy. Our star is just one of billions of stars in the galaxy, which is just one of many billions of galaxies. Our importance at the centre of things seems to have evaporated. Many scientists argue that there must be many more planets like the Earth and that some must have intelligent life on them. (Many other scientists disagree with this – it's a hot debate.)

Today, astronomers believe that the most distant stars in the Universe are 10 billion to 15 billion light years away, and that the Universe itself is 10 billion to 15 billion years old. What evidence do they have to support this idea?

The expanding Universe

The Universe is expanding. This discovery was made in the 1930s and was based on the work of an American astronomer, Edwin Hubble. He looked at the light from stars in other galaxies. He deduced that the stars (and therefore the galaxies they are in) are speeding away from us. The most distant galaxies are moving away the fastest.

Imagine such a Universe. It is full of galaxies, all rushing outwards, away from each other. Now

Figure 2h.6 Edwin Hubble discovered that distant galaxies are moving away from us. The Hubble Space Telescope is named after him.

imagine how this Universe must have been in the past. Run the movie backwards in your mind; all of the galaxies rush back together, so that they meet at a single point. That point in space and time was the start of the Universe.

SAQ

5   Imagine that all of the galaxies are observed to be moving away from us. Would this be enough to prove that, at some time in the past, they had all exploded outwards from the same point? Explain your answer.

How far is that star?

Hubble discovered that, the more distant a galaxy, the faster it was moving away from us. But how did he know how far away each galaxy was?

He relied on the work of Henrietta Leavitt who published her results in 1912. She made detailed observations of variable stars. These are stars whose brightness varies from day to day, in a regular way. By finding a pattern which related a star's brightness and the rate at which it varied, she was able to work out how far a variable star was from the Earth. This meant that astronomers could work out how far away other galaxies were.

Figure 2h.7 Henrietta Leavitt, the American astronomer who found a method of measuring the distance to other galaxies.

The Big Bang

Most physicists and astronomers now accept Hubble's picture of the start of the Universe. It began with a gigantic explosion. All of the matter and energy which is now in the Universe exploded out of a single point – this was the **Big Bang**. In the first seconds of the life of the Universe, it was incredibly dense and incredibly hot. In the billions of years which have followed, it has expanded and cooled. Stars and galaxies have formed, and planets have appeared, with life on at least one of them.

From initial temperatures of billions of degrees, space has cooled as it has expanded. Much of the energy of the Big Bang is now in the form of radiation, spread thinly throughout space, so that the average temperature is only about −270 °C (three degrees above absolute zero). Physicists have detected this background radiation, known as the **microwave background** because it consists of low-energy microwaves. It comes to us from all directions in space. This is one of the main pieces of evidence that the Universe really did come into existence as a result of a Big Bang.

How will it all end? No one is sure about this. Figure 2h.8 shows two possibilities. As the Universe expands, it gradually slows down. The gravitational attraction of the galaxies for each other causes them to decelerate. This may not be enough to stop them from moving apart for ever. However, if there is enough matter in the Universe, the galaxies may eventually slow down,

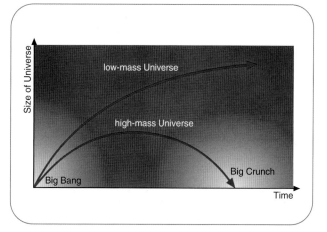

Figure 2h.8 The fate of the Universe depends on the amount of mass it contains. If there is sufficient, its expansion may stop and it will contract to a Big Crunch. If there is less mass, it will continue to expand for ever.

go into reverse and head back towards one another. The Universe will end in a Big Crunch.

None of us will live to see this happen. However, the work of the next generation of physicists will almost certainly allow us to know which is the likely fate of the Universe.

What Hubble saw

Edwin Hubble looked at the light coming from distant galaxies. For this, he needed a very powerful telescope. He noticed that the light from distant stars was redder than that from nearby stars. He realised that this '**red shift**', or change in frequency, occurred because the stars were speeding away from us.

The red shift which Hubble noticed is related to the Doppler effect. You may have noticed that, when an emergency vehicle approaches you and then goes past, the note of its siren drops as it moves away – this is the Doppler effect. In a similar way, the light waves coming from a distant star are stretched out as the star moves away from us (Figure 2h.9). Their wavelength is longer, so they are closer to the red end of the spectrum.

What Hubble discovered was that almost all galaxies appear to be moving away from us; their light is red shifted. What is more, the further away the galaxy, the faster it is moving.

From his measurements (and from more recent, more accurate ones), we can work back and deduce the age of the Universe, the time when all of the galaxies would have been together. The general agreement is that it is roughly 14 billion or 15 billion years old.

SAQ

6 Most galaxies are moving away from us but a few nearby galaxies are moving towards us; their light is shifted towards the blue end of the spectrum. Draw a diagram like Figure 2h.9 to show why the wavelength of light from an approaching star is blue shifted.

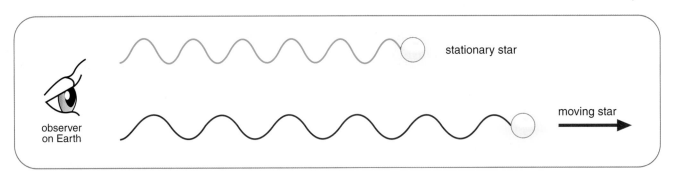

Figure 2h.9 Light waves reach us from a distant star. If the star is moving away from us, the waves are stretched out and the light looks redder.

Summary

You should be able to:

- state what stars form from

- describe the end of medium- and high-mass stars

H - summarise the life history of a star

- explain the properties of black holes

- outline the Big Bang theory of the origin of the Universe

- describe the motion of the galaxies and the microwave background radiation

H - explain how Hubble deduced that most galaxies are moving away from us and how this leads to an estimate of the age of the Universe

Questions

1 When a star dies, it may become a white dwarf, a neutron star or a black hole. What factor determines which of these it becomes?

2 Draw a diagram to represent the stages in the life of a medium-mass star, such as the Sun.

H 3 Imagine that Edwin Hubble had discovered that the light from distant galaxies was not red shifted but that it was the same as light from stars in our own galaxy.

  a What would this tell you about the Universe?

  b Could you deduce anything about the age of the Universe?

Answers to SAQs: Biology

Item B1a Fit for life

1 Photosynthesis; in chloroplasts in palisade and spongy mesophyll cells in the leaves of plants.
2 They both involve oxidation; release energy; produce carbon dioxide; produce water.
3 Aerobic respiration takes place in cells but burning does not; aerobic respiration occurs in a series of small steps, combustion does not; aerobic respiration causes a much smaller temperature rise than burning.
4 Stamina and possible cardiovascular efficiency.
5 Your answer should mention both positive and negative points about the two methods you have chosen.
6 Systolic, because this is the highest pressure.
7 You feel faint if brain cells don't get enough oxygen. If blood pressure is low, then blood may not be pushed sufficiently upwards, so that not enough blood (which carries oxygen) reaches the head.

Item B1b What's for lunch?

1 Carbohydrates, fats and proteins.
2 $70 \times 0.75 = 52.5\,g$
3 Young people are still growing, and protein is needed for growth.
4 Proteins have many functions in the body other than providing energy, for example making new cells, as enzymes, as haemoglobin, as keratin and so on.
5 Answers will vary.
6 $\dfrac{57}{1.6^2} = 22.3$

   This is well within the range of 20 to 24, so her body mass is about right.
7 They may identify with television personalities, models and other celebrities who tend to have low body weights. (You may be able to think of other reasons, too.)
8 a mouth, oesophagus, stomach, duodenum, ileum (small intestine), colon, rectum (large intestine), anus.
   b liver, pancreas (these are the only ones shown on the diagram, but there are others such as kidney, bladder).

Item B1c Keeping healthy

1 Any of the diseases caused by micro-organisms.
2 a They have decreased.
   b Yes, very similar, although the decrease is less.
   c The smaller decrease for women could be due to increases in the number of women who smoke – smoking increases the risk of getting many different kinds of cancer.
   d This could explain the relatively low mortality rate in women, where breast cancer is one of the commonest cancers.
3 a 30%
   b by about 7%
   c In both men and women, survival rates for people diagnosed with cancer improved (increased) between 1985 and 1994. The improvement has been greater for men than for women. However, survival rates were higher for women than for men in both 1985–1989 and 1990–1994.
   d More men than women get lung cancer, which has a relatively low survival rate.
4 The mosquitoes that carry malaria could live in Britain if the temperature rises.
5 Answers will vary.
6 a After about 2 days (when the temperature starts to rise); they did not feel ill straight away because it took time for the virus to begin to reproduce.
   b The white blood cells that 'matched' the virus may not have come into contact with the virus straight away; it took time for these cells to multiply and secrete the antibody.
   c The antibodies helped to destroy the virus.
7 The viral numbers would probably have stayed very low. The antibody numbers would have begun higher, or at least begun to rise much more quickly after infection by the virus.
8 Active, because your own immune system has made the antibodies.
9 a antibiotic e
   b diffusion

Item B1d Keeping in touch

1 The head is in the best position to come into contact with the environment first as you move forward, and – because it is the highest part of your body – to have the best chance of receiving information (e.g. as light, sound, smell) first.

2 The blood would absorb light, so the image on the retina would be faint, distorted or wrongly coloured.

3 A computer screen is quite close to the eyes, so the ciliary muscles are contracted, making the lens more rounded. The contraction of the muscles can tire the eyes. Looking up and into the distance allows them to relax, as they focus the eyes on a more distant object.

4 When looking at a nearby object, you are usually looking down (for example at a book). The lower part of the lens in bifocals helps to bend the light rays more, which is necessary when focusing on a close object. The upper part is used when looking into the distance, when light rays do not need bending as much.

5 Reflex actions are very fast and help you to get out of a potentially dangerous situation. Voluntary actions are slower but have much more variety, so you can make choices about what you do.

6 The vesicles of transmitter substances are only present on one side of the synapse.

Item B1e Drugs and you

1 a The incidence of lung cancer in men rose from 1951 to about 1978 and then fell.

   b The number of men who smoked fell fairly steadily from 1951 to 2002.

   c Smoking causes lung cancer, but it takes several years for the cancer to develop. Even though smoking fell from 1951 onwards, lung cancer did not start to fall until nearly 30 years later.

   d Women have always smoked less than men. However, the rate at which smoking decreased in women from 1951 was much less than that for men (indeed, it actually increased from 1955 to 1967) and by 2002 the numbers of women smoking were almost the same as the numbers for men.

   e Men may have responded to the rise in deaths from lung cancer by cutting down smoking. Women may have felt that this was not so important for them, because deaths from lung cancer in women were always much smaller than in men.

2 a Deaths were highest in 1979. They almost halved by 1989 and then continued to fall steadily until 1995, since when they have fluctuated at levels between 460 and 560 deaths per year.

   b The high figure for deaths in 1979 matches the high figure for the percentage of drivers over the limit then. Since then, the percentages have fluctuated between 15% and 21%, and it is difficult to see a clear link between the two figures. However, the high figure of 560 deaths in 2002 is mirrored by the high figure of 21% over the limit.

   c Breath tests were carried out on a much larger scale after 1979, and this can explain the improved figures for deaths and drivers over the limit since then. Drivers are now much more aware of drinking and driving, and only 4% of people now being given breath tests are over the limit, compared with 34% in 1979. This probably explains most of the reduction in the number of people over the alcohol limit who were involved in fatal accidents. It is interesting that the percentage of drivers killed who were over the limit is much higher than the overall percentage of drivers testing over the limit – showing that driving when over the blood alcohol limit greatly increases the chance of being killed in an accident.

Item B1f Staying in balance

1 An elderly person may move around less, so they generate less body heat. They may not want to spend money on heating, so keep their house colder in winter.

2 The raised hair traps air. Air is a poor conductor of heat, so it decreases the rate of heat loss by conduction. Heat lost by radiation from the skin warms the trapped air. The trapped air cannot easily move, so it is harder for convection currents to form.

3

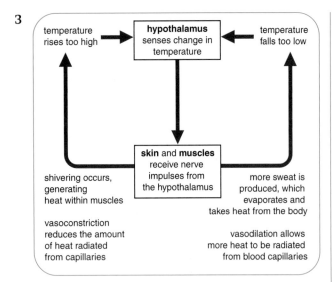

4 The liver.

5

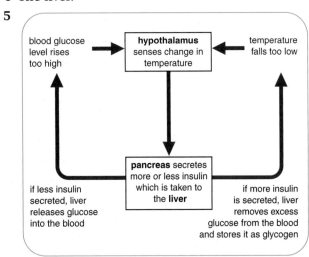

Item B1g Who am I?

1 gene, chromosome, nucleus, cell
2 64 (4^3)
3 It is sexual reproduction, because it involves gametes and fertilisation.

Item B1h New genes for old

1 The radiation could harm her fetus, causing mutations in its cells. Because the fetus is still growing, the cell that is harmed could divide many times over, so the mutation would be present in many cells in the baby's body. In particular, the radiation could harm developing gametes in her fetus's ovaries or testes, which could cause mutations that could be carried on into the next generation.

2 a BB, Bb or bb
  b BB and bb are homozygous; Bb is heterozygous.
  c BB and Bb give brown eyes, bb gives blue eyes.
3 No. The Punnett square simply shows the chance of a child having these different genotypes. The couple can have any number of children, or perhaps none at all.
4 The chance is still the same: one in four. The chance is the same each time they have a child, no matter what children they have already had.
5 a ff
  b f
  c F
  d FF or Ff – so none will have cystic fibrosis.
  e

	F
f	Ff

6

Parents	father	mother
genotype	Hh	Hh
gametes	H h	H h

	H	h
H	HH	Hh
h	Hh	hh

A child with HH or Hh will be normal, while a child with genotype hh will have sickle cell anaemia. Each time the couple have a child, there is a one in four chance that it will have sickle cell anaemia.

7 a For Amy to have cystic fibrosis, she must have two recessive alleles for it. So her parents must both have been heterozygous, with one dominant allele (so they did not have cystic fibrosis) and one recessive one. Both Sal's and Jo's ancestors must have had this recessive allele but, by chance, no one in the family had previously had two recessive alleles, except Sal's great great granduncle.
  b It is no one's fault that they have a particular allele – you cannot help your genes and did not choose them. Moreover, this allele must have come down on both sides of the family, not just Sal's side.
  c *no answer for this*

Item B2a Ecology in our school grounds

1 Grass approximately 68%; daisies approximately 15%; dandelion approximately 8%; plantain approximately 3%; bare ground approximately 6%. (These are only very approximate and any answers within a reasonable range are acceptable.) It is difficult to do this calculation because the square is large in comparison with the plants within it. You can make it easier by dividing the square up into smaller pieces.

2 Seven or eight, because the graph shows that, on this occasion, no more species were found after seven quadrats had been used.

3 Always use the same net. Always sweep for the same distance. Always sweep at the same speed. Always sweep in front or to the side of you, into undisturbed vegetation.

4 Each quadrat had an area of $0.25\,m^2$, which is about 1/160 of the total area. This could mean that the results are not very reliable, because only a small portion of the total area has been sampled.

5 Quite a large number of woodlice were recaught on the second occasion, which suggests that quite a large proportion of them were caught and marked on the first occasion. However, there are many things that we don't know – such as whether the marked ones were more likely to be caught and killed by a predator.

6 a estimated population size
$$= \frac{(22 \times 21)}{12} = 38 \text{ or } 39$$

   b She could have caught and marked more water boatmen.

Item B2b Grouping organisms

1

Plants	Animals
cells are surrounded by a cell wall made of cellulose	cells do not have a cell wall
cells may contain chloroplasts, which contain chlorophyll	cells do not contain chloroplasts
cells usually contain a large vacuole with cell sap	large vacuoles with cell sap not present (though there may be small, temporary vacuoles)
feed by photosynthesis	cannot photosynthesise; feed on other living things
cannot usually move around	can usually move around

2 It can move around, like an animal. It can photosynthesise, like a plant.

3 a Mammals have fur and feed their young on milk.
   b Fish have wet scales and gills.
   c Reptiles have dry scales and lay eggs.
   d Amphibians have a moist, permeable skin.
   e Birds have feathers and a beak.

4 Like chimpanzees, we can stand upright and use our hands for grasping and using tools. Chimpanzees differ from us in having fur and not using language. (You can probably think of many more similarities and differences.)

5 The greenfinch and the goldfinch are the most closely related, because they are in the same genus.

6 They are all streamlined, with smooth bodies. This helps to reduce friction while moving through water. They all have a tail which helps with propulsion and steering.

7 Seals and whales have small amounts of hair on their bodies, which is a mammalian characteristic. Their young develop in the mother's uterus, attached by a placenta. After giving birth, the mothers feed their young on milk. Sharks have none of these features. Sharks have a scaly body, and breathe through gills rather than lungs. They also have fins, which mammals do not have.

Item B2c The food factory

1 The branching shape provides a large surface area that can absorb sunlight. It also means that the plant has a large surface area to volume ratio, making it easier for carbon dioxide to diffuse into the photosynthesising cells.

2 Reactants: carbon dioxide and water
Products: glucose and oxygen

3 glucose + oxygen → carbon dioxide + water
The reactants in the respiration equation are the products in the photosynthesis equation, and vice versa.

4 Starch is not soluble, so could not be carried in solution in the phloem vessels. Sucrose, like all sugars, is soluble in water.

5 a As light intensity increases, so does the rate of photosynthesis, until a point (B) at which the rate begins to level off.

 b As light intensity increases, more light energy is received by the chlorophyll, which means that photosynthesis can happen faster.

6 It is warmer and there is more light.

7 Plant cells will be the same temperature as their environment, which will usually be a lot less than 37 °C. Their enzymes are adapted to work best at the lower temperature inside the plant cells.

8 The line on the graph should go up as temperature increases. Beyond a temperature of about 25 °C, the line should drop sharply until it hits the x-axis.

9 a It is not light, because even if you give the plant more light it cannot photosynthesise any faster. It must be something else, probably carbon dioxide.

 b It is carbon dioxide, because if carbon dioxide concentration increases then so does the rate of photosynthesis.

 c Light or temperature. You could provide the plant with more light or a higher temperature when the CO₂ concentration is 0.02%; if the rate of photosynthesis increases, then that was the limiting factor.

10 Light

11 Oxygen, because there is more photosynthesis than respiration.

Item B2d Compete or die

1 They would try to live near a water supply, so you would get large populations near water and smaller ones away from water. However, if there is not enough water to go round at the water supply, then some wildebeeste may move away to try to find water elsewhere.

2 The less water there is, the fewer plants will be able to get enough water for survival, so the population size will decrease.

3 a The more food they are given, the smaller the percentage of larvae that die. (Notice that the x-axis shows the number of larvae per mg of food so, as you move to the right, each larva gets less food.)

 b If less food is given per larva, there is not enough to go round and the larvae compete for it. Some will not get enough to stay alive.

4 A parasite does not usually kill its host, whereas a predator does kill its prey. A parasite usually lives in close association with its host, whereas a predator does not.

5 a About 4 weeks

 b Some may have been immune to the myxomatosis virus so they did not get the disease. Some may not have come into contact with the virus or been infected with it. Some may have had the disease but not died from it.

 c The population went back up a little way in the second year and then went down again. There are several possible reasons for this. For example, perhaps the virus almost disappeared when the rabbit population fell so low in month 12, so the rabbit population started to go up again until they were once again infected and killed. Perhaps the rabbits that were immune to the virus were able to breed after month 12, and their offspring survived for a while before the virus changed (mutated) and was able to infect even these resistant rabbits.

6 They could be parasites, because they do not kill the fish they feed on. But they do not live in close association with them, so this is more like a predator.

7 The roots of the bean plants contain nodules with nitrogen-fixing bacteria, which can increase the quantity of nitrogen-containing compounds in the soil. The crop that grows the next year can absorb some of these compounds and use them to make proteins.

Item B2e Adapt to fit

1 They are most likely to compete for food.
2 It has claws to hold prey, and a sting to kill prey.
3 They can roll into a ball, so the predator is confronted with nothing but prickles.
4
  - The fennec has large ears, giving it a large surface area to volume ratio which helps it to lose heat by radiation. The arctic fox has smaller ears, which help to conserve heat.
  - The fennec has thinner fur than the arctic fox. The thick fur traps air and reduces heat loss from the body.
  - The arctic fox has white fur, for camouflage against the snow. The fennec fox has sandy fur, for camouflage against the sand.

5

Insect-pollinated flower	Wind-pollinated flower
brightly coloured petals which attract insects	no petals, or petals are dull green or brown
anthers enclosed within the petals, so insects brush past them and pick up pollen	anthers hang outside the flower, so they easily catch the wind and pollen can be blown away
filaments usually quite stiff	filaments can bend easily so the anthers can move in the wind, helping pollen to be blown away
stigma enclosed within the petals, so insects brush against it and deposit pollen	stigma hangs outside the flower, making it easier for it to catch pollen on the wind
stigma usually flat and sticky, able to trap pollen from insects' bodies	stigma feathery, increasing its surface area and making it more likely it will come into contact with pollen
nectar produced, which attracts insects to the flower	no nectar produced

Item B2f Survival of the fittest

1 Soft parts are much more likely to be crushed by sediment accumulating on top of them, so they have no chance to become fossilised.
2 The south-west winds carried polluted air towards the east. This increased the darkness of tree trunks in the east, making it more likely that the black moths would survive.

3 Stop using warfarin as a poison for a while. The selective advantage should then swing over to the non-resistant rats. After a while, rat populations will probably contain mostly warfarin-sensitive rats, so we could use the poison again.
4 If food was in short supply, any giraffe that by chance had a slightly longer neck and could reach higher into trees to browse would have a slightly greater chance of survival. If this happened over several generations, then most of the giraffes would probably have descended from relatively long-necked ones and would have inherited the genes for long necks.

Item B2g Population out of control?

1 a They increased. As people become more affluent, they can afford to buy cars. Cars have become relatively cheaper, as production processes become more automated and efficient.
  b You may be able to pick out a number of exceptions. For example, sales have not increased very much in South Korea or India. In South Korea and India, many people cannot afford to buy or run a car. Sales have fallen in Germany and Japan in recent years. This may be because most people in Germany and Japan already have a car, or because the government is successfully persuading more people to use public transport.
  c Car sales have increased so it is likely that consumption of petrol (a fossil fuel) will also have increased.
2 a They have increased.
  b From burning fossil fuels.
  c Carbon dioxide emissions from developing countries were less than those from developed countries until about 1984. Now the opposite is true.
3 a USA, Russia, Germany, Japan, UK, France, China, Indonesia, India
  b About 10 times greater.
  c In the USA, most people have enough money to buy a car. In India, many people are poor and cannot afford one; if they do have one, they will not use it as much because they cannot afford fuel. So there is less CO_2

emitted from vehicles. In the USA, everyone uses electricity, and fossil-fuel-burning power stations release more CO_2 than those in India where fewer people use electricity. There are more industries that burn fossil fuels in the USA than in India.

   d Overall, carbon dioxide emissions per person are greater in developed countries than in developing ones.

4 a Rio de Janeiro, Beijing, Moscow and Istanbul.

   b i The figures don't tell us any reasons, but we could suggest that: the shape of the landscape around the city tends to trap air rather than letting it move away; more fossil fuels, especially coal (which contains a lot of sulfur) are burned there.

     ii Answers may vary.

    iii Answers may vary.

5 They push their tail through the surface, into the air. Oxygen can diffuse into it.

6 a River A, because it contains none of the species with a high biotic index.

   b We don't really need to know the numbers of organisms of each species, only whether it was found there or not.

Item B2h Sustainability

1 a The birds have no adaptations to avoid predation, because they have not evolved in a place where there were predators. They are now endangered by introduced predators that kill and eat them.

   b All the introduced predators could be removed from New Zealand. However, this is probably impossible. The birds could be placed on an island where no predators are present – this has actually been done on several islands around New Zealand. Some of the birds could be bred in captivity.

2 In the changed conditions, they could not run as fast as the horses with fewer toes. They were more likely to be caught and killed by predators. Their small size would also have been a disadvantage in avoiding predators.

3 Benefits: tourists might visit to see the wolves; wolves could keep down populations of grazers such as red deer, so that overgrazing was reduced.

Problems: wolves might kill domestic animals, such as sheep or pet cats; they might kill other wild animals that are endangered, such as wild cats; they might compete with other predators, such as foxes.

4 They can help to make people aware of the problems of threatened species, so that they may help by providing donations or putting pressure on firms not to sell products that may be harming habitats. They can allow researchers to find out more about the animals and how they live, so they can decide which kinds of habitats are most important for their survival in the wild.

5 a About 1100 m.

   b Only a few minutes.

   c About 50 minutes to one hour.

   d This is when the whale was moving fastest.

   e It seems to move fastest when it is deeper under the water, and slowest when it is close to the surface.

6 People will care more about whales and help to put pressure on governments and other organisations to try to conserve them. They may give money to whale conservation projects. Local people can make money from the whales, so may try to conserve them by preventing whale hunting or by making sure that the sea remains a good place for whales to visit – quiet and unpolluted.

7 a Fewer fish can be caught, so fewer people will be able to fish. Fewer people will get jobs associated with fishing, such as cleaning and packing fish, transporting them or selling them.

   b Fish may be in relatively short supply, so prices are likely to rise.

8 Habitats are more likely to survive, so species have a place to live. Climate change may be slowed down, making it less likely that species will become extinct because their habitat becomes too wet, dry, hot or cold, or because plants or other animals that an animal eats are killed by climate change. Pollution will be less, so species are less likely to be killed by toxic substances or changes to the composition of the atmosphere or the water where they live.

Answers to SAQs: Chemistry

Item C1a Cooking

1. 1. potato, chips; 2. meat, burgers; 3. flour etc, doughnuts.
2. Tough – it has the wrong texture; hard to digest; poor flavour; contains micro-organisms.
3. Appearance: egg white changes from clear and colourless to opaque and white; yolk changes from orange to yellow.
   Texture: both egg white and yolk change from liquid to solid.
4. New substances are formed; the change is permanent; an energy change is involved.
5. a group of atoms joined together.
   b the molecules change shape.
   c the meat becomes more tender, colour changes, flavour changes.
6. a the ones on the right, because they have risen.
   b the one on the left, because it hasn't risen.
   c baking powder.
7. a four.
   b six.
   c sodium, hydrogen, carbon, oxygen.
   d two hydrogen, one oxygen.
   e one carbon, two oxygen.
8. a it decomposes.
   b sodium carbonate, water, carbon dioxide.
   c sodium hydrogencarbonate.
   d bubble it into limewater.
9. a one sodium, one hydrogen, one carbon, three oxygen.
   b two sodium, two hydrogen, two carbon, six oxygen.
   c there are more atoms in the products than there are in the reactants.
10.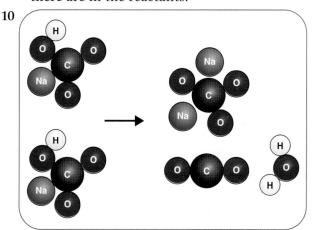

Item C1b Food additives

1. a Soup: water. Smarties: milk chocolate.
   b Soup: colour, flavour enhancers, antioxidants. Smarties: emulsifier, colours.
   c Soup: the colours are carotenes; the flavour enhancers are disodium guanylate and disodium inosate; the antioxidants are alpha-tocopherol, ascorbic acid and natural tocopherols.
   Smarties: the emulsifier is lecithin; the colours are E171, E104, E124, E110, E122, E133, and E120.
2. a Because they both get on with Emma.
   b Emma is the emulsifier. She is able to 'mix' with Oliver (oil) and with Walter (water). In this way Oliver and Walter (oil and water) are able to 'mix'.
3. a To make it look more appealing, to make it look more like a familiar 'orange' colour.
   b Tartrazine can cause itching and hyperactivity in young children.
4. a The meat should have been packed in breathable film.
   b The person should have bought dried soup and left it in the packet until they wanted to eat it.
   c The person should have bought coffee in a self-heating can (or bought a thermos flask!).

Item C1c Smells

1. The perfume has irritated her skin. Perfume should be non-irritant.
   The perfume has poisoned her. Perfume should be non-toxic.
   The perfume has reacted with water in his sweat and its smell has become unpleasant. Perfume should be unreactive.
2. They don't evaporate easily.
3. a ethanoic acid + ethanol
                    → ethyl ethanoate + water
   b A catalyst is a substance that changes the rate of a chemical reaction while remaining unchanged itself.
   c The reaction is slow.
   d Safety glasses, no naked flames, great care with spills (wipe up immediately, rinse off

hands), re-stopper all bottles quickly due to fumes, possible use of fume cupboard.

4 a solute; c dissolved;
   b solvent; d solution.

5 a propanone;
   b water;
   c ethanol;
   d no, the table says 0 g dissolve in 50 cm^3 of water;
   e propanone.

6 a The intermolecular forces between water molecules and alcohol molecules are similar to the intermolecular forces in the two pure substances.
   b The intermolecular forces between water molecules and oil molecules are weaker than the intermolecular forces in the two pure substances.
   c The intermolecular forces between nail varnish molecules and nail varnish remover molecules are similar to the intermolecular forces in the two pure substances.
   d The intermolecular forces between water molecules and ester molecules are weaker than the intermolecular forces in the two pure substances.

Item C1d Making crude oil useful

1 a Coal, oil, gas.
   b Coal from fossilised trees, oil from fossilised sea creatures, gas from fossilised trees and fossilised sea creatures.
   c They are being used up faster than they are being re-formed.
   d Oil spillages poison wildlife and pollute beaches.

2 a The fractionating tower is hotter at the bottom and cooler at the top. The hydrocarbons in the crude oil separate according to their boiling points.
   b No.
   c e.g. refinery gas, gasoline, diesel oil.

3

This is a strong covalent bond   This is a weak intermolecular bond

This is not broken when the hydrocarbon boils   This is broken when the hydrocarbon boils

4 a Reactant is undecane, products are octane and ethene.
   b Undecane is $C_{11}H_{24}$, 35 atoms in one molecule; octane is C_8H_{18}, 26 atoms in one molecule; ethene is C_2H_4, 6 atoms in one molecule.
   c Undecane and octane are alkanes, ethene is an alkene.
   d Has a C=C double bond.
   e To make a polymer (plastic).
   f As petrol.
   g The products of the cracking reaction sell for higher prices than the reactants would have done.

Item C1e Making polymers

1 a compound of carbon and hydrogen only.

2 a C_2H_6.
   b One ethane molecule consists of two carbon atoms and six hydrogen atoms.
   c i eight;   ii four;   iii one.
   d i eleven; ii four;   iii one.
   e

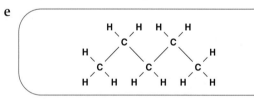

3 a A F G H J L;   f A F G H J L;
   b B C D E I K;   g A F G;
   c all;   h B K;
   d B C D E I K.   i molecular and displayed.
   e all;

4 a A F G H J L;   b B C D E I K.
   c has only single bonds/has the maximum number of hydrogen atoms per molecule for a molecule with this number of carbon atoms.

d has at least one double bond/has less than the maximum number of hydrogen atoms per molecule for a molecule with this number of carbon atoms.

5 a alkene; **b** unsaturated;

c because the hydrocarbon decolourised the bromine water.

6 a i ethene, poly(ethene);

ii propene, poly(propene).

b polymer, monomer, addition.

c poly(butene), butene, polymerisation, pressure, catalyst.

7 a
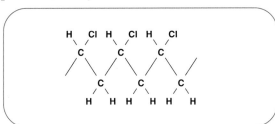

b
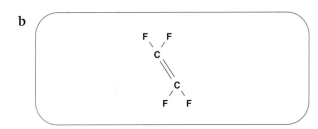

8 no change/bromine not decolourised; polymers are saturated/have no C=C double bonds.

Item C1f Designer polymers

1 a The ways in which a substance behaves.

b Easy to melt and mould, easy to colour, no sharp edges.

c Will form a thin fibre if melted and forced through a spinneret.

2 a The thermoplastic – diagram **a** – no crosslinks; the thermoset – diagram **b** – has crosslinks.

b Polymer A, is a thermoplastic, easily melted and made into thin sheets, stretchy.

c Thermosets have high melting points and are rigid.

d Thermoset, high melting point desirable with hot cookware in use, rigidity will improve damage resistance.

3 a light, packs easily, hard wearing, can be brightly coloured.

b breathable/lets sweat escape/doesn't feel clammy.

4 a Because the membrane is waterproof; because the pores in the membrane are too small for rain droplets to get through.

b Because the pores in the membrane allow molecules of water vapour to get through.

c So the water vapour can get through the nylon as well as the membrane.

d A material made by combining thin layers of two or more materials.

e Membrane is too fragile or not strong enough.

5 Incineration can produce toxic fumes and wastes a valuable resource. In order to recycle them, waste polymers must be sorted, which can be expensive. Since plastic bottles don't smash in recycling banks the way glass bottles do, the recycling banks get full quickly.

Item C1g Using carbon fuels

1 a advantages – cheap, readily available, easy to store; disadvantages – causes pollution, long start-up time; fuel-oil-powered cars are not expected as the start-up time is unreasonable.

b check it is always turned off properly to avoid leak; make sure the sitting room is well-ventilated.

c yes – cheap to buy, unusual; no – the car's fuel tank needs to be adapted, you can't fill up at every service station.

2 a water.

b helps water vapour to condense.

c carbon dioxide.

3 a blue. **b** yellow.

c more heat produced by blue flame; yellow flame produces carbon which makes gauze or glassware dirty; blue flame less likely to release carbon monoxide into lab.

4 $C_2H_4 + 3O_2 \rightarrow 2CO_2 + 2H_2O$

Item C1h Energy

1 a an exothermic reaction releases energy.
  b any burning reaction (and other examples).
  c any explosion (and other examples).
  d any burning reaction involving a flame (and other examples).
  e an endothermic reaction absorbs energy.
  f recharging a battery or sodium hydrogencarbonate reacting with citric acid.
2 a exothermic.
  b 184 kJ.
3

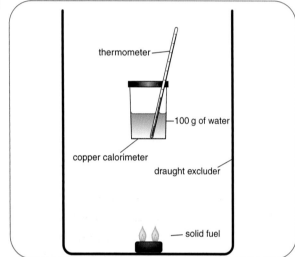

4 a hexane.
  b greater mass of fuel burnt; less water in calorimeter; lower starting temperature of water.
5 24 360 J, 24.36 kJ (method is $200 \times 4.2 \times 29$).
6 12 800 J (method is 9600/0.75).

Item C2a Paints and pigments

1 a Pigment, binder, solvent.
  b Pigment is hiding the colour already on the object and giving it its new colour; binder is sticking the pigment to the surface and will leave a gloss finish; solvent is making the paint easier to apply.
2 a It is a mixture that won't settle out.
  b The pigment particles are too small to settle out.
  c In water paints the solvent is water; in oil paints the solvent is oil.
  d Water paint.
3 a When making plastics the monomers are unsaturated; in oil paints the molecules of oil solvent are unsaturated.

  b When making thermoset plastics, cross-linking bridges form between the polymer molecules; when oil paints dry, cross-linking bridges form between the molecules of oil solvent.
  c When making plastics the monomer molecules form long chains, which is called polymerisation; when oil paints dry the molecules of oil solvent become joined together, which is called polymerisation.
4 a Thermochromic pigments are different colours at different temperatures. Phosphorescent pigments store light energy and then release it.
  b Thermochromic pigments can warn someone when an object is hot. Phosphorescent pigments are used as a safer alternative to radioactive paint.

Item C2b Construction materials

1 Five answers from:
  brick – clay, tiles – clay, glass – sand, aluminium – aluminium ore, cement – limestone/clay, mortar – sand, steel (in foundations) – iron ore.
2 a They strengthen it.
  b It is made of two materials (steel and concrete) combined together.
3 a CaO;
  b CO_2;
  c $CaCO_3$;
  d five.
  e Yes. On each side of the arrow there is one calcium atom, one carbon atom, and three oxygen atoms.
4 a Heat energy is absorbed by this reaction.
  b Heat energy is released by this reaction.
5 a i The gravel deposit was discovered.
    ii It was investigated to find out the amount and quality of gravel that was there.
    iii Roads or railways were built to give access to the area.
    iv The gravel was extracted and taken away by road or rail.
    v When the gravel deposit was finished (or no longer profitable) the work ceased.
    vi The pit was flooded or allowed to flood naturally.

vii The environment has naturally evolved into what it is today (or this evolution was assisted by an environmental project).

b Wildlife – fish and water birds; recreation – fishing, sailing, walking and birdwatching.

c making concrete.

6 a Granite is an igneous rock which forms when magma rises into the Earth's crust where it cools and sets. Limestone is a sedimentary rock which forms when layers of shells from dead sea creatures are compressed and stuck together. Marble is a metamorphic rock which forms when great heat and pressure act on limestone.

b Granite is hardest, limestone is softest.

Item C2c Does the Earth move?

1 The crust is the outer layer of the Earth and is made of plates; the mantle lies under the crust and consists of very hot rock; the core is in the centre of the Earth and is mostly iron.

2 Volcanoes, earthquakes and tsunamis happen most often at the edges of the plates.

3 The crust plus the region of the mantle directly beneath the crust.

4 a just over 90 years old.

b no-one could explain why the plates move.

c four answers from:

  i the coastlines of Africa and South America fit together.

  ii the layers of rock on the coasts of Africa and South America are the same.

  iii the same fossils are found on the coasts of Africa and South America.

  iv plate tectonics can explain how rift valleys form.

  v satellites can observe plate movements.

5 a Igneous rock forms when lava or magma cools and solidifies.

b The soil is fertile.

c Rapid cooling produces small crystals, e.g. basalt, rhyolite; slow cooling produces large crystals, e.g. gabbro, granite.

Item C2d Metals and alloys

1 a An element is made of one type of atom only. A compound is made of at least two types of atom joined by chemical bonds.

b three.

c copper, carbon, oxygen.

d five.

2 a (thermal) decomposition.

b copper carbonate
                → copper oxide + carbon dioxide

c $CuCO_3 \rightarrow CuO + CO_2$

d reactant is copper carbonate; products are copper oxide and carbon dioxide.

3 a copper oxide + carbon
                → copper + carbon dioxide

b $2CuO + C \rightarrow 2Cu + CO_2$

c reactants are copper oxide and carbon; products are copper and carbon dioxide.

4 This is a revision question.

5 a impure copper.

b positive electrode.

c pure copper.

d negative electrode.

e DC.

f copper sulphate solution.

g A solution or molten solid that conducts electricity. If the electric current is DC an electrolyte will undergo permanent change.

6 a Recycling conserves supplies of copper ore.

b Different copper alloys cannot just be melted and mixed; the scrap has to be sorted.

7

Alloy	Constituents	Property	Use
Solder	tin + lead	low melting point	joining electrical components
Brass	copper + zinc	sonorous	musical instruments
Amalgam	mercury + others	expands on solidification	dental fillings
Steel	iron + carbon	strong	reinforcing concrete
Bronze	copper + tin	easy to cast	statues
Nitinol	nickel + titanium	returns to original shape if bent	spectacle frames

Item C2e Cars for scrap

1 e.g. glass – strong, transparent; metal – malleable, strong; dashboard plastics – easy to mould, cheap, soft; instrument plastics – transparent, shatterproof; rubber – flexible.

2 a looks bad, weakens structure.

b oxygen and water.

c e.g. wet roads, roads with salt on, acid rain.

3 a steel sticks to a magnet, aluminium doesn't.
 b 2.9 times.
 c

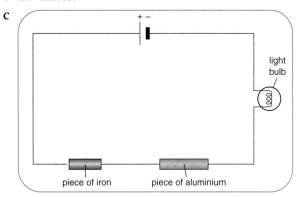

piece of iron   piece of aluminium

4 A layer of aluminium oxide forms but it is impervious and tenacious, so it protects the aluminium.

Item C2f Clean air

1 a 700 km.
 b no.
 c carbon dioxide.
 d plants.
 e less carbon dioxide, more oxygen, more nitrogen.

2 a ammonia + oxygen → nitrogen + water
 b $4NH_3 + 3O_2 → 2N_2 + 6H_2O$

3 a respiration, combustion.
 b animals, e.g. humans, horses.
 c photosynthesis.
 d plants, e.g. oak tree, flowers.

4 a Combustion removes oxygen and adds carbon dioxide. Deforestation stops the removal of carbon dioxide and the addition of oxygen. If the trees removed are burned, this removes oxygen and adds carbon dioxide. Respiration removes oxygen and adds carbon dioxide.
 b Less oxygen, more carbon dioxide.

5 a car exhausts.
 b acid rain, photochemical smog.
 c car exhausts.
 d poisonous to humans and other animals.
 e car exhausts, power stations.
 f acid rain.

6 Use electricity from 'green' sources; use less electricity; FGD in power stations; drive more efficient cars; drive less; drive cars fitted with catalytic converters.

Item C2g Faster or slower (1)

1 a cooking, rusting, combusting/exploding.
 b rusting–cooking–exploding.

2 Solid particles move from side to side (vibrate); liquid particles move slowly through the body of liquid; gas particles are free to move throughout the gas at high speed.

3 a Acid particles are closer together/more acid particles per cm^3 of solution.
 b Acid particles are further apart/fewer acid particles per cm^3 of solution.
 c Acid particles collide with calcium carbonate particles more often.
 d Acid particles collide with calcium carbonate particles less often.

4 In hotter acid the acid particles move faster and so collide harder with the calcium carbonate particles.

5 a i $32 cm^3$;
  ii $18 cm^3$.
 b The acid in experiment 1 was more concentrated (because the acid in experiment 2 had more water added to it).
 c In experiment 1 the acid particles were closer together so they collided with the calcium carbonate particles more often. In experiment 2 the acid particles were further apart so they collided with the calcium carbonate particles less often.
 d The line on the graph for experiment 1 goes up more steeply. The line on the graph for experiment 1 becomes parallel to the bottom axis first, showing that this experiment finished first.
 e All the acid was used up.

6 a

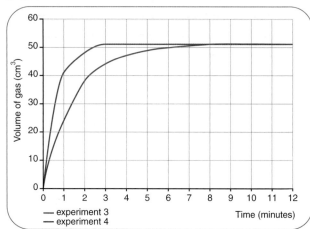

— experiment 3
— experiment 4

b experiment 3 – 31 or 32 cm³;
experiment 4 – 45 cm³.

c All the acid was used up.

d It decreased.

7 a The line on the graph for experiment 4 goes up more steeply. The line on the graph for experiment 4 becomes parallel to the bottom axis first, showing that this experiment finished first.

b The acid was hotter in experiment 4.

c In hotter acid the acid particles move faster and so collide harder with the calcium carbonate particles.

8 a between 7 and 8 cm³ per minute.

b between 3 and 4 cm³ per minute.

c 0 cm³ per minute.

Item C2h Faster or slower (2)

1 Grinding increases the surface area of the marble and so the acid particles collide with the calcium carbonate particles more often.

2 a

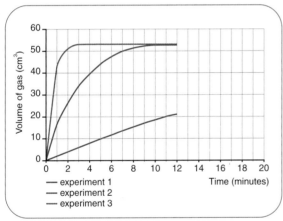

b Experiment 3 was fastest, 1 was slowest.

c The line on the graph for experiment 3 goes up most steeply. The line on the graph for experiment 3 goes parallel to the bottom axis first, showing that this experiment finished first.

d

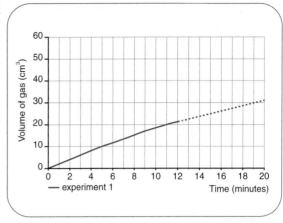

See the extrapolated line in the above graph. Your volume of gas at 20 minutes should be between 29 cm³ and 33 cm³.

3 a concentration of acid; temperature of acid; total mass of marble.

b repeat the readings.

4 a 26 or 27 cm³.

b 18 or 19 minutes.

c experiment 1 – 2 cm³ per minute;
experiment 2 – between 8 and 9 cm³ per minute;
experiment 3 – 3 cm³ per minute.

5 a hydrogen peroxide → water + oxygen

b $2H_2O_2 \rightarrow 2H_2O + O_2$

6 a In more concentrated hydrogen peroxide the particles are closer together so they collide with each other more often. In hotter hydrogen peroxide the particles are moving faster so they collide with each other harder.

b a substance that increases the rate of a reaction without being changed itself.

c manganese dioxide.

7 a The catalyst is not used up so each piece of catalyst does its job over and over again.

b $2NO + 2CO \rightarrow N_2 + 2CO_2$ (the reaction between nitrogen oxide and carbon monoxide).

c weigh it at the start, filter it out at the end, dry it and reweigh it.

d speeds up only one reaction.

8 a a very rapid reaction with gaseous products.

b TNT → nitrogen + water + carbon monoxide + carbon.

c nitrogen, water vapour, carbon monoxide.

9 Icing sugar has a much larger surface area so that the oxygen molecules in the air collide with the sugar molecules much more often.

Answers to SAQs: Physics

P1a Heating and cooling

1 It is one of the hottest objects in the photo.
2 Heat flows from hotter to colder: e.g. from food to room, from room to iced drink.
3 a 35 °C
  b 30 °C
4 Water; non-metals.
5 Metal B.
6 2600 J
7 125 °C
8 China has a greater s.h.c.; other factors: mass, ease with which heat leaves surface.
9 Boiling.
10 3 300 000 J
11 2.5 kg
12 The energy is needed to break the bonds between atoms. On solidification, energy must escape from the metal.

P1b Keeping houses warm

1 Reasons 2 and 3.
2 a All except those in **b**.
  b Cavity walls; reflective foil.
3 3 years.
4 0.25 = 25%

P1c How insulation works

1 Heat conducts out easily (and they may corrode); PVC, plastic (or wood).
2 There is a solid path for heat to conduct through; but much less is lost by convection.
3 The clothing acts as insulation on the radiator, so that less heat reaches the room by radiation.
4 Diamond; stiff bonds between atoms allow heat vibrations to pass quickly.
5 Cold, denser air sinks from freezing compartment; warmer air rises up to replace it.
6 Reflective film on inner surface of outer pane.
7 Heat escapes faster if the temperature difference is greater; so if the water is kept hot during the day, more heat escapes and must be replaced.

P1d Cooking with radiation

1 a dull black;
  b shiny, light.
2 a No convection because solid.
  b Something containing liquid, e.g. stew, curry.
3 Plastic: pass through; water: absorbed; aluminium: reflected; glass: pass through
4 Less energy wasted in heating the oven itself.
5 Twice as much mass to be heated.
6 Microwaves absorbed by the ground above.
7 Microwaves spread out.
8 Microwaves are further from the head.

P1e Communicating with infra-red radiation

1 Remote controls for TV etc, garage doors, wireless connections (and cooking, heating).
2 Direct the beam at a window so that it reflects to the TV set; show that you can change channels.
3 Need to know which device is being controlled.
4 They are equal.
5 Angles of incidence and reflection are equal.
6 Through air: put detector in path of signal; along electrical cable: attach detector which detects current (by its magnetic field); optical fibre: nothing comes out of the fibre.

P1f A wireless world

1 Eyes: light; cooking: infra-red, microwaves.
2 No, sound is not electromagnetic radiation.
3 Radio waves cannot penetrate metal car body.
4 Radio waves, light, ultraviolet, gamma rays.
5 Refraction.
6 Useful: signals reach distant points on Earth; bad: difficult to send radio signals straight out into space.
7 Lower frequencies are bent most.
8 They reflect off the ionosphere and return to the ground beyond the skip zone.
9 Microwaves; can pass through atmosphere.
10 Refraction: bending of waves when passing from one material to another; diffraction: bending when passing through a gap or around edge.

P1g Light waves

1 4 wave crests; 4 m; 2 m.
2 The particles of the water go up and down, they don't travel along.
3 Lowest frequency = longest wavelength = radio waves. Highest frequency = shortest wavelength = gamma rays.
4 10 cm/s
5 3000 Hz

P1h Earth waves

1 P-waves travel faster than S-waves.
2 S-waves.
3 Seismometer or seismograph.
4 5 h
5 White surface reflects sunlight back into space, so less warmth reaches Earth.
6 Temperature varies over surface of Earth; throughout the day; seasonally.

P2a Solar cell technology

1 Sunlight is intermittent (and none at night). Energy stored in battery can be used later.
2 No need to install cables from distant power station.
3 Power.
4 Rugged; less likely to break or go wrong in use.
5 Probably not enough roof area to supply electricity needs of everyone living below.
6 Powering fridges, phones, calculators, water pumps, spacecraft ...
7 A good absorber of infra-red (heat) radiation).
8 Glass absorbs infra-red radiation.
9 Position wind turbine in current of cold air flowing down from mountains.
10 Sunlight, which is renewable, causes convection currents (wind) which turn turbine.
11 Spoils the view/looks ugly.

P2b Generating electricity

1 Move magnet faster; stronger magnet; magnet closer to coil; more turns of wire.
2 Bigger (peak) voltage, higher frequency.
3 Twice as many cycles in same time.

4 a turbine;
   b generator;
   c boiler.
5 0.47 = 47%
6 250 J
7 Losses are less (for same power transmitted along similar cables).

P2c Fuels for power

1 More can be grown to replace that which is used.
2 a wind;
   b fossil fuels;
   c uranium, plutonium;
   d sugar cane.
3 100 J
4 36 W
5 2 A
6 8 kWh; 64 p
7 10 kWh; 80 p
8 Non-renewable.
9 Uranium, plutonium.
10 a false;
   b true;
   c false;
   d true.

P2d Nuclear radiations

1 Radon.
2 Less atmosphere to absorb cosmic rays.
3 13%
4 Medical; weapons tests; air travel; TV sets; work; power station discharges.
5 Gamma, beta, alpha.
6 Only gamma passes through lead; all three pass through open window; etc.
7 a Lead absorbs all radiation.
   b To keep the source as far from body as possible, so that radiation is weaker.
   c The user is exposed to radiation for less time.
8 Beta not readily absorbed by smoke.
9 Alpha and beta absorbed by thick walls of pipe.
10 Thin plastic does not absorb gamma radiation.
11 The radiation must be able to pass through the ground to the surface, to be detected.

P2e Magnetic Earth
1 a Left-hand end is north.
  b Strongest at ends; field lines closest together.
2 No oceans or continents; desert surface.
3 Earth rocks have been changed by erosion and new formation (think of sedimentary, metamorphic, igneous).
4 a Man-made (though women make them, too).
  b The Moon.

P2f Exploring the solar system
1 A sentence with initial letters MVEMJSUNP.
2 Meteor, comet, planet, star, galaxy.
3 a Centripetal force from centre of Earth; directed towards centre of Sun.
  b Gravity.
4 Oxygen, water, food, fuel, computers, landing module, rover vehicle, radios etc etc.
5 A measure of distance; the nearest star beyond the Sun is at a distance of 4 light-years.
6 Approx 10^{13} km.
7 The craft and its instruments can withstand higher/lower temperatures, higher radiation levels etc.

P2g Threats to earth
1 Erosion by wind and rain; movement of tectonic plates; inundation by sea; etc.
2 Gravity.
3 Most quickly when closest to Sun; most slowly when furthest.
4 Arrows always from centre of comet directed towards centre of Sun; longest when closest to Sun.
5 Asteroid, comet.
6 Explosion at side of NEO, to provide sideways force.

P2h Beginnings and endings
1 Star, red giant, supernova, black hole.
2 Greater than mass of Sun.
3 Thermonuclear fusion.
4 The outward pressure of moving particles balances the inward pull of gravity.
5 No; they might oscillate back and forth; or they might have exploded from a large area.
6 Waves from an approaching source are squashed together.

Glossary: Biology

absorption   the movement of nutrient molecules through the walls of the intestine and into the blood

abundance   the number of individuals there are in a population of a particular species in a habitat

accommodation   the adjustment of the shape of the lens and eyeball, so that light is focused precisely on the retina

acquired characteristics   characteristics that develop during an organism's lifetime, caused by the environment and not by genes

active immunity   immunity in which the body has made its own antibodies

adaptation   a feature of an organism that helps it to live in its habitat

addiction   being unable to manage without a drug

aerobic respiration   the release of a relatively large amount of energy from the breakdown of glucose, by combining it with oxygen

allele   a particular variety of a gene

amino acid   a molecule containing carbon, hydrogen, oxygen and nitrogen; there are 20 different types of amino acid that link together in long chains to form proteins

anaerobic respiration   the release of a relatively small amount of energy by the breakdown of glucose, without combining it with oxygen

animal   a multicellular organism whose cells don't have cell walls, and which feeds by eating other organisms

antibiotic   a drug that kills bacteria without harming other cells

antibodies   chemicals produced by white blood cells that help to destroy pathogens

antigens   chemicals that are part of a foreign organism; they are recognised by the immune system and attacked

asexual reproduction   reproduction involving only one parent, in which the offspring are genetically identical to their parent and to each other

axon   a long strand of cytoplasm along which nerve impulses travel

bacterium   a single-celled organism whose cells have cell walls, but no nucleus; bacterial cells are much smaller than animal cells or plant cells

balanced diet   a daily intake of food containing all the different nutrients in the correct proportions

bases   molecules that are part of a DNA molecule; there are four bases, A, C, T and G

benign   a benign tumour is one which does not shed cells that start new tumours in other parts of the body

bile   a liquid secreted by the liver; it helps to emulsify fats

bile duct   a tube that carries bile from the gall bladder into the small intestine

binocular vision   being able to see an object with both eyes at once, which allows judgement of distance

binomial   a two-word name that is given to a species of organism

biodiversity   the biological variety in a habitat, for example the number of different species living in it

blood pressure   the pressure exerted by blood on the walls of the blood vessels

body mass index (BMI)   mass in kg divided by (height in m)^2; BMI should lie between 20 and 24

bronchitis   inflammation of the bronchi

cactus   a plant that is adapted for living in deserts

cancer   a disease resulting from the uncontrolled division of cells in one or more parts of the body

captive breeding   breeding animals in zoos and other establishments, in order to increase the numbers in species that may be endangered

carbohydrates   sugars and starches

carcinogen   a substance that can cause cancer

cellulose   a polysaccharide made of many glucose molecules linked together, and which forms the cell walls of plants

central nervous system (CNS)   the brain and spinal cord

chlorophyll   a green pigment found in chloroplasts in plant cells that absorbs energy from sunlight

chromosome   a coiled thread of DNA found in the nucleus of cells

cilia   tiny hair-like extensions from a cell, which can beat steadily in unison

ciliary muscle   a ring of muscles in the eye, whose contraction alters the shape of the lens

cirrhosis   damage to the liver, often caused by alcohol

class   a group into which organisms are classified, for example the five classes of vertebrates

Class A drug   a drug that is known to cause serious harm; penalties for its use or distribution are harsh compared with Class B and Class C drugs

clear-felling   cutting down all the trees in a wood at the same time

climate change   changes in the Earth's climate, caused, for example, by global warming

clomiphene   a hormone used in fertility treatment

clone   an organism that is genetically identical to other organisms

community   all the different organisms, of all the different species, that live together in a habitat

competition   organisms compete for a resource if they both need it and it is in short supply

contraception   preventing conception, and therefore pregnancy

coppicing   sustainable use of trees by cutting them down to the ground every so many years; the trees regrow from their bases (coppice stools) and can be cut again and again

cornea   the thick, transparent layer at the front of the eye

cystic fibrosis   an illness caused by a recessive allele that prevents mucus being formed correctly

decomposers   organisms that feed on decaying leaves or other waste material

deficiency disease   an illness caused by insufficient quantities of a particular nutrient in the diet

depressant   a drug that slows down the working of parts of the brain

diabetes   an illness in which the body cannot control its blood glucose level

diastolic pressure   the blood pressure measured when the heart is relaxing

diet   the food eaten in one day

digestion   breaking down large pieces of food into small ones (physical digestion) and large molecules of food into small ones (chemical digestion)

digestive system   all the organs that are involved in the digestion of food

distribution   where a species is found

DNA   deoxyribonucleic acid; a substance found in chromosomes which carries a code used by the cell when making proteins

dominant allele   an allele whose effect is seen even if another (recessive) allele is present

dopamine   a transmitter substance found in the brain

double-blind test   a test of a new drug in which neither the researchers nor the patients know whether they are being given the drug or a placebo

drug   a substance that changes the chemical processes in the body

ecosystem   living organisms and their environment, interacting with one another

effector   a part of the body that responds to a stimulus

emphysema   an illness in which the alveolar walls are damaged, making gaseous exchange difficult

emulsification   breaking an insoluble oil into tiny globules, which can be suspended in a water liquid

endangered species   a species that is threatened, and may become extinct

endocrine gland   a gland that secretes hormones

enzymes   proteins that act as catalysts in the body

essential amino acid   an amino acid that cannot be made in the human body and so must be eaten in the diet

evolution   the gradual change in species over time

exponential growth   growth of a population where the number doubles at regular intervals

extinct   no longer in existence

fertiliser   a chemical added to soil that provides plants with nutrients and therefore helps them to grow faster and larger

fertility treatment   medical treatment that can help an infertile couple to have a child

first class proteins   proteins that contain essential amino acids

fitness   the ability of the body to perform exercise; it has several aspects, for example strength, stamina and flexibility

fossil   the remains of an animal or plant that lived long ago, turned to stone

FSH (follicle-stimulating hormone)   a hormone produced by the pituitary gland that helps in egg production and release

fungus   an organism whose body is made up of long threads called hyphae, and which feeds by digesting organic matter around it; its cells have cell walls but they are not made of cellulose

GABA   a transmitter substance in the brain

gall bladder   a small organ lying within the liver, where bile is stored

gamete   a sex cell, such as an egg or a sperm, that will fuse together to form a zygote

gene   a length of DNA that codes for the production of a protein, or for a particular characteristic

genetic code   coded instructions for making proteins, carried in DNA

genetic diagram   a way of describing what happens during a cross between two organisms; it often contains a Punnett square

genotype   the genes that an organism has

genus   a group of similar species

geographical isolation   the separation of two populations by a geographical barrier

glucose   a type of sugar

glutamate   a transmitter substance in the brain

glycogen   a polysaccharide that acts as an energy store in liver and muscles

habitat   the place where an organism lives

haemoglobin   the red pigment in red blood cells that transports oxygen

hallucinogen   a drug that produces distorted pictures and ideas in a person's brain

heat stroke   symptoms caused by having a body temperature well above normal; metabolism is badly affected

herbicide   a chemical that kills weeds

heterozygous   possessing two different alleles of a gene

homeostasis   the maintenance of a constant internal environment in the body

homeothermic   able to keep the internal body temperature constant

homologous chromosomes   two 'matching' chromosomes that carry the same genes

homozygous   possessing two similar alleles of a particular gene

hormone   a chemical, secreted by an endocrine gland, which is transported in the blood and affects organs in other parts of the body

host   an organism in which a parasite lives and feeds

hunting   killing wild animals

hybrid   an organism produced as a result of reproduction between individuals belonging to two different species; hybrids are usually infertile

hyphae   a long thread that forms part of the body of a fungus

hypothalamus   a part of the brain that senses and controls body temperature

hypothermia   having a temperature well below normal

immune response   the response of the immune system to invasion of the body by pathogens or other foreign cells

immune system   cells and tissues that react to the invasion of the body by pathogens and help to destroy them

immunity   you are immune to a disease if your body can destroy the pathogens before they make you ill

incidence rate   the number of new cases of a disease occurring per 100 000 people in a given period of time

infectious disease   a disease caused by a micro-organism, which can be passed from one person to another

insect pollination   transfer of pollen from anther to stigma, carried out by an insect

insulin   a hormone secreted by the pancreas that reduces blood glucose level

iris   the coloured part of the eye, which can adjust the size of the pupil

kwashiorkor   a type of malnutrition caused by lack of protein in the diet

lactic acid   a substance produced during anaerobic respiration in mammals

leaf litter   a layer of decaying leaves, such as is found on the ground in a wood

leguminous plants   plants belonging to the pea and bean family

lens   a transparent structure found in the eye; it refracts light rays and helps to focus an image onto the retina

LH (luteinising hormone)   a hormone produced by the pituitary gland that helps in egg production and release

limiting factor (for photosynthesis)   a factor that limits the rate of photosynthesis – for example, a low temperature, a low concentration of carbon dioxide, a low light intensity

lipase   an enzyme that digests fats

liver   a large organ lying just beneath the diaphragm; it has many functions, including the secretion of bile

long sight   being unable to focus on near objects

lysozyme   an enzyme found in tears and other body fluids, which destroys bacteria

malaria   an infectious disease caused by protozoans, which are injected into the body by mosquitoes

malignant a malignant tumour sheds cells which grow into new tumours in other parts of the body

mark−release−recapture a technique that can estimate the number of animals in a habitat

mass extinction the extinction of large numbers of species at the same time

memory cells white blood cells that have been produced in response to a pathogen, and that remain in the body for some time afterwards

menstrual cycle the approximately monthly cycle in women in which egg production and release alternates with menstruation

metabolic reaction a chemical reaction that takes place in a living organism

millimetres of mercury (mmHg) a unit for measuring blood pressure; 1 mm Hg is equivalent to 133 Pa

mitosis a type of cell division in which the daughter cells are genetically identical to each other and to the parent cell

monocular vision being able to see an object with only one eye at a time

mortality rate the number of deaths occurring per 100 000 people in a given period of time

motor neurone a nerve cell that transmits impulses from the CNS to an effector

MRSA methicillin-resistant *Staphylococcus aureus* − a bacterium that cannot be killed by most antibiotics, in particular methicillin

mucous membrane a layer of cells lining the trachea and other parts of the respiratory and digestive system, which contains goblet cells making mucus

mutation an unpredictable change in an organism's genes or chromosomes

mutualism a relationship between two organisms of different species, in which they live closely together and both get benefit from this relationship

mycelium a mass of threads (hyphae) that forms the body of a fungus

natural selection the survival of only the best-adapted organisms; only they reproduce and pass on their advantageous features to their offspring

nectar a sugary liquid produced in some flowers, to attract insects for pollination

negative feedback a control system in which the increase in level of a parameter brings about actions that cause its level to decrease

nerve a group of nerve fibres

nerve impulses electrical signals transmitted along nerve fibres

nervous system the body organs that are involved with the production and transmission of nerve impulses − the brain, spinal cord and nerves

neurone a nerve cell

niche the way a species fits into an ecosystem

nicotine the addictive drug in tobacco

nitrogen fixation the conversion of chemically inert nitrogen gas, N_2, to a more reactive form such as nitrate, NO_3^-, or ammonium ions, NH_4^+

obesity having a body weight that is much greater than it should be

oestrogen a hormone secreted by the ovaries which helps to control the menstrual cycle

optic nerve the nerve that carries impulses between the eye and the brain

oral contraceptive a pill that prevents pregnancy

ovule structures found inside the ovary of a flower which contain the female gametes

oxygen debt extra oxygen that the body requires after vigorous exercise, to help to break down lactic acid formed during anaerobic respiration

painkiller a drug that helps to relieve pain

pancreas a gland lying close to the stomach; it makes pancreatic juice, which helps in digestion, and also the hormone insulin, which reduces blood glucose levels

parasite an organism that lives in close contact with another organism, feeds on it and does it harm

particulates tiny pieces of carbon and other materials that can damage the lungs if breathed in

pascal (Pa) a unit for measuring pressure; one pascal is equivalent to a pressure of one newton per square metre

passive immunity immunity caused by the acquisition of antibodies from another organism

pathogen an organism that causes disease

performance enhancer a drug that increases athletic performance

peripheral nervous system the part of the nervous system other than the brain and spinal cord, made up of nerves and receptors

pesticide a chemical that kills pests, such as insects or fungi

phenotype the features of an organism, partly caused by genes and partly by environment

phloem tubes tubes that carry sucrose and other substances from one part of a plant to another

photosynthesis the production of glucose and oxygen from carbon dioxide and water, using light energy

pitfall trap a trap into which insects may fall as they move about over the ground

placebo a substance that has no measurable effect on the body, used to compare against a drug in a drug trial

plant a multicellular organism whose cells have cellulose cell walls, and which feeds by photosynthesis

pollen grains tiny particles that contain the male gametes of a plant

pollination the transfer of pollen from the anther of a flower to a stigma

pollution causing damage to the environment, for example by the addition of a toxic chemical

polysaccharide a carbohydrate made of many monosaccharide (sugar) molecules linked together

pooter a tube through which you can suck to capture insects and other small organisms

population a group of organisms of the same species, all living in the same place at the same time, and that can interbreed with one another

predator an animal that kills and eats other animals

prey an animal that is killed and eaten by another animal

progesterone a hormone secreted by the ovaries and placenta, which helps to maintain the uterus lining and also pregnancy

protease an enzyme that digests protein

protein a substance whose molecules are made of long chains of amino acids

Punnett square a diagram that shows the possible genotypes resulting from the fusion of gametes

pupil the circular gap in the centre of the iris, through which light passes

quadrat a square area that is studied in ecology, as a sample of a larger area

RDA recommended daily allowance; RDAs have been calculated for the different nutrients in our diet

receptor an organ or cell that detects a stimulus

recessive allele an allele whose effect is seen only if a dominant allele is not also present

red–green colour blindness (deficiency) the inability to distinguish red from green

Red List a list of endangered species prepared by the International Union for Conservation of Nature and Natural Resources

reflex action a fast, automatic response to a stimulus

reflex arc the pathway along which a nerve impulse travels during a reflex action

rehabilitation learning to manage without a drug and to rebuild a life

respiration the release of energy from food substances such as carbohydrates; it happens in every living cell

retina the part of the eye that contains receptors and is sensitive to light

root nodules small swellings on some plants' roots which contain nitrogen-fixing bacteria

secondary sexual characteristics features that develop during puberty, such as deep voice in men, breasts in women

secretion making and releasing a chemical

sex hormones hormones such as testosterone and oestrogen that are involved in reproduction

sexual reproduction reproduction involving gametes, usually from two different parents; the offspring are genetically different from each other and their parents

shivering rapid, uncoordinated contraction and relaxation of muscles, which generates heat

short sight being unable to focus on distant objects

small intestine the longest part of the digestive system, where digestion is completed and digested food is absorbed

social drug a drug that is so widely used that it has become acceptable

speciation the formation of a new species

species a group of organisms that share the same features and that can breed together successfully

species diversity the number of different species living in a habitat

spinal reflex a reflex action in which the nerve impulse passes through the spinal cord

starch a kind of carbohydrate; a polysaccharide

stimulant a drug that increases activity in some parts of the brain

stimulus a change in the environment that is detected by the body

stomach a part of the digestive system in which food can be kept for several hours; protein digestion begins here

stroke brain damage caused by a burst blood vessel or a blockage of a blood vessel supplying the brain

sucrose a sugar made of two sugar units linked together

suspensory ligaments a ring of fibres that supports the lens and holds it in the centre of the eye

sustainable resource something that is used by humans – for example fish, trees – that can renew itself as we use it and that we could therefore continue to use over a long period of time

synapse the place at which two neurones meet; there is a small gap between them called the synaptic cleft

systolic pressure the blood pressure measured when the heart is contracting

tar a component of tobacco smoke that can cause cancer

target organ an organ that is affected by a hormone

testosterone a hormone secreted by the testes which is responsible for the development of secondary sexual characteristics and sperm production

tolerance needing to take more and more of a drug to get the desired effect

toxins poisons

transect a line along which organisms are identified and counted, sometimes inside quadrats placed at intervals along the line

transmitter substance a chemical which carries a nerve impulse across a synapse

tuberculosis (TB) an infectious disease caused by a bacterium

Tullgren funnel a funnel into which soil or leaf litter is put, to help to trap decomposers and other small organisms

tumour a lump caused by uncontrolled cell division

vaccination being given a harmless form of a pathogen, which makes you immune to the disease that it can cause

vasoconstriction narrowing of arterioles taking blood to the skin, reducing blood flow

vasodilation widening of arterioles taking blood to the skin, so more blood flows near the body surface

vector an organism that transfers a pathogen from one person to another, for example mosquitoes are vectors for malaria

vesicle a very small 'space' inside a cell, surrounded by a membrane

vitamin an organic substance needed in very small quantities in the diet

VO$_2$ max the maximum rate at which the body can use oxygen before having to change to anaerobic respiration

voluntary action an action that is under the control of the will

warfarin a poison used to kill rats, and also in medicine to reduce the chance of blood clots forming and causing a heart attack

whale a large marine mammal

white blood cells cells found in the blood that defend the body against pathogens

wind pollination transfer of pollen from anther to stigma, carried out by an insect

withdrawal symptoms unpleasant symptoms that a person suffers if they stop taking a drug to which they are addicted

zygote the cell formed when two gametes fuse together at fertilisation

Glossary: Chemistry

acid rain   rain with a pH below 4.5

active packaging   food packaging that contributes in some way to preserving or heating the food

addition polymer   a polymer in which the monomer is an alkene

addition polymerisation   the formation of a polymer in which the monomer is an alkene

additives   extra chemicals added to food while it is being blended or packaged

alkane   a hydrocarbon compound in which all the hydrogen and carbon atoms are joined to their neighbours by single covalent bonds

alkene   a hydrocarbon compound in which two of the carbon atoms are joined to each other by a double covalent bond

alloy   a mixture of a metal with one or more other substances (usually one or more other metals)

amalgam   an alloy of mercury and other metals such as tin, silver and copper

antioxidants   food additives which prevent chemicals in the food reacting with oxygen from the air

artificial   man-made, not natural

atmosphere   the air; the layer of gases surrounding the Earth

atom   one of the tiny particles that all matter is made of, consisting of a positive nucleus orbited by one or more negative electrons

baking powder   a powder containing sodium hydrogencarbonate which gives off carbon dioxide when heated. When it is present in food the food will 'rise' during cooking

balanced symbol equation   a description of a chemical reaction using the chemical formula of each substance involved and having the same number of each type of atom on either side of the arrow

basalt   a fine-textured igneous rock containing small crystals, made from an iron-rich lava

binder   a component in paint that helps the paint to stick onto the surface and also provides a protective finish

biodegradable   able to rot or decompose by natural means, i.e. by the action of micro-organisms

bond   a way of holding two atoms or molecules together. Can be strong or weak. (see **intermolecular forces, covalent bonding, ionic bonding**)

brass   an alloy of copper and zinc

breathable   allowing gases through. This term is particularly used for clothing materials that allow water vapour through and therefore don't feel clammy

breathable film   clear film for packaging meat which allows gases in and out

bronze   an alloy of copper and tin

catalyst   a substance that changes the rate of a chemical reaction while remaining unchanged itself

cellulose   a rigid material contained in plant cell walls

cement   a construction material made by heating together limestone and clay. Cement is mixed with sand and water to make mortar, which binds bricks together to make walls

chemical change   a change in which new substances are formed, e.g. combustion

chemical reaction   a change in which new substances are formed

chemical test   a way of identifying something which involves a chemical change, e.g. burning (not a physical change, e.g. melting)

chemicals   the substances everything is made of

collision   when two or more objects hit each other

colloid   a mixture that won't separate out. Usually consists of very tiny solid or liquid particles/droplets suspended in another liquid

combustion   a chemical reaction in which oxygen and a fuel react, releasing useful energy

complete combustion   used to describe the burning of a hydrocarbon to form carbon dioxide and water

composite   two or more materials combined together

compound   a substance made from more than one type of atom (and therefore more than one type of element) chemically bonded together. Can be decomposed

concentration   a measure of how much solute is dissolved in each dm^3 (litre) of solution

concrete   a strong, hard substance used for building, made by mixing cement, water and gravel

constituents   the parts that a material or finished article is made of

core   the central layer of the Earth that contains iron

corrosion   being eaten away by a chemical reaction

covalent bonding   a way of holding atoms together that involves the sharing of pairs of electrons. This type of bonding is strong

cracking   breaking large hydrocarbon molecules into smaller ones. Cracking produces more of the petroleum fraction and also produces alkenes that are used to make polymers

cross-linking bridges   covalent bonds that hold polymer chains together by strong forces, resulting in a rigid (thermosetting) polymer that doesn't soften when heated

crude oil   a complex mixture of hydrocarbons found in the Earth's crust and originally formed by the fossilisation of dead sea creatures

crust   the solid outer layer of the Earth

decolourised   has lost its colour. Specifically when an alkene causes the orange colour of bromine to disappear

decompose   to break down into simpler substances

deforestation   making forests smaller by cutting down trees

degassing   the release of dissolved gases from another substance

dehydrate   to remove water from a substance

denaturing   causing a protein molecule to change its shape, usually by heating

digest   to break down large molecules of food into small ones

dissolving   the mixing that happens when a solute is put into a liquid called a solvent, e.g. when stirring salt into water

double covalent bond   a covalent bond in which the bonding atoms share two pairs of electrons. Shown in displayed formulae as a double line, e.g. C=C

ductile   can be drawn out easily into wires

dust explosion   an explosion in which the two reactants are oxygen from the air and a fine combustible dust such as flour (also called 'powder explosion')

dye   a coloured substance used to colour other substances

electrolysis   a chemical change caused to a substance by passing a DC electric current through it. Used to purify copper

element   a substance made from one type of atom. Cannot be decomposed

emulsifier   an additive that enables water and oil to mix

endothermic reaction   a chemical reaction that removes energy from the surroundings, usually in the form of heat, making the surroundings colder

energy value   the amount of energy released when one gram of a fuel burns

ester   a substance made by reacting an acid with an alcohol. Used as a solvent or synthetic perfume

evaporating   turning from a liquid to a gas (or vapour), a physical change

excess   more than the minimum necessary

exothermic reaction   a chemical reaction that releases energy to the surroundings, usually in the form of heat, making the surroundings hotter

explosion   an extremely rapid exothermic reaction with gaseous products

fair test   a comparative experiment in which only one factor (known as the input variable) is changed

finite   fixed in amount. When used to describe a resource, 'finite' means it will run out if we keep using it

flavour   what food tastes like

flavour enhancers   substances that improve the taste of savoury food

flue gas desulfurisation   removal of sulfur dioxide from chimney gases. Often used at power stations

formula unit   the group of atoms represented by the formula of a compound, e.g $CaCO_3$. This term is most often used when the bonding in the compound is ionic

fossil fuels   fuels, mainly hydrocarbons, that are the preserved remains of dead plants and animals: coal, oil and natural gas

fractional distillation   separating a mixture, e.g. crude oil, on the basis of the different boiling points of its components

fractions   the partly purified substances produced when distilling crude oil

fuel   a substance that releases useful amounts of stored energy when it burns

fuel economy   the amount of fuel a machine needs to do a particular job

gabbro   a coarse-textured igneous rock containing large crystals, made from an iron-rich lava

graded bedding   the formation of rock layers with heavy particles below lighter particles

gradient the slope of a graph

granite a coarse-textured igneous rock containing large crystals, made from a silica-rich lava

hydrocarbon a compound of carbon and hydrogen only

hydrophilic literally means 'water loving', used to describe a substance or part of a molecule that will dissolve in water but not in oil

hydrophobic literally means 'water hating', used to describe a substance or part of a molecule that will dissolve in oil but not in water

igneous rock a type of rock formed when lava, either within the Earth's crust or emerging from a volcano, solidifies

impervious will not allow other substances to penetrate

incomplete combustion the burning of a hydrocarbon in which it forms carbon monoxide and water, or carbon (soot) and water. Caused by inadequate oxygen supply

insoluble will not dissolve (in a particular solvent) e.g. gravel is insoluble in water

intelligent packaging food packaging that contributes in some way to preserving or heating the food

intermolecular forces the bonds that hold molecules together. Intermolecular forces are never strong; they can vary between weak and very weak

ionic bonding a way of holding the atoms in a compound together. Involves the attraction between positively charged ions and negatively charged ions. This type of bonding is strong

irreversible change a change in which products cannot go back to being reactants

irritant a substance that causes irritation, itching or reddening of skin

laminate a material made from thin layers of two or more other materials stuck together

land fill disposal of waste by burial in pits

lava molten rock (magma) when it comes to the Earth's surface

lithosphere the Earth's crust and outer mantle taken together

magma molten or semi-molten rock in the Earth's mantle

malleable can be moulded into new shapes without breaking

mantle the layer of the Earth between the crust and the core. The mantle is composed of magma. The outer mantle is rigid but the inner mantle can flow

metamorphic rock a type of rock formed from a different rock type that was subjected to heat and pressure

micro-organisms bacteria, viruses, fungi and other tiny living things

molecule the group of atoms represented by the formula of a compound, e.g. H_2O, or an element, e.g. O_2. This term is most often used when the bonding in the compound or element is covalent

monomer a substance made of small molecules that can join together to make large molecules (see **polymer**). Alkenes can act as monomers

nitinol an alloy of nickel and titanium which always returns to its original shape if bent

non-renewable a resource that is either not being re-formed at all or is being re-formed much more slowly than it is being used up is described as non-renewable

oil paint a paint in which the solvent is an oil or liquid hydrocarbon. Gloss paints are oil paints

oil slick oil floating on the sea. Kills marine life and seabirds

opaque does not let light pass through it

ore a rock or mineral containing enough of a pure metal or of a metal compound to be worth mining it

oxidation a description of a reaction in which oxygen is gained by one or more of the reactants

permanent change a change in which products cannot go back to being reactants

phosphorescent used to describe pigments that can absorb and store energy when light shines on them. When the light is switched off they release the energy slowly in the form of light

photochemical smog a persistent mixture of smoke and fog caused by the action of sunlight on certain chemicals

photosynthesis a process by which green plants use light energy to change carbon dioxide and water into glucose and oxygen

physical change a change in the properties of a substance, e.g. melting, boiling

pigment a component in paint that provides the new colour and also hides the original colour of the surface to be painted

plate tectonics the theory that the Earth's crust is broken into slow-moving plates

pollutant a harmful substance present in an environment in amounts larger than is natural

poisonous harmful if swallowed. Possibly leading to death

polymer a substance consisting of very large molecules made by joining together small molecules called monomers. Plastics and artificial fibres are synthetic polymers. Proteins are natural polymers

polymerisation the formation of a polymer from its monomer

products the substances formed in a chemical reaction

properties the ways in which a substance behaves

pumice a type of porous igneous rock

rancid a description of a fat or oil that has taken on an unpleasant taste due to reacting with oxygen from the air

rate the speed at which something happens

rate of reaction the speed at which reactants are used up or products are created

raw uncooked (food), unprocessed (other material)

reactants the substances used up in a chemical reaction

reaction see chemical reaction

recycle to reclaim materials from a used object and use them to make new objects

reduction a description of a reaction in which oxygen is lost by one or more of the reactants

reinforced concrete concrete that has been strengthened by the addition of steel bars

respiration a chemical reaction happening in cells, in which oxygen and glucose (or some similar fuel) react to form carbon dioxide and water, releasing useful energy

rhyolite a fine-textured igneous rock containing small crystals, made from a silica-rich lava

saturated used to describe a hydrocarbon compound in which all the hydrogen and carbon atoms are joined to their neighbours by single covalent bonds. If shaken with bromine water such a compound will not decolourise the bromine water

sedimentary rock a type of rock formed from tiny particles called sediments that have been glued together

shared pair two electrons that orbit two atoms at the same time; a single covalent bond

single covalent bond a covalent bond in which the bonding atoms share one pair of electrons. Shown in displayed formulae as a single line, e.g. C–C

sodium hydrogencarbonate a compound of sodium, hydrogen, carbon and oxygen having the formula $NaHCO_3$. It is added to plain flour to make self-raising flour. See **baking powder**

solder an alloy of tin and lead

soluble will dissolve (in a particular solvent), e.g. salt is soluble in water

solute a substance that can dissolve in a liquid called a solvent, e.g. when stirring salt into water the salt is the solute

solution a liquid mixture that will not separate out, made from a liquid called a solvent with a substance called a solute dissolved in it. e.g. when stirring salt into water the salty water is the solution

solvent a liquid that other substances may dissolve in, e.g. when stirring salt into water the water is the solvent

specific catalyst a catalyst that only speeds up one reaction or only speeds up a very few reactions

stainless steel an alloy of iron, nickel and chromium. Stainless steel doesn't rust

steel any iron-based alloy. Mild steel is over 99.75% iron, the remainder being carbon

subduction when one tectonic plate (usually oceanic) is pushed beneath another tectonic plate (usually continental)

successful collisions collisions in which the colliding particles hit each other hard enough to change into products

surface area the total amount of surface of a solid object or a number of solid objects. If a single piece is ground into smaller pieces, its surface area increases

synthetic man-made, not natural

tangent a line drawn on a line graph, just touching a curved line, in order to measure the gradient of the curve at that point

tectonic plates the pieces into which the Earth's crust is broken. Tectonic plates move due to convection currents in the mantle

tenacious   holding on firmly to a surface or object

texture   the feel of a material

thermal decomposition   the breakdown of a substance into two or more simpler substances, caused by heating

thermochromic   used to describe a pigment or dye that changes colour as its temperature changes

TNT   a high explosive, full name trinitrotoluene

toxic   poisonous

unsaturated   used to describe a hydrocarbon compound in which two of the carbon atoms are joined to each other by a double covalent bond. If shaken with bromine water such a compound will decolourise the bromine water

unsuccessful collisions   collisions in which the colliding particles do not hit each other hard enough to change into products and simply bounce off each other

volatile   able to evaporate easily

water paint   a paint in which the solvent is water. Emulsion paints are water paints

word equation   a summary of a chemical reaction in which the names of all the reactants are written to the left of an arrow, and all of the products are written to the right of the arrow

Glossary: Physics

absorbed   when radiation is taken in by an object

alpha radiation   radiation from a radioactive substance (in the form of particles)

alternating current (AC)   electric current which flows first one way, then the other, in a circuit

amplified   when the size of a signal has been increased

amplitude   the greatest displacement of a wave, measured from its undisturbed level

analogue signal   a signal which is coded by changing the shape of a wave

artificial satellite   a man-made spacecraft in orbit around the Earth or elsewhere in the solar system

asteroid   a rock in space; most orbit between Mars and Jupiter

beta radiation   radiation from a radioactive substance (in the form of particles)

Big Bang theory   the idea that the Universe started by exploding outwards from a single point

biomass   a material, recently living, used as a fuel

black hole   a massively dense object, often a dead star, whose gravity is so strong that light cannot escape

centripetal force   a force which holds an object in a circular orbit; it acts towards the centre of the orbit

comet   an icy, dusty 'snowball' whose orbit brings it close to the Sun every now and then

conduction   occurs when a substance transmits heat or electric current without the substance moving

convection   occurs when heat energy is carried by a flow of liquid or gas

cosmic rays   high-energy radiation from space

crest   the highest point on a wave

critical angle   the smallest angle of incidence which gives total internal reflection

decommissioning   dismantling and safe disposal of a power station at the end of its useful life

diffraction   occurs when a wave spreads out as it travels through a gap or past an edge

digital signal   a signal which is coded as a series of 1s and 0s (or ons and offs)

direct current (DC)   electric current which flows in the same direction all the time

dynamo effect   generation of electricity when a coil moves near a magnet

efficiency   the fraction of energy which is converted into a useful form

electromagnetic spectrum   the family of radiations similar to light

electrons   electrically charged particles which flow to form an electric current

emitted   when radiation is given out

endoscope   medical device for seeing inside a patient's body

fossil fuel   a material, formed from long-dead material, used as a fuel

frequency   the number of vibrations or waves in a second

galaxy   a group of billions of stars in space, held together by gravity

gamma radiation   electromagnetic radiation from a radioactive substance

generator   a device such as a dynamo, used to generate alternating current (AC)

global warming   the increase in the Earth's average temperature caused by an increase in the greenhouse effect

greenhouse effect   warming of the Earth caused by absorption of infra-red radiation in the atmosphere

infra-red   invisible radiation beyond the red end of the spectrum

in phase   describes two waves with the same frequency which are in step with each other

insulating materials   materials which conduct heat energy very slowly

interference   occurs when two waves, whose frequencies are close together, overlap so that they are difficult to separate

intermolecular bonds   the weak bonds between a molecule and its neighbours

ionisation   occurs when a particle (atom or molecule) becomes electrically charged by losing or gaining electrons

ionising radiation   radiation, for example from radioactive substances, which causes ionisation

ionosphere   electrically charged layer high up in the atmosphere

kilowatt-hour (kW h)   a unit for the energy transferred by an electrical appliance

kinetic energy (KE)   the energy of a moving object

laser   a device for producing a narrow beam of light of a single colour or wavelength

light-year   the distance travelled by light in a year (about 10^{13} km)

longitudinal wave   a wave in which the vibration is forward and back, along the direction in which the wave is travelling

magnetic field   the area around a magnet or electric current in which a magnet will feel a force

main sequence   describes the long period of normal life of a star

meteor   a dust particle which enters the Earth's atmosphere; appears as a 'shooting star'

microwave background   radiation which is spread throughout space, left over from the Big Bang

microwave radiation   electromagnetic radiation similar to radio waves, used for transmitting mobile phone signals

multiplexing   the interleaving of several signals so that they can be sent along the same cable or optical fibre

National Grid   the system of power lines, pylons and transformers used to carry electricity around the country

near-Earth object (NEO)   an object in space whose path may bring it close to the Earth

neutron star   a late stage in the life of a massive star

non-renewable resource   a material which, when used up, will not naturally be replaced

nuclear fuel   a substance used to fuel a nuclear power station

optical fibres   long strands of glass along which light carries signals

passive solar heating   occurs when buildings are designed to be heated by the Sun's rays

payback time   the time taken for the savings from installing insulation, for example, to cover the initial cost

penetrating power   how far radiation can penetrate into different materials

photocell   a device which transfers the energy of sunlight directly to electricity

planet   a large, solid object in orbit around the Sun

planetary nebula   ring of material thrown out into space when a star collapses to form a white dwarf (has nothing to do with planets)

plutonium   a metallic element used as a nuclear fuel

power lines   cables used to carry electricity from power stations to consumers

protostar   early stage in the life of a star, as it begins to form and shine

P wave   primary seismic wave which travels as a longitudinal wave

radiation   energy spreading out from a source carried by particles or waves

red giant   a medium-mass star which has expanded and cooled, towards the end of its life

red shift   the change in wavelength of light from a star or galaxy, caused by its movement

reflected   when radiation bounces off an object

refraction   the bending of light when it passes from one material to another

renewable energy resource   a material, used as a source of energy, which is automatically replaced

seismic wave   wave travelling through the Earth, caused by an earthquake

seismometer   an instrument for detecting and recording seismic waves

shadow zone   area of Earth's surface where seismic waves are not detected

shock wave   see **seismic wave**

solar flare   outburst of hot material from the surface of the Sun

specific heat capacity (s.h.c.)   a measure of how much energy a material can hold

specific latent heat (s.l.h.)   a measure of how much energy is needed to melt or boil a material

star   a ball of hot, glowing gas, held together by its own gravity

supernova   explosion of a massive star, before it dies

S wave   secondary seismic wave which travels as a transverse wave

temperature   a measure of how hot or cold something is

thermogram   an image in which different temperatures are represented by different colours

thermonuclear fusion   the process in stars, in which small atomic nuclei join together and release energy

total internal reflection (TIR) occurs when 100% of light reflects back inside a solid material

transformer a device used to change the voltage of an AC electricity supply

transverse wave a wave in which the vibration is at right angles to the direction in which the wave is travelling

trough the lowest point on a wave

turbine a device which is caused to turn by moving air, steam, water etc.

ultraviolet invisible radiation beyond the violet end of the spectrum

Universe all the matter and energy of which we can ever know

uranium a metallic element used as a nuclear fuel

visible light electromagnetic radiation which can be detected by our eyes

visual pollution occurs when something spoils the look of an area

voltage the 'push' of a battery or power supply in a circuit

wave speed the speed at which a wave travels

wavelength the distance between neighbouring crests (or troughs) of a wave

white dwarf late stage in the life of a medium-mass star

Physics formulae

Module P1 Energy for the home

You should be able to state and use the following equations:

◆ Efficiency of an energy transfer (page 268):

$$\text{efficiency} = \frac{\text{useful energy output}}{\text{total energy input}}$$

◆ Speed of a wave (page 294):

wave speed = frequency × wavelength

◆ Energy required to raise the temperature of a material (page 260):

energy required = mass × specific heat capacity × increase in temperature

◆ Energy required to cause a change of state (page 263):

energy required = mass × specific latent heat

Module P2 Living for the future

◆ You should be able to calculate the electrical power rating of an appliance using the equation (page 324):

power = voltage × current

◆ You should be able to use these equations in the context of a power station to calculate energy input, energy output or waste energy output and efficiency (page 318):

fuel energy input = electrical energy output + waste energy output

$$\text{efficiency} = \frac{\text{electrical energy output}}{\text{fuel energy input}}$$

◆ You should be able to state and use the equation (page 324):

energy transferred = power × time

Periodic Table

Key

relative atomic mass
atomic symbol
name
atomic (proton) number

1	2											3	4	5	6	7	8
				1 **H** hydrogen 1													4 **He** helium 2
7 **Li** lithium 3	9 **Be** beryllium 4											11 **B** boron 5	12 **C** carbon 6	14 **N** nitrogen 7	16 **O** oxygen 8	19 **F** fluorine 9	20 **Ne** neon 10
23 **Na** sodium 11	24 **Mg** magnesium 12											27 **Al** aluminium 13	28 **Si** silicon 14	31 **P** phosphorus 15	32 **S** sulfur 16	35.5 **Cl** chlorine 17	40 **Ar** argon 18
39 **K** potassium 19	40 **Ca** calcium 20	45 **Sc** scandium 21	48 **Ti** titanium 22	51 **V** vanadium 23	52 **Cr** chromium 24	55 **Mn** manganese 25	56 **Fe** iron 26	59 **Co** cobalt 27	59 **Ni** nickel 28	63.5 **Cu** copper 29	65 **Zn** zinc 30	70 **Ga** gallium 31	73 **Ge** germanium 32	75 **As** arsenic 33	79 **Se** selenium 34	80 **Br** bromine 35	84 **Kr** krypton 36
85 **Rb** rubidium 37	88 **Sr** strontium 38	89 **Y** yttrium 39	91 **Zr** zirconium 40	93 **Nb** niobium 41	96 **Mo** molybdenum 42	[98] **Tc** technetium 43	101 **Ru** ruthenium 44	103 **Rh** rhodium 45	106 **Pd** palladium 46	108 **Ag** silver 47	112 **Cd** cadmium 48	115 **In** indium 49	119 **Sn** tin 50	122 **Sb** antimony 51	128 **Te** tellurium 52	127 **I** iodine 53	131 **Xe** xenon 54
133 **Cs** caesium 55	137 **Ba** barium 56	139 **La*** lanthanum 57	178 **Hf** hafnium 72	181 **Ta** tantalum 73	184 **W** tungsten 74	186 **Re** rhenium 75	190 **Os** osmium 76	192 **Ir** iridium 77	195 **Pt** platinum 78	197 **Au** gold 79	201 **Hg** mercury 80	204 **Tl** thallium 81	207 **Pb** lead 82	209 **Bi** bismuth 83	[209] **Po** polonium 84	[210] **At** astatine 85	[222] **Rn** radon 86
[223] **Fr** francium 87	[226] **Ra** radium 88	[227] **Ac*** actinium 89	[261] **Rf** rutherfordium 104	[262] **Db** dubnium 105	[266] **Sg** seaborgium 106	[264] **Bh** bohrium 107	[277] **Hs** hassium 108	[268] **Mt** meitnerium 109	[271] **Ds** darmstadtium 110	[272] **Rg** roentgenium 111							

Elements with atomic numbers 112-116 have been reported but not fully authenticated

*The Lanthanides (atomic numbers 58-71) and the Actinides (atomic numbers 90-103) have been omitted.

Cu and Cl have not been rounded to the nearest whole number

Group 8 is usually called Group O; see *Additional Science Class Book*, Item C3c

Index

Bold page references are to Higher material.

absorbers, **273**, 275, 276, 300, **307**, **308**
absorption of food, 14
absorption of signals, 285
abundance, 66–7, 94, 95
accommodation (eye), **30**
acetylcholine, 35
acid in digestion, 13–14
acid rain, 124, 228, 235
acquired characteristics, **118**
acrylic paints, 201
active immunity, 22
active packaging, 158–9
adaptations, 83, 102–8, 132
addiction, 38–9, 42
addition polymerisation, 178, **179**
additives, 155–60
address code, 280
adhesives, 164
aerials, 277, **278**, **288–9**
aerobic respiration, 1–2, 8, 92, 125
age effects, 5, 8, **32**, 56
agility, **3**
AIDS, 16
air composition, 233
air density, **272**, **287**
air heating, 261, 269, **272–3**
air pollution, 115, 122–5, 127, 188, 233
alcohol, as fuel, 322
alcohol effects, 5, 37, 38, 41–3
alcoholism, 42–3
algae, 75, 125, 127
alimentary canal, 8, 13
alkanes, 169–70, 174–7
alkenes, 169–70, 175–6, 178
alleles, **61–3**
alloys, 218, 221–4
alpha (α) radiation, 59, 331, **332**, 334
alternating current (AC), 315–16
altitude, 1, **287**
aluminium, 205, 218, 221, 229–30, 252
aluminium cars, 179, 227, 228–30
aluminium foil, 271
aluminium oxide, 228–9
alveoli, 20, 40, 41
amalgam, 221
amino acids, 9, 10, 11, 13, 14, **99**
ammonia gas, **232**
ammonium ions, **99**
Amoeba proteus, **78–9**
amphibians, 75, 80, **81**, 113, 133, 137
amplification, **280**
amplitude, 290, 291
amylase, 13, 54
anaemia, 8, 16, 60, **62–3**
anaerobic respiration, 2, **4**
analogue signals, 280, **280–1**, **282**, 286
angle of incidence, 281
angle of reflection, 281
animal cells, 76–7

animal competition, 94–8
animal counting, 68–71
animal extinction, 351
animal kingdom, 75, 76
animal rights activists, 165
animal testing, 25, 165
anodes, 220
antelopes, 103
antibiotics, 16, 23–4, 38
antibodies, 21–2
antigens, 21–2
antioxidants, 155, 157
Archaeopteryx, **81**
arteries, **2**, 4–5, 48
arterioles, **48**
arthritis, 10, 11, 257
artificial ecosystems, 72–3, 137
asbestos, 266–7
asexual reproduction, 55
asteroid impact, 117, 290, 340, 350, 351, 353
asteroids, 344, 350, **351–2**, 353
astronauts, 339, 341, 346–7, **348**
astronomy, 340, **341–2**, 344–5, 353, 356–60
Athlete's foot, 19
atmosphere, **99**, 232–8, 287, 301–3, 348
atom splitting, **327**
atom vibration, 239, **271**, **272**
atoms, 149, 151, **152**, 174–7, 331
Aurora Borealis, **342**
autism, 23
average rate of reaction, **245–6**
axons, 33, **34**

backbones, 79
background radiation, 330, 341, **342**
bacteria, 19–24, 38, 40–1, 60, 75, **77**, 113
bacteria in food, 13, 99, 148
bacteria in plants, **99**
bacteria in sewage, 125
baking powder, 90, 151–3
balanced diet, 8–10, **11**
balanced symbol equations, 1, **89**, **152–3**, **190–1**, **191**, **207**, **236**
bar magnets, 338, 339
basalt, 215, **216**
bases in DNA, 54, 55, 60
batches of pulses, **282**
bats, 83–4, 136
batteries, 193–4, 305, 315
bends (diving), **140**
beta (β) radiation, 331, 332, 334
Bible stories, **112**, **117**
Big Bang, 359–60
Big Crunch, 359–60
bighorn sheep, 117
bile, 14
bile duct, 14
binders (paint), 200
binocular vision, 32–3
binomials, 82–3

biodegradable polymers, 186
biodegradable structures, 184
biodiversity, **73**
biomass fuels, 322
biotic index, 127
bird navigation, 338
birds, 75, 77, 80, 84, 105, 113
birth rates, **126**
black holes, 344, 357, **358**
black surfaces, **273**, 275, 308
black widow spider, 35
blast furnace, 261
blind spot, 29
blindness, 28
blood, **2**, 4–6, 12, 14, 19, 21, 48
blood alcohol limit, 43
blood capillaries, 48, 60
blood pressure, 4–6, 39, 43, 157
blood sugar levels, 11, 49–50
blood vessels, 5, 6, 11, 39, 42, **50**
body growth and maintenance, 8, 9
body mass, 10, 11, 58
body mass index (BMI), 11
body temperature, 46–9
body weight, 5, 10, 11–12
boiling, **169**, **263**
boiling points, 168–9
boiling water, 259, 262, 317, 320
bombs, 244, **326**, **327**, **336**
bond breaking, **194**, **262**
bond making, **194**
bonds, **156**, **169**, 218
bones, 346–7
bowel diseases, 8, 23
brain, 1, 5–6, 28–9, 30, 33, 37–8
brain cells, 6, 49
brain development, 60
brain and mobile phones, 277
brass, 221
bread mould (*Mucor haemalis*), **78**
breast milk, 22
breathable materials, 159, 184, 185
breathing, 1–2, 60
    see also respiration
breeding, 83, 136
brick buildings, 205
brick walls, 265, 269, 284
brightness of stars, 345, 356, 359
broad-leaved trees, 143
broadcasting, 285–7, 294
bromine water, **177–8**
bronchi, 20, 40, 41
bronchitis, 39, 40–1, 124
bronze, 222, 224
building materials, 205–10
Bunsen burners, 189–90
burning, 1, 193, 195–6, 232–3, 245
    of fossil fuels, 122, 124, 215, **233**, 235, 265, 303, 317, 322
    see also combustion
butane, 174
butene, 175

butterflies, 116
cables, 313, 318–19, **320**
cacti, 105, 114
calcium carbonate, 206–7, 236, 240–5, 249
calcium oxide, 206–7
calcium sulfate, 236
calorimeters, 195
camels, 105–6
camouflage, 84, 104, 115
cancer, 16–17, **17–18**, 39, **40**, 51, **326**
    bowel cancer, 8
    breast cancer, 11, **17**
    lung cancer, 17, 39
    skin cancer, 125, 300
cancer treatment, 137, 330–1, 335
cannabis, 37, 38, 39
capillaries, **48**, 60
captive breeding programmes, 136
car fuels, 188, 189, 191, **302**
carbohydrates, 8–10, 13, 49, 89, 122, 149, **150**, 182
carbon
    in steel, 221, 228
    in tobacco smoke, **40**
carbon dioxide
    in the atmosphere, **91**, 122–3, 215, 232–3, 302–3
    from baking powder, 151–3
    from carbon monoxide, 235–6, 251
    from cement making, 206–7
    from combustion, 188–91, 232–3
    from copper carbonate, 219
    in heat trapping, 236
    in photosynthesis, 77, 88–92, 122, 232, **233**
    in reaction rate measurement, 240–5
    in respiration, 1, 2, 41, 46, **92**, 232–3
    in spacecraft, 346
    test for, 151, 189
carbon dioxide emissions, 133, **144**, 180, **325**
carbon fuels, 188–92
carbon monoxide, 39–40, 189–90, 191, 234, 235, 251
carcinogens, **40**
cardiac muscle, 4
cardiovascular efficiency, **4**
cars, 180, 191, 220, 221, 227–31, **273**
casting, 222
catalyst reuse, 251
catalysts, 170, 178, 250–1
catalytic converters, 235–6, 251
cathodes, 220
cats, **83**, 84, 97
cavity walls, 265–6, 269–70
cell division, 17, **40**, 56, **61**
cell surface membrane, 76, **78**, **79**
cell walls, **24**, 76, 77, **78**, 89, **150**
cells (biology)
    in animals, 76–7
    and diseases, 16, 18, 21, 28, **40**, 300
    and genes, 54–6
    in the nervous system, 33
    in plants, 76–7, 88, 89–90, 91, 92
    in respiration, 1–2, 6, 8, 49
cellulose, 77, 89, 90, **150**
Celsius scale of temperature, 195, 258, **259**

cement, 205, 206–7
Census of Marine Life Project, 72
central heating systems, 219, 220, 260, 270–1, **273**
central nervous system, 33, 34
centripetal force, **345**
cerebellum, **37**, **38**, 80
CFCs (chlorofluorocarbons), 125, **301–2**
chain molecules, 174, 178, 182
charged particles, **342**
chemical bonds, **156**, **169**, 218
chemical change, 149, **150**
chemical digestion, 13
chemical reactions, 147, 149, 152–3, 158, 178, **194**, 228, 239–56
chemical test for carbon dioxide, 151, 189
chlorophyll, 77, **79**, 88, 89
chloroplasts, 76, 77, **78**, **79**, 88, 89
chromosomes, 16, 54–5, **61–2**, 63, **78**
cilia, 20, 40
ciliary muscle, 29, **30–1**
circulatory system, 5–6, 39, 116
cirrhosis of the liver, 43
classification of organisms, 75–7, **77–9**, 79–80, **81**
classification of vertebrates, 79–80
clay, 205, 206–7
cleaners of parasites, 98
clear-felling, 143
climate, 236, 260, 302–3, 351
climate change, 131–2, 133, 215, 265, 303
clomiphene, **51**
clones, 55
clothing, 47–8, 182, 183–4, 201, 267, 300, 333, 336
coal-fired power stations, 124, 235, **310**, 317, **318**, 322
coatings, 271, **308**
cocaine, 38
codes, 280, **282**
coils of wire, 314, 315, 319, 339
collapsing stars, 356–7
collisions with Earth, 350–1, **353–4**
collisions of electrons, **272**
collisions of particles, 239, 240–1, 249, 250, 252
colloids, 200
colossal squid (*Mesonychoteuthis hamiltoni*), 72
colour additives in foods, 155, 157
colour blindness, 16, 29, 60, 61
colour pigments, 200, 201–2
combustion, 1, 188–90, **190–1**, 232–3, **233–4**
    see also burning
combustion products, 188
comets, 290, 340, 344, 350, 352, **353**
command code, 280
common polypody fern (*Polypodium vulgare*), 76
communicating with infra-red radiation, 279–83
communications systems of satellites, 341, **342**
communities of organisms, 71–3
compass needle, 212, 338

competition, 94–7, **118**, 132
composite materials, 205
compounds, 151, 218
computers, 279, 284, 353
concentration, **90**, 240–1, 242–3, 244–5
conception, 51
concrete, 205–6
condensation, 262, 317, **351**
conduction, 47, 222, 227, 269, **271–2**, **276**
conductivity, 220, 222, 227, 228
conductors, **271–2**
conifers, 75, 113, 143
conservation, 71, 135–7, 137–42, 230
construction materials, 205–10
continental plates, 211, **213**
contraception, 51
convection currents, **213**, 245, 269–70, **272–3**, 309
convergent evolution, 85
cooking, 147–54, 188, 239, 275–8, 324
cooking methods, 147–8
cooling, 48–9, 193, 215, 236, 257–64
cooling towers, 313, 317, **318**
copper, 218–21, 227, **228**, 230
copper carbonate, 219
copper ore, 218–19, 220
copper oxide, 219, **228**
copper sulfate solution, 220
coppicing, 143
core (Earth), 211, 212, 232, **298–9**, 339
core (star), **357**
cornea, 29, **30**, 31, **32**
corrosion, 228–9
cortex, **37**, **38**
cosmetics, 165, 300
cosmic rays, 330, 341, **342**, 347
cost savings, 180, 261, 265, 266
costs in car manufacture, **229**
costs of energy, 324–5
costs in power stations, **323**, 326–8
coughing, 40–1
covalent bonds, 169, 170, 175–6, **176–7**, **182–3**
cracking (crude oil), 169–70, 197
craters, 350, 351
creation, **112**, **117**, **118**
crests of waves, 290, 294
critical angle, 281
crop yields, **73**
cross-linking bridges, **183**, **201**
crude oil, 167–73, 174, 188, 191, 230
crust, see Earth's crust
crystals, in rocks, 215
cyanobacteria, 232
cystic fibrosis, 16, 60, **61–2**
cytoplasm, 19, 76, **77**, **78**, 90

daily energy requirements, 8
Darwin, Charles, **117–18**, 119
data collection, 242, 243
decolourising bromine water, **178**
decommissioning nuclear power stations, **326**, **328**
decomposers, 69
decomposition, 152–3, 207, 219, 250–1, 252
decompression chambers, **140**
deficiency diseases, 8, 9, 16

deforestation, **234**, 303
degassing, **232**
degrees of temperature, 195, 258, **259**
dehydration, 47, 158
denaturation, 47, 91, **150**
dendrites, 33, **34**
density of air, **272**, **287**
density of black holes, **358**
density of metals, 228, 229
depressants, 37, **38**, 42
deserts, 84, 94, 105, 310
designer polymers, 182–7
desulfurisation, 236
diabetes, 10, 11, 16, 49–50
diarrhoea, 21, 99
diastolic pressure, 5
diesel, 168, 191, 234–5
diet, 5, 8–12, 16, 17, 58
dieting, **12**
diffraction, **277–8**, **288–9**
digestion, 12–14, 60, 148, **150**
digestive system, 1, 12–14, 60, 97, 99
digital radio and TV, **281**, 286
digital signals, **280–1**, **282**
dinosaurs, 110, 117, 131, 132, 351
direct current (DC), 305, 315
disc diffusion, 24
diseases, 16–20
displayed formulae, 174–6
dissolved substances, 163, 240
distribution, 94, 95, 97, 103
diving, **140**
DNA, 16, 17, 54, 55, 59–60, 77, 125, 182, 300
DNA fingerprinting, 336
dodos, 131, 132
dogs, 82, 83, 97, 136
domestic appliances, 189–90, 275–6, **323–4**
dominant alleles, **61–3**
dopamine, **37**
Doppler effect, 360
dose of radiation, 330, 332, 347
double-blind tests, **25**
double covalent bonds, 169, 170, 175–6, **176–7**, **201**
double glazing, 265, 266, 269, **308–9**
Down's syndrome, 56
drink driving, 42, 43
drug classification, 38, 39
drug habit, 38
drug misuse, 37, 38–43
drugs (medicines), 19, 23, 24–5, 37–8, 59, 137
drugs trials, 25
drying of paint, 200, **200–1**
ductile metals, 221
dust
    of asbestos, 266
    in the atmosphere, 215, 303, 350
    in space, 344, 350, **351**, 357
dust explosion, 252–3
dyes, 157, 201–2
dynamo effect, 314, 315, **342**

E numbers, 155
ear, 28
Earth, threats to, 350–5
Earth age, **112–13**, **118**, 232

Earth core, 211, 212, 232, **298–9**, 339
Earth magnetism, 212, 338–43
Earth movement, 211–17
Earth structure, 232, 298–9, 340, 350
Earth temperature, 215, 236, 302
earth waves, 297–304
earthquakes, 212, 290–1, 297–8, **298–9**, 350
Earth's crust, 205, **209**, 211–12, **213**, 214, **298**, 340
    oil prospecting, 167, 168
    ore extraction, 220, 221, 229–30
earthworm (*Lumbricus terrestris*), 66, 77, 104
ecology, 66–74, **95**
ecosystems, 72–3, **95**, 137
edible mushroom (*Agaricus campestris*), 78
education programmes, 136
effectors (nervous system), 34
efficiency of light bulbs, 267–8
efficiency of power stations, **318**
egg cooking, 147, 148–50
egg fertilisation, **50–1**, 56, 58, 60, **62**, **107**
eggs of animals, 80, **108**
electric charge, 331, **332**, **342**
electric currents, **295**, 305, **307**, 314–16, 319, **320**, **324**
    in the Earth, 338, 339
    in space, **342**
electric heaters, **273**
electrical conductivity, 220, 222, 227, 228
electrical impulses to the body, 33–5
electrical impulses to the brain, 1, 28, 29, 33, **34**
electrical power, **323–4**
electrical signals, 280, **295**
electricity consumption, 314, 324–5
electricity costs, 324–5
electricity generation, 313–21
    by fuel cells, 191
    by nuclear power, **325–8**
    by solar cells, 305–9
    by wind power, 309–11
    in space, 271
electricity meters, 324, **325**
electricity supply, 306–7, 314, 318–20, **323–4**
electrolysis, 220
electromagnetic radiation, 276, 284, 290, **295**
electromagnetic spectrum, 276, 286, 299
electromagnetic waves, 285, 286, 292
electromagnets, 319, 339
electrons, **177**, **272**, **307**, **332**
elements, 218
emitters, **273**, 275
emphysema, 39, 40, 40–1, 124
emulsification, **14**, **156**
emulsifiers, 155–6
emulsion paints, 200
endangered species, 131–43
endocrine glands, 49
endoscopes, 281–2
endothermic reactions, 193–4, **194**, 207

energy, 193–9
    from glucose, 1, 2, 8, 89, 258
    input, **318**
    output, **318**
    storage in plants, 89–90
energy change, 149
energy efficiency, 267–8, **318**
energy-efficient houses, 265, 269
energy flow, 258, **271–2**, **318**, **320**
energy loss, 265, 269–70, **320**
energy recovery, 261
energy retention, 269–70
energy saving, **144**, 265, 269
energy source choice, 167, **323**
energy transfer, **320**, **323**, 324–5
energy transformation, **323**, 324
energy value of fuel, 188, **196**
engineering polymers, 180, 227
environmental concerns, 143–4, 230, 311, **323**, **325–6**
environmental effects, 58, 116–17, 167–8, 171, 207
enzymes, 1, 13, 14, 20, 47, 54, **78**, 91, **251**
epicentre, 297
erector muscles, **48**
esters, 162–3, 164
estimation (organism population), 70–1
ethane, 174–5, 176
ethanoic acid, 162–3, **196**
ethanol, 162–3, 195
ethene, 175–6, 177, 178
evaporation, 47, 48, 105, **233**
evaporation of perfume, **161–2**
evaporation of solvents, 163, 164
evaporation in space, 352, **354**
evolution, 75, **81**, 85, 112, **118–19**, 232
exercise, 3–4, 5, 8, 11, 40, 346–7
exhaust fumes, 234, 235
exothermic reactions, 158, 193, **194**, 207, 245, 252–3
expanding Universe, 358–9, 360
expansion of air, **272**
experiment results, 70, 242–6
explosions, 193, 197, 245, 252–3
exponential growth, 121
extinction of species, 117, 131–3, 136, 142, 351
eye damage, 300
eyes, **11**, 28–33

fair testing, 195
farming, **73**, 215, **234**, 235, 310
faster reactions, 240–3, 245, 249, 250–3
fatigue, 3
fats in foods, 8, 10–14, 122, 155, 156
fats in plants, 89, **90**
fatty acids, 10, 13, 14
fault detection, 335
feathers, 80, **81**, 84, 105, 167
feedback, **48**
ferns, 75, 76, 113
fertile offspring, 83, **119**
fertilisation, 56, 58, 60, **62**, **107**
fertilisers, 73
fertility treatment, 51
fibre in food, 8

fibre optics, 281–2, 306
fibreglass, 265, 269, 270
fibres of asbestos, 266–7
fibres of textiles, 182, 227
field studies, 66–73
filament (flower), **107**
filament lamps, 267, 268
finite resources, 167
firebombs, 244
fires, 197, 223–4, 245, 350
first class proteins, 9
fish, 72, 75, 80, **81**, 98, 104, 113
fish in polluted water, 124, 125–6, 133
fish populations, 142–3
fishing, 142–3, 235
fitness, 1–7, 346–7
fitness programmes, **3–4**
Five Kingdom system, 76
flames, 189–90, 245
flammable substances, 197
flavour enhancers, 155, 157
flexibility, **3–4**
flour, 148, 151, 253
flowering plants, 75, 76, 88, 113
flowers, 76, 88, **107–8**
flue gas desulfurisation (FGD), 236
foam polymers, 182
foam in wall cavities, 265–6, 269, 270
focusing light, 29–33, **309**
food, 8–15, 17, 20, 49–50, 122, 330
    competition for, 94, 95–7
food additives, 155–60
food energy, 258
food supplements, **11**, 17, 99
food supply, 8–9, **126**
forests, 88, 133, 143, **234**, 303
formula units, 149, 151–2, **153**
formulae, **152**, **190**
fossil fuel burning, 122, 124, 215, **233**, 235, 265, 303, 317, 322
fossil fuel supplies, 121, 167, 191
fossils, **81**, 82, 110–13, 117, 131, **212**, 351
foxes, 84–5, 106, 114
fractional distillation, 168–9, 174
fractions, 168–9, 169, 188
free electrons, **272**
freezing water, 106, 262
frequencies of currents, **316–17**
frequencies of waves, **276**, **278**, 285–6, **287**, 292, 293–4, **295**
frequent collisions, 240–1, 249
FSH (follicle-stimulating hormone), **51**
fuel burning, 193, 195–6, 232–3, **234**, 235
fuel choice, **323**
fuel economy, in cars, **229**
fuel energy, **318**
fuel oil, 168, 169, 170, 188
fuel rods, 322, **326**
fuels, 188–92
    for power generation, 322–9
    for spacecraft, 346, 347
fungi, 19, 23, **78**, 113, 127, 133

GABA (transmitter substance), **38**
galaxy, **342**, 344, 347, 356, **357**, 358–60
gall bladder, 14
gametes, 56, 60, **107**

gamma (γ) radiation, 284, 286, 330–5, **342**
gas clouds, **351**, 356
gas collection, 240, 242, 243–6, 249
gas-fired power stations, 317, 322
gas pressure, 241
gases, 239
gasohol, 322
Geiger counters, 330, 331, 335
gender (sex), 8, 58, 63
generating electricity, 313–21
generators, 314–15, 317, 322, **342**
genes, 16–17, **40**, 54–65, 114–16, **118**
genetic code, 54
genetic diagrams, **62–3**
genetic mutations, 16–17, **301**
genetics, 60–3, **118**
genotypes, **61**, **62**
genus, 82
geographical isolation, **119**
geology, 211–16, **340**, 351
glands, 33, 34, 48, 49, **51**, 80
glass, 205, 227, **272**, 276, 284, 300, **308**, **336**
glass optical fibres, 281–2
global population growth, 121, **127**, 133, **234**
global warming, 102, 122, 133, 180, 215, 236, 265, 302, **325**
glucose
    as energy store, 49–50
    in photosynthesis, 89–90, **99**
    in respiration, 1–2, 8, 10, 47, 92, 258
glutamate, **38**
glycerol, 10, 13, 14
glycogen, **49**
goblet cells, 20, 40
Gore-Tex®, 183–5
graded bedding, **216**
gradient of graphs, **245–6**
grain elevators, 253
granite, 208, **209**, 215
graphs, 242–3, 245–6
gravitational potential energy, **357**
gravitational pull, 340, 344, **345**, **353**, 357, **358**, 359
gravity, 344, **345**, 348, **351**, **353**, 357
Greek medicine, 257
greenhouse effect, **144**, 236, 302
grid of power, 318–20, **342**
grizzly bears, 102
grouping organisms, 66, 75–87
Gulf stream, 236
gullet (oesophagus), 13, 39

habitat creation, 137
habitat destruction, 133, 143
habitat protection, **73**, 134, 135
habitats, 66, 70, 71, 83–4, 102
haemoglobin, 8, 40, 54, 60, **62–3**
hair, 48, 50, 54, 80
Hall–Héroult process, 230
hallucinogens, 37
hand-grip monitor, **3**
harder collisions, 240, 241
hazardous materials, **326**, 333, 336
health, 3, 9, 10–12, 16–27, **327**
heart, **2**, 3, **4**, 5, **12**, 56, 80
heart attack, 11, 39, 43

heart disease, 3, 10, 11, 39, 43, 157
heart rate, 2
heat conduction, 47, 222, 227, 269, 270–2, **276**
heat energy, 189–90, 207, 257–64, 317, 322
    of burning fuel, 193, 197
    of light bulbs, 267–8
heat energy measurement, **195–6**
heat flow, 258, **271–2**
heat inputs, 47
heat loss, 48, 269–70
heat outputs, 47
heat stroke, 47
heat trapping, 236
heating, 188, 257–64
heating food, 147–50
heating houses, 265–8, 308
height, 9, 11, 58
herbicides, 73
heroin, 37, 38, 39
hertz (Hz), 285, 294, **316**
heterozygous people, **61–2**
hip replacements, 185
HIV, 16
holes in the ozone layer, **301**
homeostasis, 46–7
homologous chromosomes, **61**
homozygous people, **61**
hormones, 49–51
horses, 59, 82, **83**, 111–12, 322
hosts, 19, 88, 97
hot air heating, 261, 269, **272–3**
house heating, 265–8
house insulation, 265–6, 269–70
household waste, 125
human effects on the atmosphere, **233–6**
human effects on extinctions, 131, 132
Human Genome Project, 55
human population, 121–30, 133, **143–4**, **234**
human species (*Homo sapiens*), 80, 81, 83
hunting, 131, 132–3, 137, 141, **142**
hybrids, **83**
hydrocarbon compounds, 167, 168–9, 174, **201**
hydrocarbons, 174–6, **176–8**, 188–91, 197
hydrochloric acid, 13, 20, 240–5, 249
hydrogen atoms, **177**
hydrogen fuel, 191, **302**, 356
hydrogen gas, **351**, **357**
hydrogen peroxide solution, 250–1
hydrogen pop test, 252
hydrophilic substances, 156
hydrophobic substances, 156
hyphae, 19, **78**
hypothalamus, **48**
hypothermia, 42, 46, 47

Ice Age, 111, 131
ice packs, 262
ice in the sea, 102–3, 106
ice in space, 344, 350, **351**, 352, **354**
igneous rocks, **209**, 215, **216**
images (eye), 28, 29–30
immune response, 22

immune system, 16, 17, 21, 23, 99
immunisation, **22-3**
immunity, 22-3
impervious layers, 228-9
incidence rate, **17-18**
incinerators, 125, 184, 186
indicator species, 127
infection, 16, 20, **24**, 38
infectious diseases, 3, 18-23, **126**
infra-red radiation, 270-1, **273**, 275,
      276, 279-86, **295**
   from the Sun, 299, 305, **308-9**
   and global warming, 122
infra-red sensors, 279
inheritance, 60, **62-3**, 63, **118**
input energy, **318**
insects, 68, 69, **73**, 75, **78**, 105, **107-8**,
      110, 113, 132
insoluble substances, **90**, 164
insulating materials, 182, 265, 269,
      270
insulation, 105, 265-74
insulators, **271**, 319
insulin, 16, 49-50
intelligence, 58, 60
intelligent packaging, 158-9
interference (signals), **278**, **282**, 285
intermolecular bonds, **262**
intermolecular forces, **164-5**, **169**,
      182-3
intestines, 13, 14, 99
invertebrates, 125, 132, 133
ionisation, 331, **332**
ionising radiation, 59, **326**, 331, 341
ionosphere, 287
ions, 77, **99**, **332**
iris (eye), 29
iron, 205, **216**, 221, 228, 340, 351
   in food, 8, 16
iron core, 205, 212, 232, 319-20, 339
iron (III) hydroxide, **228**
irradiation, 330

joints, 3, 11, 185
joules (J), 195-7, 258, 259, 267, 324
joules per kilogram (J/kg), 259, **263**
joules per second (J/s), **323**

keratin, 54
keys (animal identification), 66
kidneys, 5, 6, 46
kilojoules (kJ), 195
kilojoules per kilogram (kJ/kg), **263**
kilowatt-hours (kWh), 324-5
kilowatts (kW), **323**, 324-5
kinetic energy (KE), **262**, **271**, **272**, **276**,
      309, **357**
kingdom classification, 76-7, **77-9**
kwashiorkor, 8, 9
Kyoto Agreement, **144**

labels
   of energy efficiency, 267
   of food ingredients, 155, 156
   of power rating, **323**
lactic acid, 2
Lamarck, Jean-Baptiste, **118**
laminates, **184**
landfill sites, 125, 184, 336

larvae, 80, **108**, 127, **213**, 214-16
laser eye surgery, **32**
laser light waves, 292
laser pulses, 281, **282**
latent heat, 261-2, **263**
lava, 214-15, **216**
lead, 218, 332, 333
leaf litter, 69, 88
leaves, 69, 76, 77, 88, 89, 104
lens of eye, 29, **30-1**, 31, **32**, 300
LH (luteinising hormone), **51**
lichens, 115, 127
life on Earth, 75, **112-13**
ligaments, 29, **30-1**, **32**
light
   in the eye, 29-30, 31-2
   from the Sun, 59, 215, 299-303,
      305, 307-9, 350, **354**
light bulbs, 267-8, 292, **323**, 325
light competition, 94, 95
light energy, 89, 193, 267-8
light reflection, 281-2
light refraction, 29, **30-1**, 281
light transmission, 281-2
light waves, 290-4, **295**, 360
light-years, **347**
lime water, 151, 189
limestone, 205, 206-8, **209**
limiting factors, **91**
lions, **83**, 103, 135
lipase, 13, **14**
liquid paraffin, 169-70
liquid petroleum gas (LPG), 168, 188
liquids, 239
lithosphere, 212
liver, **2**, 13, 14, 38, 49
liver cells, 18, 19, 43, 49-50, 76
liver damage, 42, 43
liverwort (*Pellia epiphylla*), 76
living cells, *see* cells (biology)
living organisms, 66-87
local authorities, **144**
loft insulation, 265, 266, 269
long-distance communication, **287**
long sight, 31, **32**
longitudinal waves, 297-8
lung disease, 266
lungs
   of animals, 80, 235
   of humans, 3, 16, 20, 40-1, 46, 60,
      61, 235
   when diving, **140**
lymph, 14
lysozyme, 20

magma, **209**, 213-15, 232
magnetic Earth, 212, 338-43
magnetic fields, 212, 315, 338, 339,
      341, **342**, 348
magnetic navigation, 338
magnetic poles, 338
magnetic storms, **342**
magnetron, 150
magnets, 314, 315, 338, **342**
main sequence stars, **358**
mains frequency, **316**
malaria, 16, 18-20
malleable metals, 221, 228
mammals, 75, 79, 80, **81**, 85, 113

mammoths, 111, 131, 132
mantle, 211, 212, **213-14**, 232, 298,
      339, 340
marble, 208-9
marble chips (calcium carbonate), 240,
      249
mark-release-recapture technique, 70
Mars, 340, 344, 346, 347-8, 350, **352**
mass, **358**, 359
mass extinction, 132
mass of water, 195, **196**, 259, **260-1**
massive stars, 356-7
maximum volume of oxygen (VO_2
      max), **4**
meat, 148-50, 155, 159, 239
medical diagnosis, 257
medical radiation sources, **327**, 330-1
medicines, *see* drugs
melanin, 59, 299, 300
melanoma, 17
melting ice, 102, 149, **259**, 262
melting polymers, 182, 184
membranes (cells), 24, 76, **78**, **79**
membranes (materials), **184**
membranes (mucus), 20, 40
memory, 58
memory cells, 22
memory metal, **222**
men, 8, **17-18**, 50-1, 55, 60, **62-3**
menstrual cycle, **50-1**
metabolic reactions, 1, 47, 54, **90**, 92
metal conduction, 222, 227, **271**, **272**
metals, 218-21
metamorphic rocks, **209**
meteors, 344, 351
meters, 5, 324, **325**
methane, 174-5, 188, 189-90, **190-1**,
      322
micro-organisms, 16, 18, 19, 22, 148
microwave background, 359
microwave ovens, 150, 276, 277
microwaves, 150, 276-7, 284, 286,
      287-9, **295**, **347**
migration, 139, 338
Milky Way, 344
millimetres of mercury (mm Hg), 5
mineral resources, 121, **126**, 207
minerals in food, 8, 16, 17
minerals in the soil, 77, 110
mining, 220, 266-7
mirrors, **309**, **353**
mitochondrion, 19, 76
mitosis, 55, 56
mixtures, **233**
MMR vaccine, 23
mobile phone chargers, 194, 320
mobile phone masts, 277, **278**, 284, **288**
mobile phones, 277-8, 280, 284-5,
      286, 293
molecular forces, **164-5**
molecular formulae, 174-6
molecule vibration, 150, 239
molecules, 149, 152, **153**, 156
   in combustion, 190-1
   in conduction, **272**, **276**
   in crude oil, 170
   in evaporation, **161-2**
   of hydrocarbons, 174-7
   in ionisation, **332**

molecules (continued)
  in living cells, **1**, 9–10, 12, **262**, **272**, **276**, **332**
  in melting, **262**
  in oil paints, **200–1**
  in plants, 89–90
monocular vision, 33
monomers, 174, 178
Moon, 230, 302, 339–40, 345, 346, 350
Morse code, 280, **282**, 293
mortality rate, **17–18**
mortar, 205, 206
mosquitoes, 19–20
moths, 115
motor neurones, 33, 34
moulding polymers, 182, 184
mountains, **213**, 310, 330
mouth, 13, **22**
MRSA, **24**
mucous membrane, 20, 40
mucus, 20, 40, 41, 60, **61**, 104
multiplexing, **282**, **286**
muscle cells, 1, 2, 40
muscles in humans, 2–4, 8, 9, **12**, 33, 34, 37, 346–7
musical notes, **295**
mutations, 16–17, **24**, **40**, 58–60, 116–17, **301**
mutualism, 98–9
myelin sheath, 33, **34**
myxomatosis, 97, 114

nail varnish, 163–4, **165**
naphtha, 168
National Grid, 313, 318–19, **342**
natural ecosystems, 72–3
natural polymers, 182
natural resources, 121, **126**, 230
natural selection, **112**, 114–19
nature conservation, 71, 135–7, 137–42
navigation, 338, 341
near-Earth objects (NEOs), 353–4
nectar, **107–8**
negative charge, **177**, **332**
negative feedback, **48**
nerves, 1, 28, 33
nervous system, 28, 33–5
nesting sites, 94, 133–6, 136
nets (organism sampling), 68
neurones, 33–4, **37**
neutron stars, 357
niche (ecology), **95**
nickel, **222**, 339, 340, 351
nicotine, 37, 39–40
nitinol, **222**
nitrates in the soil, **73**, 90, **99**
nitrogen in the atmosphere, **99**, 232, 233, **342**
nitrogen in diver's blood, **140**
nitrogen fixation, **99**
nitrogen monoxide, **236**
nitrogen oxides, 234, 235–6
noise (signal transmission), **280–1**, **286**
non-biodegradable polymers, 184
non-destructive testing, 335
non-metal conduction, **271–2**
non-renewable resources, 167, 184, 191, 220, 230

nuclear bombs, 245, **326**, **327**, 336
nuclear fission, **327**
nuclear fuels, 322, **325**, **326**, 336
nuclear fusion, **357**
nuclear power stations, **310**, 318, 322, **325–8**, 336
nuclear radiations, 330–7
nuclear weapons, **326**, **327**
nucleic acid (RNA), 19
nucleus of atom, **177**
nucleus of cell, 76, **78**, **79**
nutrients, 8, **12**, 88
nylon, 182, 183, 184

obesity, 10, 11
oceanic plates, 211, **213**
oceans, formation of, **232**
oesophagus (gullet), 13, 39
oestrogen, 50–1
off-peak power, **325**
oil-fired power stations, 317, 322
oil paints, 200
oil politics, **167–8**
oil prospecting, 167, 298
oil refineries, 169, 170, 197
oil spills, 167, 171
oils, 89, **90**, 155–6, 200
optic nerve, 29
optical fibres, 281–2, 306
optimum temperature, 47
orbits, 344, 345, 346, 352
ore extraction, 218, 221, 229–30
organism communities, 71–3
organism grouping, 66, 75–87
organism sampling, 66–71
organisms, 66–87
oryx, 132, 133, 136
oscilloscopes, 316
osmosis, **90**
ospreys, 134, 135, 136
output energy, **318**
ovaries, 49, 50, **51**, 56, 60, **62**, **107**
ovens, 275–6
oviduct, **51**
ovulation, 50–1, 56
ovules, **107**
oxidation of glucose, 1
oxidation of oil paints, **201**
oxidation reactions, **228**
oxides of nitrogen, 234, 235–6
oxygen
  aluminium reaction, 228–9
  in the atmosphere, 1, 232–3, **287**, **342**
  in blood, 6, 39–40, 41, 60
  in combustion, 188–91, 232–3, 245
  in explosions, 252–3
  from hydrogen peroxide, 250–1
  from photosynthesis, 89, 90, 232, **233**
  reactions with food, 155, 159
  for respiration, 1–2, **4**, 8, **92**, 232–3, 346
  sulfur dioxide formation, 235
  in water, 125–6, 127, 133
oxygen debt, 2
ozone depletion, 124–5
ozone layer, 124–5, 300, **301–2**

packaging, 158–9, 182
painkillers, 37, 38
paints, 200–3
palaeontology, 131
pancreas, 13, 16, 49
paper, **332**, 334
paraffin, 168, 169–70
parasites, 19, 88, 97, 98
particle collision, 239, 240–1, 249, 250, 252
particles, **177**, 200, **262**, **287**, **351**, **357**
particulates, 39, **40**
pascals, 5
passive immunity, 22
pathogens, 18–24, **40**, **126**
peak demand, **325**
peak voltage, 316
penicillin, 23, **24**
penicillin-resistant bacteria, **24**
peppered moth (*Biston betularia*), 115
performance enhancers, 37
perfumes, 161–3
periods, 50–1
peripheral nervous system, 33
pesticides, **73**, 137
petrol, 168, 169, 170, 180, 188, 191, 234–5, 322
pH, 13, 127, 235
phenotypes, **61**
phloem tubes, 89
phosphorescent paints, 202
photocells, 305, 306–7, **307–8**, 315
photochemical smog, 235
photographic film, 332, 335, 336
photosynthesis, 77, 88–92, **99**, 232, 233
photosynthesis rate, 90–1
physical change, 149, 162
physical digestion, 12
pigments, 200, 201
pitfall traps, 69
pituitary gland, 48, **51**
placebos, **25**
placenta, 22, 40, 80
planetary nebula, 356
planets, 340, 344, 345, 350, **351**, 353, 356
plankton, 139–40
planning rules, **144**
plant cells, 76–7, 88, 89–90, 91, 92
plant competition, 94–5
plant kingdom, 75, 76
plants, 55, 67–8, 76–7, 137, 302, 351
Plasmodium, 18–19
plastics, 170, 174, 180, 182, 227, 230, **271**, 276, 334
plate tectonics, 211–12, **212–14**, 232, 350
platypus, 79
plutonium, 300, 322, **325**, **326**, **327**
poison arrow frogs, 103–4
poisonous substances, 116, 167, 171, 189–90, 234
polar bears, 102–3, 106
polar ice caps, 122, 236, 303
pollen grains, **107**, **108**
pollination, 88, **107–8**
pollutants, 233, 234–6
pollution, 115, 122–6, 127, 133, 167, 171, **234–6**, 306, **310**
polymer molecules, 174, 178, 182–3

polymerisation, 178, **201**
polymers, 170, 174–87, **271**
polyoxymethylene (POM), 180
polyphenylenesulfide (PPS), 180
polypropene, 178
polysaccharides, 10, **49**
polystyrene, 170, 182, **183**
polytetrafluorethylene (PTFE), **184**, 185
polythene, 170, 178, 182, **183**
pond weed (*Spirogyra longata*), **78–9**
pooters, 68, 70
population (people), 121–30, 133, **143–4**, **234**
population (species), 70, 94, 96–7, 142
positive charge, **177**, **332**
potatoes, 89, 149, **150**
powder explosions, 252–3
power lines (cables), 313, 318–19, **320**
power ratings, **323–4**
power stations, 302, **309**, **310**, 313, 315, 317, **318**, 322, 325
power supply, 307, 314, 318–20, **323–4**, **342**
predator–prey relationships, 96–7
predators, 94, 98, 103–4, 114, 116, 132, 135
pregnant women, **11**, 22, 40
pressure, 5
pressurised gases, 243
pressurised reaction vessels, 178
prey, 96–7, 103–4, 132, 135
primary coils, 319–20
primary (P) waves, 297–8, **298–9**
probiotics, 99
products, chemical, 147, 152
progesterone, 50–1
propane, 174
propene, 175, 178
properties of metals and alloys, 222, 228–30
properties of perfumes, 161
properties of polymers, 182, 184
protease, 13
protecting habitats, 135
protecting species, 133–5
protective clothing, 300, 333, 336
protective layers, 228
protective shields, 341, 347
proteins
   in cells, 1, 21, 54, 59, 182
   in food, 8, 9–10, 12, 13, 122, 149, **150**
   in plants, 90, **99**
Protoctista, **78–9**, 113
protons, **342**
protostars, **357**
protozoa, 16, 18–19
PTFE (polytetrafluorethylene), **184**, 185
pulse (body), 4
pulses of infra-red radiation, 279–82
pumice, **216**
Punnett square, **62–3**
pupil (eye), 29, 34
purifying copper, 220
purifying crude oil, 168–70
pylons, 313, **318**, 319

quadrats, 66–8, 70
quantitative estimates, 70–1
quarrying, 207, 220

rabbits, 33, 94, 97, 114, 165
racing pigeons, 338
radar, 150, 348
radiation, 47, 48, 269, 270–1, **273**, **308**, 348
radiation cooking, 275–8
radiation dose, 330, 332, 347
radiation film badges, 332
radiation levels, 330–1
radiation sources, 330, 333
radiation in space, 270–1, **273**, **301**, 330, 341–2
radiation therapy, 335
radiators, 260, 269, 270–1, **273**
radio bands, 286, 294
radio transmission, **281**, 284–8, 294, 346
radio waves, 284–7, **287–9**, **295**, **347**
radioactive decay, 335, 339
radioactive labelling, 336
radioactive paints, 202
radioactive rocks, 330
radioactive substances, 330–7
radioactive tracing, 335
radioactive waste, **325–6**, 336
radiography, 59
radioisotopes, **327**
radon, 330
Rafflesia, 88
rainfall, 124, **233**, 350, **354**
rainforests, 88, **234**, 303
rate of chemical reactions, 239–56
rate of photosynthesis, 90–1
rats, 116, 131, 165
rattlesnakes, 35
raw food, 147–9
raw materials, 8, 88, 178, 184
reactants, 147, 152, **194**, 239–45, 249, 251
reaction rates, 239–56
   measuring, 240, 242–6
reactions, *see* chemical reactions
receptor cells (eye), 28, 29, 60
receptors (nervous system), 33, 34, 35
recessive alleles, **61–3**
recommended daily allowances (RDAs), 9–10
recycling cars, 230
recycling copper, 220–1
recycling energy, 261
recycling waste, 125, **144**, 184, 186, 230
red blood cells, 18, 19, 40, 60
red giants, 356
red–green colour blindness, 16, 29, 60, 61
red kites, 133, 134, 135
Red List (IUCN), 132, 133
red shift, 360
reduction reactions, 219, **228**
reflection, **273**, 285, **354**
reflective coatings, 271
reflective foil, 265, 269, 271, 275
reflex actions, 34
reflex arcs, 34
refracted light, 29, **30**, 281
refracted waves, 287, **298–9**
refrigerators, **301–2**, 306
rehabilitation, 39
religion, and diet, **10**, **11**

remote living, 306–7
renewable energy sources, 305, **310**, 322
reprocessing plants, **336**
reproduction, 54–6, 97, 102
reptiles, 75, 80, **81**, 85, 113
residues, 168
resistance, electrical, **320**
resource use, 121, **126**
respiration
   in animals, 92, 232–3
   in humans, 1–2, 8, 40–2, 47, 49, 193, **234**, 258
   in plants, 89, 92, 232–3
   see also breathing
respiratory passages, 20
results, recording, 70, 242–6
retaining energy, 269–70
retina, 28–30, 31
retinal implants, 28
rhyolite, 215, **216**
rift valleys, **212**, **213**
river pollution, 125–6
road accidents, 43
rock vibrations, 297–8
rockets, 230, 346
rocks, 207–9, 211–12, 214–15, **340**, **348**, 350–2
roots, 76, 77, **99**
rubbish, *see* waste disposal
running, 2, 37
rusting, 227–8, 230, 239

salt, 5, 155, 157, 228
sampling, 66–71
Sankey diagrams, 268
SARS disease, **126**
satellite broadcasts, **287–8**, 341
satellites, **212**, 341, **342**, **353**
saturated hydrocarbons, **176–7**, **179**
scales of measurement, **259**
scrap metal, 220–1, 230
secondary coils, 319–20
secondary sexual characteristics, 50
secondary (S) waves, 297–8, **298–9**
sedimentary rocks, 111, **209**
seismic waves, 297–8, **298–9**, 339
seismometers, 297
sense organs, 28–36
sensory neurones, 34–5
sewage, 125, 126, 133, **144**
sex hormones, 50–1
sex inheritance, 63
sexual reproduction, 55–6, **107**
shadow zones, **298**, **299**
shared pair of electrons, **177**
shiny surfaces, **273**, 275
shivering, 47
shock waves, 297–8
short sight, 31, **32**
sickle-cell anaemia, 16, 60, **62–3**
side effects, 25
sight, 28–33
sight correction, **32**
signal detection, 279–80, 284
signal reception, **277–8**, **281**, 285, **288**
signal transmission, **280–1**, 284–8, **295**, **347**
silica, 215, **216**

silicon, 307
single cells, 75, **77**, **78–9**, 112
single covalent bonds, 169, 170, 175–6, **176–7**
Sites of Special Scientific Interest (SSSIs), 135
skin, 28, 48, **54**, 97, 98
skin damage, 17, 20, 299–301
skyscrapers, 205, 206
smells, 88, **108**, 161–6
smoke detectors, 334
smoking, 17, 37, 39–41
sodium carbonate, 152–3, 162–3
sodium hydrogencarbonate, 90, 151–3, 194
soil, **73**, 77, 88, 90, **99**, 104, 124, 143
solar cell technology, 305–12
solar flares, 341, **342**
solar heating, 308, **308–9**
solar panels, 271, **307**
solar power stations, **309**
solar system, 340, 344–9, 350, **351**
solar water heaters, 308
solder, 221
solid state detectors, 332
solids, 239, 249
solubility, 163–4
solutes, 163, 240
solutions, **90**, 163
solvents, 37, 163–4, 200
sound waves, 280, **282**, **288**, **295**
space radiation, 270–1, **273**, **301**, 330, 341–2
spacecraft, 270–1, 339, 341, 346, 347–8
speciation, **119**
species, 66, 70, 72, 81–5, **119**
species diversity, **73**
specific heat capacity (s.h.c.), 259–60, **260–1**
specific latent heat (s.l.h.), 262, **263**
specificity of catalysts, **251**
spectrum of light, 292, 299
speed of light, 292, 294–5, **347**
speed of waves, 293–5
sperm, 50, **51**, 56, **62–3**
spinal reflexes, 34
spores, 76
sporting ability, **58**
squirrels, 95–6, 132, 134, 135
stamina, 3, **4**
starch, 8, 10, 12, 13, 54, 76, 89, 90
stars, **342**, 344, 345, 356–60
    death of, 356
    distance of, 359
    formation of, **357–8**
    radiation from, **357**
starvation, **12**, 215
steam, 262, **309**, 317, 322
steel, 205–6, 218, 219, 221, 259, 261, 285
    in cars, 180, 227, 228, 229
steelworks, 261
sterilisation, 334
steroids, 37, 38
stimulants, 37, 38, 39
stimuli, 28, 34
stomach, 12, 13, **54**
stone buildings, 205, 208, 235
storage of energy, 89–90

storage of radioactive materials, 333, **336**
stress, 5
stroke, 6, 39, 43
subduction, **213–14**
substations, 313, 319, **342**
successful collisions, **241**
sucrose, 89
Sudan Red, 157
sugar levels, 11, 49–50
sugars, 10, 13, 14, **78**, **89**, **99**, 232
sulfur, 235, 252
sulfur dioxide, 124, 127, 215, 234, 235, 236, 252
sulfuric acid, 124, 162–3
Sun, 344, **351**, 352, **353**, 356, **358**
sun protection factor (SPF), 301
Sun radiation, 265, 270–1, **273**, 299–301, 305–12, 330, 341
Sun surface temperature, 299, 341
sunburn, 17, 300
sunlight, 59, 215, 299–303, 305, 307–9, 350, **354**
sunlight for photosynthesis, 77, 88, 89, 90, **91**
sunscreens, **273**
suntan lotion, 300–1
supernovae, **342**, 356–7, 358
surface area, **14**, 249, 252, 253
surface radiation, **273**, 275
survival, 94, 102–4, 105, 110–20, 133
suspensory ligaments, 29, **30–1**, **32**
sustainability, 131–46
sustainable development, 143–4
sustainable resources, 142–4
sweating, 47, 48, 183–4, 227
sweeping technique, 68
swifts, 105
switching equipment, 319, **342**
symbols, 1, 89, **152–3**, **190–1**
symptoms, 21
synapses, **34–5**, **37**
synthetic polymers, 182
systolic pressure, 5

tangents to graphs, **245–6**
tar in tobacco smoke, 39, **40**
target organs, 49
tectonic plates, 211–14, 232, 350
teeth, 13, 80, **81**, 138
telecommunications, 306, 341
telephones, 280, 281, **282**, **295**, 305
telescopes, 345, 352, 353, 356, 360
temperature
    of the air, 102, 302
    of the body, 46–9
    of the Earth, 215, 236, 302
    in photosynthesis, 90–1
    and reaction rates, 241, 243–5
    of the seas, 106, 236, 260
    in space, 348
    of stars, **357**, 359
    of the Sun, 299, 341
temperature control, 47–9
temperature difference, 258, **271–2**
temperature measurement, 47, 257, 258, 259
temperature rise, 195, 259, **260–1**, 262
temperature scale, 195, 258, **259**

tenacious layers, 228–9
testes, 49, 50, 56, 60, **61**
testing
    on animals, 25, 165
    of theories, **340**
testosterone, 50
text messaging, 293
textile fibres, 182, 227
texture of food, 149, 156
theory testing, **340**
thermal decomposition, 152–3, 207, 219
thermal imaging, 48, 257, 279
thermochromic paints, 201–2
thermograms, 257, 265, 270
thermonuclear fusion, **357**
thermoplastics, **183**
thermosets, **183**
thickness measurements, 334
tidal waves, 290–1
tigers, **83**, 103, 135
tiles, 205
time gaps, **286**
timebase setting, **316**
timing rates of reaction, 240, **245–6**, 249
tissue, 4, 6, **17**, 25, 40, 49
TNT (trinitrotoluene), 252
tobacco, 39, 40
tolerance (drugs), 38
total internal reflection (TIR), 281–2, **287**
toxins, 21–2, 35, 38, 42, 43, 188
transects, 68
transformers, 319–20
transmitter substance, 35, **37–8**
transverse waves, 291, 297–8
trees, 59, 72, 76, 88, 95, 143, 302
troughs of waves, 290
tsunamis, 212, 290–1, **354**
tuberculosis (TB), 16, **22**
Tullgren funnels, 69
tumours, **17**, 300, 335
tuning in radio and TV, 285–6
turbines, 309–11, 315, 317, 322
TV remote control, 279–80
TV transmission, **281**, 284, 285–6
twins, 51, 54, 58

ultraviolet (UV) radiation, 17, 59, 124–5, 284, 286, 299–301, 305, **308**
unbalanced force, **345**
underfloor insulation, 265, 269
units of electricity, 324–5
Universe, 344, 347, 356, 358–60
unsaturated hydrocarbons, **177–8**, **179**, **201**
unsuccessful collisions, **241**
uranium, 322, **325**, **326**, **327**, 330
urine, 46, 125
uterus, 40, **50–1**

vaccination, **22**, 23
vacuoles, 76, **78**
vacuum, 269, **273**
variable stars, 359
variation, 58–9, 60, **61**

vasoconstriction, **48**
vasodilation, **48**
vegans, **11**
vegetarians, **10–11**
venoms, 35
Venus Express spacecraft, 270–1
vertebrates, 79–81, 132
very fast reactions, 252–3
vesicles, **34–5**
vibration of atoms, 239, **271**, **272**
vibration of molecules, 150, 239
vibration of particles, 239, **262**
viruses, 16, 19, 20, 21, 97, 125
visible light, 284, 286, 299, 305, **308**
visual pollution, **310**
vitamins, 8, **11**, 16, 17, 147
VO₂ max, **4**
volatile substances, **162**
volcanoes, **212–14**, 214–15, **216**, 232,
    234, 290, 291, 303, 350
voles, 96–7
voltage, 305, **307**, 316, 319–20, **324**
volume of gas collected, 240, 242–6,
    249
voluntary actions, 34

warfarin, 116
warmth, 122, 193, 265, 302–3, **308**, 346
waste disposal, 125, 184, 230, **326**, 335,
    336

waste energy, 261, 265, 268, 317–18
waste materials, **143**, 322, **323**, **325–6**,
    336
water
    and corrosion, 228, 229
    in deserts, 105–6
    from combustion, 188–91
    from respiration, 1, 2
    for photosynthesis, 88, 89, 90
    as solvent, 163–4
water balance, 46
water paints, 200
water pollution, 125–6, 127, 167
water specific heat capacity, 259–60
water supply, **126**, **144**, 335, **336**
water vapour, **194**, **232**, 233, 252–3
waterproof clothing, 183–4
watts (W), **323**, 324–5
wave diffraction, **277–8**
wave frequencies, **276**, **278**, 285–6,
    287, 292, 293–4, **295**
wave refraction, 287, **298–9**
wave speed, 193–5
wavelengths, 290, 292, 293–4, **295**, 360
waves, 290
weak intermolecular forces, **161–2**,
    165, **169**, **182**, **183**
weasels, 96–7
weather conditions, **233**, 236, 260, **310**
weight, 5, 9, 10, 11–12

weightlessness, 346
whale products, 141
whales, 85, 137–9, **139–40**, 141, **142**
whaling, 141, **142**
white blood cells, 21, 22, **40**
white dwarfs, 356
white surfaces, **273**
wind farms, 309, **310**, 311, 318
wind-pipe (trachea), 20, 40, **140**
wind power, 305, 309–11
wind turbines, 309–10, **310**
windows, 205, 265, 269, **308–9**
wings, 75, 80, 83, 84, 105
wireless communications, 279, 284–9
wireless hotspots, 284
withdrawal symptoms, 38–9
women, 8, **17–18**, 50–1, 60, **62–3**
wood, 72, 143, **233–4**
woodlands, 69, 72–3, 95–6, 135, 136,
    143
word equations, 1, 89, 152, 206, **228**,
    235, 240
worms, 66, 72, 77, 97, 104, 110, 113

X-rays, 59, 284, 286, 330, 335, **358**

zero temperature, **259**
zygotes, 56, 58, 60, **62**

Acknowledgements

Photographs

Cover image, Francisco Chanes / Custom Medical Stock Photo / SPL; **B1a.1**, J. Marshall / Tribaleye Images / Alamy; **B1a.2**, Imagestate / Alamy; **B1a.3**, Sean Demspey / PA / EMPICS; **B1a.4, B1a.5, B2d.12, P1e.1**, Royalty-free / Corbis; **B1a.6, B1c.1, B1d.12**, Phanie / Rex; **B1a.7**, Lehtikuva / Rex; **B1a.8**, Zephyr / SPL; **B1b.2**, J. B. Russell / Panos; **B1b.4**, Frank Siteman / Rex; **B1c.2**, James Stevenson / SPL; **B1c.9**, Andy Crump, TDR, WHO / SPL; **B1c.14**, David Copeman / Alamy; **B1c.16**, Biomedical Imaging Unit, Southampton General Hospital / SPL; **B1c.17**, Karen Kasmauski / Corbis; **B1d.1**, National Eye Institute, National Institutes of Health; **B1d.2**, M. Humayun; **B1d.18, C1d.7d, P1h.1, P1h.2**, Reuters / Corbis; **B1e.1**, Bob Rowan; Progressive Image / Corbis; **B1e.3**, Ed Kashi / Corbis; **B1e.4**, Mark Peterson / Corbis; **B1e.5**, Baumgartner Olivia / Corbis Sygma; **B1e.6**, Du Cane Medical Imaging Ltd / SPL; **B1e.10, C2f.7**, Astrid & Hanns-Frieder Michler / SPL; **B1e.11, C2d.1b, P1g.9**, Hulton−Deutsch Collection / Corbis; **B1f.2**, Photofusion Picture Library / Alamy; **B1f.3a**, Damien Lovegrove / SPL; **B1f.3b**, Paul Whitehill / SPL; **B1f.3c**, Samuel Ashfield / SPL; **B1f.5**, Getty Images News; **B1f.9, P1g.3**, AFP / Getty Images; **B1g.1**, Andrew Syred / SPL; **B1g.3**, Michael Boys / Corbis; **B1g.6**, L. Willatt, East Anglian Regional Genetics Service / SPL; **B1h.2**, Doug Allan / SPL; **B1h.4**, Colin Cuthbert / SPL; **B2a.2, C2e.3**, Leslie Garland Picture Library / Alamy; **B2a.5, B2a.11, B2d.3, B2e.12**, Eleanor Jones; **B2a.7**, Volker Steger / SPL; **B2a.8**, Adam Hart-Davis / SPL; **B2a.10**, Burkard Scientific (Sales) Ltd; **B2a.12**, Russ Hoddinott / NPL; **B2a.15**, Museum of New Zealand Te Papa Tongarewa, negative number I.006709; **B2a.16a, B2a.16b, B2b.19, B2h.13**, Geoff Jones; **B2b.9**, Dave Watts / NPL; **B2b.11, B2h.6**, Dietmar Nill / NPL; **B2b.12a**, Michael Prince / Corbis; **B2b.12b**, Bernard Walton / NPL; **B2b.14a**, Carol Walker / NPL; **B2b.14b**, Lynn M. Stone / NPL; **B2b.14c**, Juan Manuel Borrero / NPL; **B2b.15**, Chris Collins / Corbis; **B2b.17a, B2d.6b**, Colin Varndell / NPL; **B2b.17b, B2d.9**, Daniel Heuclin / NHPA; **B2b.20**, Mark Smith / SPL; **B2c.2, C1c.3**, Nick Garbutt / NPL; **B2c.3**, Dr Jeremy Burgess / SPL; **B2d.1, B2h.7**, Terry Andrewartha / NPL; **B2d.2**, Bruce Davidson / NPL; **B2d.5**, Robert Harding Picture Library Ltd / Alamy; **B2d.6a**, Niall Benvic / NPL; **B2d.11**, Peter Blackwell / NPL; **B2d.13, C1a.2e**, Cordelia Molloy / SPL; **B2d.14**, John Kaprielian / SPL; **B2e.1**, Jeff Foott / NPL; **B2e.2**, Mary McDonald / NPL; **B2e.4**, Mahipal Singh / OSF; **B2e.6**, Hilary Pooley / OSF; **B2e.7**, Barry Mansell / NPL; **B2e.13**, Stan Osolinski / OSF; **B2e.14**, Mike Brown / OSF; **B2e.15**, Hanne & Jens Eriksen / NPL; **B2e.16, B2e.17**, Daniel Cox / OSF; **B2e.18a**, Graham Hatherly / NPL; **B2e.18b**, B. & C. Alexander / NHPA; **B2e.22**, Satoshi Kuribayashi / OSF; **B2e.23**, Dennis Kunkel / OSF; **B2e.24**, Z/S Formula / OSF; **B2f.2**, John Downer / NPL; **B2f.3, C1a.1d, C2f.1**, Corbis; **B2f.4**, Silkeborg Museum, Denmark / Munoz-Yague / SPL; **B2f.8a, B2f.8b**, Stephen Dalton / NHPA; **B2f.10**, Warwick Blass / NPL; **B2f.11**, Ingo Arndt / NPL; **B2f.12**, Blickwinkel / Alamy; **B2f.13, C2c.9, P2e.9**, SPL; **B2f.14**, Wellcome Library, London; **B2g.7**, Stan Kujawa / Alamy; **B2g.9**, The Daily Telegraph 2005; **B2g.13, P1h.11, P1h.14, P2f.5, P2f.6, P2g.1**, NASA / SPL; **B2h.1**, George Bernard / SPL; **B2h.2, C2b.8b**, Jonathan Blair / Corbis; **B2h.3**, Victor Habbick Visions / SPL; **B2h.4, B2h.5**, Martin Harvey / NHPA; **B2h.8**, Kent News & Picture / Corbis Sygma; **B2h.9**, Nick Turner / NPL; **B2h.11**, Geogphotos / Alamy; **B2h.12**, Steve Kaufman / Corbis; **B2h.14a, B2h.14c**, Brandon Cole / NPL; **B2h.14b**, Doug Perrine / NPL; **B2h.14d**, Doc White / NPL; **B2h.18**, Mary Evans Picture Library; **B2h.19**, Mark Carwardine / NPL; **B2h.21**, Greenpeace / Kate Davison; **B2h.22**, Renee Morris / Alamy; **C1a.1a**, Philip Wilkins / ABPL; **C1a.1b, C1a.4a**, Maximilian Stock Ltd / ABPL; **C1a.1c**, Foodfolio / Alamy; **C1a.1e**, Food Features; **C1a.1f**, ATW Photography / ABPL; **C1a.2a**, Norman Hollands / ABPL; **C1a.2b, C2d.2b**, Sheila Terry / SPL; **C1a.2c, C1a.7b**, Gerrit Buntrock / ABPL; **C1a.2d**, Mark Dyball / Alamy; **C1a.2f**, Anthony Blake / ABPL; **C1a.4b**, Clive Streeter / Dorling Kindersley; **C1a.7a**, Plainpicture / Alamy; **C1b.1, C1b.2, C1b.3a, C1b.3b, C1b.5, C1b.7, C1h.4, C2d.2a, C2d.5, P1b.6**, Vanessa Miles; **C1b.8**, Coston Stock / Alamy; **C1c.4**, AM Corporation / Alamy; **C1c.6**, Adrianna Williams / Zefa / Corbis; **C1c.7**, Andreas Pollok / Getty; **C1c.9**, The Body Shop International plc; **C1d.1**, Richard Folwell / SPL; **C1d.2**, Matthew Polak / Corbis Sygma; **C1d.7a, C2h.5**, Bettmann / Corbis; **C1d.7b**, Vanessa Vick / SPL; **C1d.7c**, Lawson Wood / Corbis; **C1e.1**, Imagebroker / Alamy; **C1e.2, C1e.3, C1e.6a, C1e.6b, C1h.3, C2h.3**, Jeremy Pembry / Cambridge

University Press; **C1f.1a**, Diomedia / Alamy; **C1f.1b**, Elizabeth Whiting & Associates / Alamy; **C1f.1c**, The Garden Picture Library / Alamy; **C1f.1d**, Shout / Alamy; **C1f.1e**, **C1f.1f**, Ace Stock Limited / Alamy; **C1f.3a**, Blasius Erlinger / Zefa / Corbis; **C1f.3b**, **C1f.5a**, **C2a.4a**, **C2a.4b**, **C2a.4c**, **C2b.7**, David Acaster; **C1f.4**, Stock Connection / Alamy; **C1f.5b**, **P1e.10**, **P2d.14**, Tek Images / SPL; **C1f.6a**, Roger Ressmeyer / Corbis; **C1f.6b**, Nick Hawkes; Ecoscene / Corbis; **C1g.1**, **P2a.12**, David Hoffman Photo Library / Alamy; **C1g.2**, David Sanger Photography / Alamy; **C1g.3**, David Stares / Alamy; **C1h.1**, Victor De Schwanberg / SPL; **C1h.2**, Rick Doyle / Corbis; **C2a.1**, DIY Photo Library; **C2a.3a**, Hubert Stadler / Corbis; **C2a.3b**, Bill Bachmann / Alamy; **C2a.5**, Robert Llewellyn / Corbis; **C2b.1**, Justin Kase / Alamy; **C2b.2**, Maximilian Stock Ltd / SPL; **C2b.3**, Jose Fuste Raga / Corbis; **C2b.4**, Alan Schein Photography / Corbis; **C2b.5**, Jon Bower / Alamy; **C2b.6**, Robert Brook / SPL; **C2b.8a**, David Keaton / Corbis; **C2b.8c**, Doug Houghton / Alamy; **C2c.4**, Yann Arthus-Bertrand / Corbis; **C2c.7**, Douglas Peebles / Corbis; **C2c.8**, National Gallery, London, UK / Bridgeman Art Library; **C2d.1a**, Philadelphia Museum of Art / Corbis; **C2d.1c**, Fotofacade / Alamy; **C2d.1d**, **P1a.8**, Matthias Kulka / Corbis; **C2d.2c**, **C2h.6**, **P1b.7**, **P2b.2**, Andrew Lambert Photography / SPL; **C2d.3a**, **C2d.3b**, GC Minerals / Alamy; **C2d.6**, David Lees / Corbis; **C2d.7**, **C2e.2b**, Ashley Cooper / Corbis; **C2d.8**, Mediscan; **C2d.9**, Danny Lehman / Corbis; **C2d.10**, Yiorgos Karahalis / Reuters / Corbis; **C2d.11**, **P2d.2**, Pascal Goetgheluck / SPL; **C2e.1**, Transtock Inc. / Alamy; **C2e.2a**, Régis Bossu / Sygma / Corbis; **C2e.4**, Denis Balibouse / Reuters / Corbis; **C2e.5a**, **C2e.5b**, Williams Haynes Portrait Collection, Chemists' Club Archives, Chemical Heritage Foundation Collections, Philadelphia, PA, USA; **C2e.6**, George Hall / Corbis; **C2e.7**, Jeremy Walker / SPL; **C2f.4a**, Alisdair Macdonald / Rex; **C2f.4b**, Travelshots.com / Alamy; **C2f.4c**, Dr Morley Read / SPL; **C2f.5**, Mauro Fermariello / SPL; **C2f.6**, **P1c.1**, **P2a.4**, **P2a.7**, **P2c.1**, **P2c.8**, **P2d.9**, Martin Bond / SPL; **C2f.8**, Greenpeace / Morgan; **C2f.9**, David Askham / Garden Picture Library; **C2g.1a**, R. Holz / Zefa / Corbis; **C2g.1b**, John Mead / SPL; **C2g.1c**, Magrath Photography / SPL; **C2g.9**, Mary Evans Picture Library / Alamy; **C2h.4**, Malcolm Fielding, Johnson Matthey plc / SPL; **C2h.7**, Richard Hamilton Smith / Corbis; **P1a.1**, **P1b.1**, **P1c.3**, Tony McConnell / SPL; **P1a.2**, Dr Ray Clarke & Mervyn Goff / SPL; **P1a.5a**, Larry Dale Gordon / Zefa / Corbis; **P1a.5b**, David Muench / Corbis; **P1a.6**, James Holmes / SPL; **P1b.3**, Building Regulations Explanatory Booklet, Office of the Deputy Prime Minister. Crown copyright material is reproduced with the permission of the Controller of HMSO and Queen's Printer for Scotland; **P1b.4**, Jim Winkley; Ecoscene / Corbis; **P1b.5**, Philippe Gontier / Eurelios / SPL; **P1c.4**, ESA / D. Ducros; **P1c.9**, Roger G. Howard Photography; **P1d.1**, **P1f.5a**, **P2d.5b**, Martyn F. Chillmaid / SPL; **P1d.3**, Serge Kozak / Alamy; **P1d.4**, E. Klawitter / Zefa / Corbis; **P1d.5**, Carlos Dominguez / SPL; **P1e.7**, Deep Light Productions / SPL; **P1f.1**, Big Cheese Photo LLC / Alamy; **P1f.5b**, John Pettigrew; **P1f.9**, Lynette Cook / SPL; **P1f.11a**, **P2a.1**, Peter Bowater / SPL; **P1g.7**, Jim Byrne QAPhotos; **P1h.8**, Lauren Shear / SPL; **P1h.9**, Jon Hicks / Corbis; **P1h.10**, David R. Frazier Photolibrary, Inc. / Alamy; **P1h.12**, Jerry Mason / SPL; **P2a.3**, Peter Menzel / SPL; **P2a.8**, David Nunuk / SPL; **P2a.10**, **P2c.9**, Novosti Photo Library / SPL; **P2a.15**, Country Guardian; **P2b.6**, Bill Longchore / SPL; **P2b.13**, **P2b.14**, **P2b.15**, National Grid; **P2c.2**, Warren Gretz / NREL / DOE / US Department Of Energy / SPL; **P2c.3**, **P2c.4**, **P2d.7**, **P2d.8**, Andrew Lambert; **P2c.7**, Will Mcintyre / SPL; **P2c.10**, British Nuclear Group; **P2d.5a**, Klaus Guldbrandsen / SPL; **P2d.6**, Juan Silva / Getty; **P2d.10**, AJ Photo / SPL; **P2d.11**, Paul Rapson / SPL; **P2d.12**, Martin Dohrn / SPL; **P2e.1**, Jef Maion / Nomads' Land – www.maion.com / Alamy; **P2e.3**, James L. Amos, Peter Arnold Inc. / SPL; **P2e.7**, **P2f.4**, **P2g.3**, Detlev Van Ravenswaay / SPL; **P2e.8**, SOHO / ESA / NASA / SPL; **P2e.11**, Chris Madeley / SPL; **P2e.12**, Gary Hincks / SPL; **P2g.2**, David Parker / SPL; **P2g.4**, David A. Hardy / SPL; **P2g.5**, Walter Pacholka, Astropics / SPL; **P2g.7**, Eckhard Slawik / SPL; **P2g.9**, Chris Butler / SPL; **P2h.2**, **P2h.3**, NASA / ESA / STScI / SPL; **P2h.6**, Emilio Segre Visual Archives / American Institute Of Physics / SPL; **P2h.7**, Harvard College Observatory / SPL.

Abbreviations: ABPL, Anthony Blake Photo Library; NHPA, Natural History Picture Library; NPL, NaturePL.com; OSF, Oxford Scientific; Rex, Rex Features; SPL, Science Photo Library.

Picture research: Vanessa Miles.

Text extracts

page 1, from Everest: *Expedition to the Ultimate*, Reinhold Meissner (1979), page 179, Oxford University Press, ISBN 0 718 21218 5; **page 10**, from *The Impact of Small Arms on the Population: A Case Study of Kitgum and Kotido Districts*, Bruno Ocaya et al. (2001), page 26, Action for Development of Local Communities / Oxfam (http://www.smallarmsnet.org/docs/saaf10.pdf).